S0-BKQ-197

Manual of the Vascular Plants of Wyoming

Garland Reference Library of Science and Technology (Vol. 4)

Oxytropis nana Nutt., a Wyoming endemic collected by Thomas Nuttall on his journey across Wyoming in 1834

Manual of the Vascular Plants of Wyoming

Robert D. Dorn
Illustrations by Jane L. Dorn

In two volumes

Volume I

Equisetaceae to Grossulariaceae

Garland Publishing, Inc., New York & London

1977

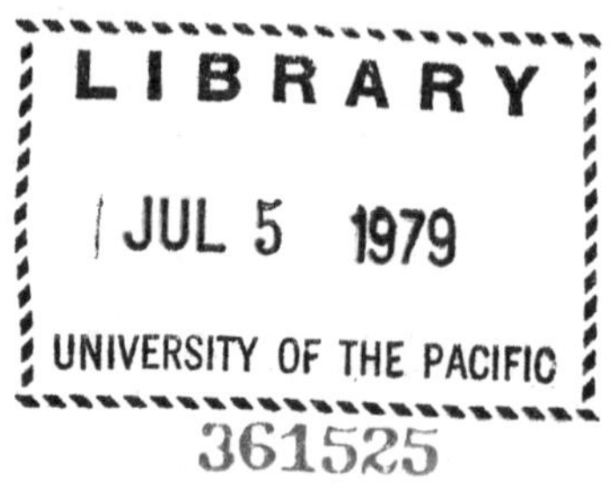

Library of Congress Cataloging in Publication Data

Dorn, Robert D
Manual of the vascular plants of Wyoming.

(Garland reference library of science and technology ; v. 4)
Includes index.
Bibliography: p.
1. Botany--Wyoming. I. Title. II. Series.
QK195.D67 582'.09'787 76-24772

ISBN 0-8240-9905-2

Printed in the United States of America

To Thomas Nuttall,
first botanist in Wyoming, whose botanical perception and expertise were unmatched by any other American taxonomist of the nineteenth century

CONTENTS

Manual of the Vascular Plants of Wyoming

Volume I

Equisetaceae to Grossulariaceae

INTRODUCTION

In November of 1971, while at the University of Wyoming, I undertook the task of writing a manual of vascular plants for Wyoming on a part-time basis. Since the project would have to be completed in less than 3 years, a number of short-cuts had to be introduced. These include brief descriptions; distributions given for Wyoming only and in broad categories; limited synonomy; and exclusion of (1) all infraspecific taxa, (2) species to be expected in the state but not yet found, (3) some annual grain field weeds, and (4) species growing only under irrigation including lawn weeds. Nearly all species of the latter two groups can be found in Hitchcock's *Manual of the Grasses of the United States* and/or in Bailey's *Manual of Cultivated Plants* (A notable exception is *Astragalus cicer* L., irrigated field, Boulder, Sublette Co.). Completion of this treatment in so short a time would not have been possible without the extensive previous work of Aven Nelson and C. L. Porter at the Rocky Mountain Herbarium. My own studies of the Rocky Mountain flora began in 1967 in Montana and included collecting and field studies in all parts of Wyoming from 1969 through 1974.

Since the primary purpose of the manual is for identification of vascular plants, the taxa are arranged alphabetically within the three major groups of Pteridophytes, Gymnosperms, and Angiosperms. Relationships can be shown in only a general way using a linear sequence. One of the more recent systems of classification is used in the summary for those interested in family relationships as far as the present state of our knowledge is concerned. Future changes will be necessary as more data become available.

A somewhat conservative treatment is used for most "difficult" groups for which detailed biosystematic studies have not been completed. This seems the best service for those who are not professional taxonomists. In addition, some studies are unconvincing in their conclusions so are not always followed. References which appear in *Vascular Plants of the Pacific Northwest*, Parts 1-5, by C. L. Hitchcock et al. are usually not included here in order to conserve space. Those interested in infraspecific taxa should refer to Hitchcock's treatment, the references he lists, the references listed herein, Harrington's *Manual of the Plants of Colorado*, or *Gray's Manual of Botany*.

All descriptions have been drawn from Wyoming material except in a few cases where such material was not available. Every species has also been run through the keys. Distributions are given in quadrants of the state except for plants which have been collected in only one or two counties, in which case the counties are listed. The localities can be narrowed down further in many cases by noting the habitat. Thus, an alpine habitat in the NE quadrant of the state can only be in the Big Horn Mountains.

I have attempted to include all names which appeared in the 1909 Manual of Coulter and Nelson for species accepted by them, and which apply to Wyoming plants. I have not attempted to include all the synonomy from Rydberg's work. In addition, synonyms based on Wyoming collections for which type specimens are on file in the Rocky Mountain Herbarium have been included.

There are numerous errors in the literature in citation of original descriptions of species. I have been able to check about 90 percent of these descriptions. For those I could not check, and for which there was disagreement in the literature, I have chosen what I thought to be the most reliable source with monographs having a high priority. In several cases involving species described from Europe, *Flora Europaea* was followed rather than the consensus in this country.

A few hints may be helpful for non-professionals, particularly students. A complete specimen is always desirable and frequently is absolutely necessary. This includes the root system (except for trees and shrubs), mature flowers, and even fruits if possible. More than one specimen is also desirable. Both fruits and flowers may be necessary for identification in such families as the Cruciferae, Umbelliferae, and Boraginaceae. In most other families the keys are based primarily on flowers. When a complete specimen is not practical, height, habit, and type of root system should be recorded. For all specimens record location, elevation, date, habitat, and flower color (sometimes changes on drying). All terms should be understood before deciding on a choice in the key. This is extremely important. A glossary is available at the end of this book. In case of doubt or if your plant lacks the character referred to, try both choices. Read both choices of a lead entirely before deciding on a choice. Check your progress with the family, genus, and species descriptions. Individuals may occasionally be found growing in habitats other than the ones stated or they may fall outside the range of the description for that species. Measurements are based on dried material. There may be slight shrinkage in drying so fresh measurements may be slightly larger. Those in parentheses are extremes rarely encountered. Numbers of parts (e. g., stamens 5 or leaflets 3) are per flower, per leaf, etc. unless otherwise indicated.

During the preparation of this treatment, no fewer than ten species were collected in the state for the first time. Many areas of the state remain almost unknown botanically and many new state records can be expected, particularly along the state's borders and especially in the SW and NE corners.

Only two new combinations have been necessary for this treatment (in Cymopterus and Haplopappus).

The illustrations were contributed by Jane L. Dorn. D. H. Knight contributed valuable suggestions for the section on vegetation types. D. J. Crawford made available some unpublished data for the treatments of Chenopodium and Populus. The unpublished work of R. Hill was drawn upon for the treatment of Mentzelia. G. Pierce, E. Nelson, E. Prather, J. Humbles, J. Landon, and C. Reeder helped search for lost specimens and brought some specimens to my attention. J. Reeder made a few suggestions on nomenclatural problems.

R. D. Dorn

KEY TO THE MAJOR GROUPS

Disseminules spores; plants never having flowers or seeds; ferns, horsetails, club mosses, and other pteridophytes
1. PTERIDOPHYTES below

Disseminules seeds; plants often having flowers; conifers, jointfir, and flowering plants

Ovules and seeds mostly borne on the surface of scales, the scales aggregated into a cone, occasionally the cone is fleshy and berry-like; evergreen trees and shrubs with needle-like or scale-like leaves; conifers and jointfir
2. GYMNOSPERMS p. 6

Ovules and seeds enclosed in an ovary or fruit; trees, shrubs, vines, and herbs; leaves only rarely as above; flowering plants 3. ANGIOSPERMS p. 7

1. PTERIDOPHYTES

Dissemination by spores; flowers and seeds lacking.

1. Stems jointed and grooved lengthwise; horsetails and scouring rushes EQUISETACEAE p. 69

1. Stems not jointed or grooved lengthwise

2. Leaves like a four-leaf clover MARSILEACEAE p. 74

2. Leaves not as above

3. Leaves grass-like in a basal tuft; sporangia at base of the leaves; plants aquatic or amphibious ISOETACEAE p. 72

3. Leaves usually not as above; sporangia at base of the leaves or not; plants terrestrial

4. Leaves simple, sessile, and scale-like; plants moss-like
   5. Spores of 2 kinds, 1-4 large megaspores in separate sporangia and many small microspores in separate sporangia (microspores usually lacking in 1 species); strobili mostly 4 sided; leaves with ligules
      SELAGINELLACEAE p. 91
   5. Spores of 1 kind, small and numerous; strobili cylindrical or not developed; leaves without ligules
      LYCOPODIACEAE p. 73
4. Leaves not with the above combination of characters; mostly fern-like plants
   6. Sporangia usually on the underside of vegetative-like leaves, the leaves sometimes slightly modified; roots not thick, fleshy, and radially spreading; stems and petioles often bearing scales POLYPODIACEAE p. 77
   6. Sporangia on a specialized branch appearing to be continuous with the main stalk or appearing to arise from near the base of a sterile leaf; roots thick, fleshy, and radially spreading; stems and petioles lacking scales OPHIOGLOSSACEAE p. 74

2. GYMNOSPERMS

Disseminules seeds, these borne on scales, the scales aggregated into cones; trees or shrubs with evergreen needle-like or scale-like leaves.

Branches remotely jointed; leaves scale-like, 2 at a node, usually united below; shrubs in arid areas of SW Wyo. EPHEDRACEAE p. 97

Branches not jointed; leaves linear or needle-like, or if scale-like, then not remote but overlapping; shrubs and trees; widespread

Leaves linear or needle-like, not opposite or whorled, sometimes in fascicles of 2-5; trees with seeds in woody or subwoody cones PINACEAE p. 98

Leaves scale-like and opposite, or needle-like and whorled; trees or shrubs with seeds in a fleshy berry-like cone CUPRESSACEAE p. 95

3. ANGIOSPERMS

Disseminules seeds, these enclosed in an ovary when immature, the ovary becoming the fruit at least in part; trees, shrubs, vines, and herbs with flowers.

Plants submerged or floating-leaved aquatics, usually limp out of water (avoid temporarily flooded terrestrial or emergent plants) GROUP I. p. 8

Plants emergent aquatics or terrestrial

Leaves mostly parallel-veined, simple, rarely reduced to sheaths; flower parts in 3's or 6's (rarely in 2's or 4's); vascular bundles scattered in stem; root system usually fibrous or bulbous; monocotyledonous herbs GROUP II. p. 12

Leaves mostly net-veined (sometimes obscure), simple or compound, rarely lacking; flower parts usually in 2's, 4's, or 5's (rarely in 3's) or the parts many; vascular bundles arranged in rings in the stem; root system various, often taprooted; dicotyledonous trees, shrubs, woody vines, or herbs GROUP III. p. 15

GROUP I. Submerged and floating-leaved aquatics.

1. Plants thallus-like, free floating at the water surface or below, mostly less than 10mm long but rarely to 25mm; stems and leaves not differentiated; duckweeds LEMNACEAE
1. Plants usually not free floating except when fragmented, usually over 10mm long; stems and leaves usually well differentiated
  2. Leaf blades round to broadly oval, usually over 10cm long, mostly entire, on long petioles which attach near a deep notch at base of blade; water lily NYMPHAEACEAE
  2. Leaf blades and petioles not as above
    3. Leaves simple and subopposite, usually with bunches of smaller leaves in their axils, at least above; leaves dilated at base, finely toothed; fruits solitary and sessile in leaf axils NAJADACEAE
    3. Leaves not as above; fruits various
      4. Leaves whorled on an elongate stem
        5. Leaves compound or at least dichotomously divided into linear segments
          6. Leaf divisions with at least 1 side minutely spiny-toothed CERATOPHYLLACEAE

6. Leaf divisions not toothed HALORAGACEAE

5. Leaves simple

7. Leaves mostly 2-4 per node HYDROCHARITACEAE

7. Leaves mostly more than 4 per node HIPPURIDACEAE

4. Leaves opposite, alternate, or in a basal clump, or rarely inconspicuous

8. Leaves in a single basal clump; rhizomes and runners lacking (incomplete specimens may incorrectly run here)

9. Leaves linear and subterete, 1-6cm long Subularia in CRUCIFERAE

9. Leaves flat, linear or not, often with a differentiated blade and petiole, often over 6cm long

10. Leaves with lance-elliptic to elliptic-oblanceolate or oblong blades 15mm or less long on slender petioles; pistils solitary Limosella in SCROPHULARIACEAE

10. Leaves usually not as above; pistils several per flower ALISMATACEAE

8. Leaves on an elongate stem or basal along a rhizome or runner

11. Leaves compound with thread-like divisions, with a lacy appearance

12. Leaves bearing scattered bladders LENTIBULARIACEAE

12. Leaves without bladders Ranunculus in RANUNCULACEAE

11. Leaves not as above

13. Stems thread-like (often appearing like leaves) and arising from a horizontal rhizome, usually in clumps less than 12cm high; leaves reduced to sheaths CYPERACEAE

13. Stems not as above; leaves various

14. Leaves opposite

15. Leaves about 1mm wide or less and mostly over 25mm long; stipules usually present; fruits in leaf axils, stalked and beaked
Zannichellia in POTAMOGETONACEAE

15. Leaves mostly wider or shorter; stipules present or not; fruits various

16. Leaves often 3 at lower nodes or at nodes with flowers; flowers and fruits on a peduncle or in a peduncled spathe HYDROCHARITACEAE

16. Leaves 2 at a node; flowers and fruits sessile or nearly so

17. Submerged leaves linear; floating leaves, if any, club-shaped to oval; fruit with 2-4 seeds; stamen 1 CALLITRICHACEAE

17. Submerged leaves linear to spatulate; fruit with many seeds; stamens 3-6

18. Opposite pairs of leaves connate-sheathing around stem CRASSULACEAE

18. Opposite pairs of leaves not connate-sheathing ELATINACEAE

14. Leaves alternate or basal (rarely fascicled and superficially appearing opposite)

19. Leaves 3-5 parted with net venation
Ranunculus in RANUNCULACEAE

19. Leaves mostly entire; venation net or parallel

20. Leaves basal, less than 12(20)cm long, occasionally with a few much shorter cauline leaves; plants often stoloniferous

21. Pistils numerous in each flower
Ranunculus in RANUNCULACEAE

21. Pistils solitary
Limosella in SCROPHULARIACEAE

20. Leaves cauline, or if apparently basal, these over 12cm long; plants stoloniferous or not

22. Leaves oval to lanceolate with net venation; stipules completely encircling stem Polygonum in POLYGONACEAE

22. Leaves or stipules not as above

23. Leaves with stipules POTAMOGETONACEAE

23. Leaves lacking stipules (sometimes auricled)

24. Inflorescence of globose heads, the male heads above the female
SPARGANIACEAE

24. Inflorescence not as above

25. Leaves few, all near base of stem; each flower and fruit subtended by a scale
Scirpus subterminalis in CYPERACEAE

25. Leaves relatively numerous and scattered along the stem; each flower and fruit not subtended by a scale POTAMOGETONACEAE

GROUP II. Monocots and exceptional dicots except submerged and floating-leaved aquatics.

1. Plants parasitic on coniferous trees; stems jointed; perianth segments mostly 3 LORANTHACEAE
1. Plants not as above
 2. Flowers unisexual and numerous, in dense cylindrical spikes at tip of stem, the male spike above the female; female spikes usually over 7cm long; perianth of bristles; plants usually over 1m high and in moist areas; cattails TYPHACEAE
 2. Flowers not as above; height and habitat various
  3. Flowers unisexual, in dense globose heads, the male heads above the female; perianth of membranous scales; leaves linear, often sheathing the stem; plants of wet places SPARGANIACEAE
  3. Flowers, if in globose heads, bisexual or with well developed sepals and petals, or rarely the plants dioecious; leaves and habitat various
   4. Flowers in the axils of 1-4 chaffy bracts; perianth of bristles or minute scales or lacking; leaves linear and sheathing the stem, rarely reduced to the sheathing base; fruit indehiscent and 1 seeded; grasses and sedges
    5. Flower subtended by 1 bract (rarely 2), the pistillate flower sometimes also enclosed in a perigynium (when flowers unisexual); stem usually solid or pithy in internodes, often triangular in cross section; leaf sheaths normally not split CYPERACEAE
    5. Flower subtended by 2 or more bracts; stem often hollow in internodes, round or flat in cross section; leaf sheaths often split GRAMINEAE

4. Flowers not as above; perianth of 6 scales or not of scales or bristles; leaves and fruits various

6. Inflorescence a spike-like raceme 2-15mm wide, often loosely flowered; perianth segments green; stamens 6, the anthers longer than the filaments; leaves linear, the sheaths open, glabrous JUNCAGINACEAE

6. Inflorescence usually not spike-like, or if so, without the other characters

7. Perianth of 6 separate, similar, chaffy, often brownish scales; inflorescence never an umbel; rushes JUNCACEAE

7. Perianth with sepal-like or petal-like parts or lacking, or if chaffy, the flowers in umbels

8. Pistils usually 10 to many, arranged in a ring or a globose head ALISMATACEAE

8. Pistils solitary (flowers rarely unisexual)

9. Plants prostrate, usually on mud flats, the perianth lacking although 1 or 2 floral bracts may be present; flowers and fruits solitary in leaf axils CALLITRICHACEAE

9. Plants not as above

10. Ovary superior; stamens more than 3 (3 in small alpine or amphibious annuals)

11. Plants with sepals and petals well differentiated, the petals blue or purplish to rose; leaves linear or linear-lanceolate and sheathing, alternate; roots fibrous, bulb lacking; plants often of sandy areas or prairies COMMELINACEAE

11. Plants not with the above combination of characters

12. Plants annual, to 10cm high, with opposite leaves and many ovules per ovary, the flowers less than 2mm long and axillary ELATINACEAE

12. Plants not as above

13. Inflorescence a fleshy spike subtended by a white or yellow spathe; perianth lobes and stamens 4 ARACEAE

13. Inflorescence not as above; perianth parts and stamens rarely 4

14. Ovules 1 per ovary; fruit an achene; flowers sometimes unisexual; venation of leaves usually obscure as a result of small size, hairs, or obscure veins, or else prominently net-veined; bulbs lacking, taproot often present POLYGONACEAE

14. Ovules many per ovary; fruit a capsule or berry; flowers bisexual; venation of leaves usually distinctly parallel; bulbs often present, taproot lacking LILIACEAE

10. Ovary inferior; stamens 3 or fewer

15. Stamens 1 or 2; flowers irregular ORCHIDACEAE

15. Stamens 3; flowers regular

16. Leaves whorled at least in part, scattered along stem RUBIACEAE

16. Leaves not whorled, equitant and near base of plant IRIDACEAE

GROUP III. Dicots except submerged and floating-leaved aquatics.

1. Plants trees, shrubs, or woody vines, woody throughout SERIES A Below

1. Plants herbaceous or woody only at base (woody plants with flowers will also run here)

  2. Calyx or corolla or both lacking (calyx-like involucre sometimes present but this subtends several to many flowers) SERIES B p. 34

  2. Calyx and corolla present, rarely intergrading and the parts of each numerous (corolla sometimes deciduous in older flowers)

    3. Petals separate to base (rarely united below ovary around a carpophore, or with a single blue or purple petal) SERIES C p. 44

    3. Petals united at least at base SERIES D p. 56

SERIES A. Woody dicots.

1. Plants vines or twining shrubs

  2. Leaves and branches alternate (ignore tendrils and flower stalks)

    3. Tendrils present VITACEAE

    3. Tendrils lacking

      4. Leaves toothed CELASTRACEAE

4. Leaves entire although sometimes with a pair of lobes at base SOLANACEAE

2. Leaves and branches opposite or whorled

5. Leaves compound *Clematis* in RANUNCULACEAE

5. Leaves simple

6. Leaves lobed or toothed and on long petioles *Humulus* in CANNABACEAE

6. Leaves entire or nearly so, sessile or nearly so *Lonicera* in CAPRIFOLIACEAE

1. Plants trees or shrubs, not twining although sometimes prostrate

7. Leaves palmatifid with mostly 3-5 linear, spinulose-tipped segments, also with axillary fascicles of often simple linear leaves *Leptodactylon* in POLEMONIACEAE

7. Leaves not as above

8. Leaves and branches opposite (rarely subopposite) or whorled

9. Leaves compound

10. Pith of older (and often younger) stems over half the diameter of stem; leaves pinnately compound; fruit a berry; shrub *Sambucus* in CAPRIFOLIACEAE

10. Pith of older stems usually less than half the diameter of stem; leaves pinnately or palmately compound; fruit a samara or achene; tree or shrub

11. Plants shrubs to 3dm high; sepals blue, 15-50mm long *Clematis* in RANUNCULACEAE

11. Plants trees or shrubs mostly over 3dm high; sepals not blue, less than 15mm long

12. Leaflets mostly 3 or 5, the margins usually lobed or with a few large teeth; bundle scars 3 ACERACEAE

12. Leaflets mostly 5 or 7, the margins entire to toothed; bundle scars more than 3 OLEACEAE

9. Leaves simple

13. Leaves somewhat silvery or gray from scales at least beneath (sometimes dotted with brown scales also)

14. Leaves often dotted with brownish scales; plants usually 1m or more high, the branches sometimes spine-tipped *Shepherdia* in ELAEAGNACEAE

14. Leaves lacking brownish scales; plants usually much less than 1m high, the branches not spine-tipped *Atriplex* in CHENOPODIACEAE

13. Leaves not silvery or gray from scales

15. Leaves, or some of them, with 3-5 primary lobes

16. Leaves usually indented at base, all lobed and with secondary lobes or coarse teeth; fruit a samara ACERACEAE

16. Leaves rounded or pointed at base, usually somewhat evenly toothed, the terminal leaves often not lobed; fruit a drupe *Viburnum* in CAPRIFOLIACEAE

15. Leaves lacking lobes (rarely sinuately lobed)

17. Leaves conspicuously pubescent on both sides, usually more so beneath

18. Leaves regularly toothed

<u>Jamesia</u> in HYDRANGEACEAE

18. Leaves entire or nearly so, rarely sinuately lobed

19. Leaf blades 5-25mm long, with 3 main veins arising at or near base

<u>Philadelphus</u> in HYDRANGEACEAE

19. Leaf blades often over 25mm long, with definite pinnate venation

20. Leaf tips usually acute or acuminate; leaf hairs usually attached near their middle

CORNACEAE

20. Leaf tips usually obtuse or rounded; leaf hairs attached at their base CAPRIFOLIACEAE

17. Leaves glabrous or with a few scattered hairs especially on margins (rarely with minute branched hairs beneath)

21. Leaves toothed

22. Leaves mostly over 4cm long; leaf tip acuminate <u>Viburnum</u> in CAPRIFOLIACEAE

22. Leaves mostly less than 4cm long; leaf tip not acuminate, rarely short cuspidate

23. Flowers and fruits in the leaf axils or merely paired if terminal; leaf blades mostly less than 3 times as long as wide

24. Branchlets glabrous or glabrate; plants often upright; flowers axillary CELASTRACEAE

24. Branchlets pubescent; plants creeping; flowers terminal <u>Linnaea</u> in CAPRIFOLIACEAE

23. Flowers and fruits in a terminal raceme; leaves mostly more than 3 times as long as wide <u>Penstemon</u> in SCROPHULARIACEAE

21. Leaves entire or wavy-margined or rarely sinuately lobed

25. Year-old stems red; flowers in a terminal inflorescence; leaves usually with hairs attached near their middle with the 2 ends free CORNACEAE

25. Year-old stems not red; flowers various; leaves without hairs as above

26. Leaves somewhat leathery, 0.3-3cm long, with very minute branched hairs beneath, the base of each opposite pair appearing to be continuous but below the articulation point <u>Kalmia</u> in ERICACEAE

26. Leaves not as above

27. Flowers in a terminal raceme; leaves mostly over 3 times as long as wide <u>Penstemon</u> in SCROPHULARIACEAE

27. Flowers usually axillary (merely paired if terminal); leaves mostly less than 3 times as long as wide CAPRIFOLIACEAE

8. Leaves and branches alternate (rarely in fascicles or densely clustered at base)

28. Leaves compound

29. Leaflets with spiny-toothed margins, the spines often about 1mm or more long BERBERIDACEAE

29. Leaflets lacking spiny-toothed margins

30. Leaves divided into many mostly linear segments *Artemisia* in COMPOSITAE

30. Leaves not as above, often with rather broad leaflets

31. Plants bearing spines or prickles ROSACEAE

31. Plants lacking spines or prickles

32. Leaflets mostly 3-7

33. Leaflets toothed or lobed or with wavy margins (poison-ivy keys here) ANACARDIACEAE

33. Leaflets entire *Pentaphylloides* in ROSACEAE

32. Leaflets mostly 9 or more

34. Leaflets entire, with midrib prolonged to a short bristle *Amorpha* in LEGUMINOSAE

34. Leaflets toothed, the midrib not prolonged

35. Stems somewhat soft, the pith 1/2 to 2/3 their diameter; fruit covered with reddish or stellate hairs; leaflets somewhat glaucescent beneath; stamens 10 or less ANACARDIACEAE

35. Stems hard, the pith less than 1/3 their diameter; fruit not covered with reddish or stellate hairs; leaflets usually not glaucescent beneath; stamens mostly 15-20 *Sorbus* in ROSACEAE

28. Leaves simple (rarely with a pair of nearly distinct basal lobes)

36. Leaves scale-like, mostly about 1mm long and overlapping, resembling a juniper TAMARICACEAE

36. Leaves not as above

37. Plants with spines, thorns, prickles, or spine-tipped branches

38. Leaves rarely over 3mm wide, 5-20 times longer than wide, entire

39. Twigs, at least the youngest, somewhat tomentose _Tetradymia_ in COMPOSITAE

39. Twigs not tomentose _Sarcobatus_ in CHENOPODIACEAE

38. Leaves mostly over 3mm wide, usually less than 10 times longer than wide, entire or not

40. Leaves silvery or gray on one or both surfaces from scales or scale-like hairs

41. Leaves about the same color on both sides; shrubs mostly less than 1m high CHENOPODIACEAE

41. Leaves greenish above, silvery beneath; trees or shrubs mostly well over 1m high _Elaeagnus_ in ELAEAGNACEAE

40. Leaves not silvery or gray from scales, sometimes so from definite hairs

42. Leaves 3-5 parted, the divisions often with 3 linear subdivisions _Artemisia_ in COMPOSITAE

42. Leaves not as above

43. Leaves entire

44. Leaves prominently palmately veined, lighter beneath than above, pubescent *Ceanothus* in RHAMNACEAE

44. Leaves obscurely veined, about the same color on both sides, pubescent or not

45. Plants tomentose at least on young branchlets; spines divergent from main branches *Tetradymia* in COMPOSITAE

45. Plants not tomentose; spines often merely the tips of main branches (see also *Lycium* in SOLANACEAE) *Grayia* in CHENOPODIACEAE

43. Leaves toothed or lobed

46. Leaves 3-5 palmately lobed, with somewhat rounded teeth; venation palmate; prickles sometimes present on internodes GROSSULARIACEAE

46. Leaves usually not lobed (rarely pinnately lobed), with usually pointed teeth; venation pinnate; prickles not present on internodes ROSACEAE

37. Plants not as above

47. Leaves linear, evergreen, 1 or 2 grooved beneath, resembling needles of a fir tree; mostly alpine or subalpine *Phyllodoce* in ERICACEAE

47. Leaves not as above; alpine or not

48. Leaves or their much divided segments mostly 3mm or less wide, 5-20 times longer than wide

49. Leaves succulent, nearly cylindrical (rarely flattened), glabrous or farinose; plants mostly in alkaline areas — Suaeda in CHENOPODIACEAE

49. Leaves not as above; habitat various

50. Leaves, or many of them, with 3 lobes at tip, silvery or gray hairy on both sides — Artemisia in COMPOSITAE

50. Leaves not as above

51. Plants of wet areas; bud scales solitary; willows — Salix in SALICACEAE

51. Plants mostly of dry areas; bud scales usually several or buds obscure

52. Leaves toothed; flowers solitary in leaf axils — Calylophus in ONAGRACEAE

52. Leaves usually entire or lobed (or the margins scabrous or ciliate); flowers terminal or axillary

53. Leaves and stems with long spreading hairs along with short, usually stellate hairs; flowers axillary — CHENOPODIACEAE

53. Leaves, stems, and flowers not combined as above

54. Leaves gray or silvery on both sides from minute scales

Atriplex in CHENOPODIACEAE

54. Leaves not gray or silvery, or if so, then from distinct hairs

55. Flowers in a terminal, branched inflorescence (sometimes in heads); venation on upper leaf surface obscure

56. Leaves somewhat green above, silvery or gray beneath, not filiform

Eriogonum in POLYGONACEAE

56. Leaves equally green, silvery, or gray on both sides, sometimes filiform COMPOSITAE

55. Flowers sessile or nearly so and axillary (sometimes in heads); venation on upper leaf surface usually prominent, often reticulate

57. Leaves with rolled margins, entire or rarely 3 lobed at tip, usually lighter beneath from hairs ROSACEAE

57. Leaves not as above COMPOSITAE

48. Leaves mostly over 3mm wide and less than 10 times longer than wide

58. Leaf margins entire or nearly so

59. Leaves pale beneath and with yellowish resinous dots — *Ledum* in ERICACEAE

59. Leaves not as above

60. Leaf margins usually rolled, hairy beneath

61. Plants of dry areas; bud scales several

62. Stems and leaves with many distinct stellate hairs — *Ceratoides* in CHENOPODIACEAE

62. Stems and leaves without stellate hairs, the hairs often dense and tangled

63. Flowers in a terminal, branched inflorescence; venation on upper leaf surface obscure — *Eriogonum* in POLYGONACEAE

63. Flowers sessile and axillary; venation on upper leaf surface usually prominent and reticulate — *Cercocarpus* in ROSACEAE

61. Plants of wet areas; bud scales solitary; willows — *Salix* in SALICACEAE

60. Leaf margins usually flat, often not hairy beneath

64. Crushed leaves with a sage odor
Artemisia in COMPOSITAE

64. Crushed leaves lacking a sage odor

65. Plants mostly of wet areas; bud scales solitary; flowers unisexual, borne in catkins; willows
Salix in SALICACEAE

65. Plants of wet or dry areas; bud scales more than 1 (rarely normally obscure); flowers not as above

66. Plants with stems and leaves covered with minute gray-mealy scales, mostly woody near base only
CHENOPODIACEAE

66. Plants not as above

67. Leaf blades mostly 15mm or less long, glandular-puberulent; SW desert area
Brickellia in COMPOSITAE

67. Leaf blades longer or not glandular or the plants not in the SW desert area

68. Leaves with 3 prominent, somewhat parallel veins arising from nearly the same point near base of blade

69. Leaves lighter beneath
Ceanothus in RHAMNACEAE

69. Leaves equally green on both sides
Chrysothamnus in COMPOSITAE

68. Leaves not as above

70. Leaves ovate to cordate, some usually with a pair of basal lobes; plants over 2dm high SOLANACEAE

70. Leaves not as above or else the plants smaller

71. Leaves sharply acute at tip; plants of the plains and foothills, rarely in the mountains

72. Leaves white-tomentose beneath, less so and greenish above, mostly lanceolate to oblanceolate, rarely linear
Eriogonum in POLYGONACEAE

72. Leaves not white-tomentose beneath, or if so, then linear or equally tomentose above (also see *Lycium* in SOLANACEAE) COMPOSITAE

71. Leaves often obtuse or rounded at tip; plants often of the mountains

73. Leaves either glandular-puberulent or canescent or tomentose, about equally green (or silvery) on both sides COMPOSITAE

73. Leaves not glandular-puberulent nor canescent or tomentose, or if so, distinctly lighter beneath

74. Leaves white-tomentose beneath, less so and greenish above; deserts and plains *Eriogonum* in POLYGONACEAE

74. Leaves not as above; often in the mountains

75. Plants with linear-lanceolate to elliptic leaves which are silver-scaly on both sides but sometimes more so beneath; calyx 4 parted; fruit fleshy or drupe-like, 1 seeded ELAEAGNACEAE

75. Plants without the above combination of characters ERICACEAE

58. Leaf margins toothed or lobed, sometimes slightly so

76. Crushed leaves with a sage odor; many leaves with 3 lobes or teeth at tip *Artemisia* in COMPOSITAE

76. Crushed leaves lacking a sage odor; leaves lobed or not

77. Plants alpine, mostly less than 8(10)cm high

78. Leaves white-tomentose beneath, green above, shallowly pinnately lobed or crenate *Dryas* in ROSACEAE

78. Leaves not as above

79. Plants with unisexual flowers in catkins; bud scales solitary *Salix* in SALICACEAE

79. Plants not as above ERICACEAE

77. Plants not as above

80. Plants with densely glandular-puberulent leaves and stems, the leaf blades mostly ovate and 15mm or less long, green on both sides; rare in SW desert areas *Brickellia* in COMPOSITAE

80. Plants not as above

81. Leaves with 3 prominent, somewhat parallel veins arising from nearly the same point near base; leaves not lobed, longer than wide

Ceanothus in RHAMNACEAE

81. Leaves not veined as above or else the leaves lobed or the blades as wide as or wider than long

82. Plants trees with whitish bark; petioles pubescent; leaves usually doubly toothed and nearly twice as long as wide; flowers borne in catkins; Black Hills

Betula in BETULACEAE

82. Plants not with the above combination of characters

83. (moved to left margin)

83. Plants mostly of wet areas (rarely in upland forests); bud scales solitary; flowers unisexual, borne in catkins; willows

Salix in SALICACEAE

83. Plants not with the above combination of characters

84. Plants upright shrubs 0.5-3m high with finely toothed leaves, the teeth terminated by, or reduced to, short bristles (not sessile glands) which may be gland-tipped, the leaves lighter beneath than above; young twigs glandular-pubescent; NW Wyo. ERICACEAE

84. Plants not as above; widespread

85. Twigs usually roughened with blister-like resinous bumps; leaves often glandular-dotted; flowers in catkins; plants often of moist areas Betula in BETULACEAE

85. Twigs usually not as above; leaves, flowers, and habitat various

86. Leaf blades, or many of them, asymmetrical at base, 1 side extending lower than the other; twigs of year often hairy; petioles usually hairy; fruit a samara or drupe; E Wyo. ULMACEAE

86. Leaf blades symmetrical at base or nearly so (except rarely when lobed); twigs, petioles, and fruit various; widespread

87. Leaves cordate or subcordate at base, usually acute at tip, definitely longer than wide, not lobed but often doubly toothed; twigs often hairy; fruit a nut or nutlet; Black Hills BETULACEAE

87. Leaves usually not as above; twigs and fruit various; widespread

88. Plants trees bearing catkins; buds often sticky with resin; leaves elliptic or lanceolate to deltoid or suborbicular, not lobed; aspen and cottonwoods Populus in SALICACEAE

88. Plants trees or shrubs with or without catkins; buds usually not sticky; leaves never deltoid unless lobed

89. Leaves 5cm or more long, pinnately lobed or parted, not toothed or with broad rounded teeth; fruit an acorn; oaks FAGACEAE

89. Leaves often shorter, mostly palmately lobed or not lobed (rarely with a single pair of basal lobes), toothed or not; fruit not an acorn

90. Plants with doubly toothed leaves which are longer than wide; buds usually stalked; flowers in catkins or cone-like structures

Alnus in BETULACEAE

90. Plants not as above

91. Plants depressed evergreen shrubs to 7cm high with ovate to orbicular or rarely elliptic leaf blades 5-25mm long, the teeth usually tipped with minute spines ERICACEAE

91. Plants not as above

92. Plants with mostly elliptic to oval leaves, some over 4cm long, finely toothed to near base; flowers mostly 2-5 in axils of leaves; petals lacking; fruit a 3 seeded berry; spur shoots never present; bud scales pubescent; plants usually in moist areas of NW Wyo. Rhamnus in RHAMNACEAE

92. Plants not as above

93. Petals lacking; fruit a samara; leaves elliptic to lanceolate, doubly toothed

Ulmus in ULMACEAE

93. Petals present or rarely lacking; fruit not a samara; leaves various

94. Stamens usually 5; ovary 1 celled; ovules or seeds several to many; fruit a berry; leaves usually 3-5 palmately lobed GROSSULARIACEAE

94. Stamens either more than 5 or plants without the above combination of characters

95. Leaves sessile or nearly so, linear to oblanceolate, less than 8mm wide, mostly serrulate, usually pubescent; flowers solitary and axillary, the ovary inferior; plants of rather dry, open areas at lower and middle elevations *Calylophus* in ONAGRACEAE

95. Leaves and flowers not with the above combination of characters

96. Buds covered by 2 outer symmetrical scales, these glabrous; leaves not lobed; ovary inferior, the fruit a berry; plants of the mountains ERICACEAE

96. Buds with more than 2 outer scales, these often pubescent; leaves lobed or not; ovary inferior or superior, the fruit various; habitat various

97. Leaves mostly ovate to cordate, some with a single pair of basal lobes *Solanum* in SOLANACEAE

97. Leaves not as above ROSACEAE

SERIES B. Dicots lacking a calyx or corolla or both.

1. Plants parasitic on coniferous trees; stems jointed; perianth segments mostly 2-4 LORANTHACEAE
1. Plants not as above
  2. Flowers in dense heads subtended by an involucre, the head usually appearing like a single flower (rarely 1 flower per head); ovary inferior; stamens united by their anthers (rarely free); flowers sometimes unisexual COMPOSITAE
  2. Flowers not in heads or lacking the other characters
    3. Plants woody trees or shrubs
      4. Leaves compound
        5. Plants shrubs to 3dm high; sepals blue, 15-50mm long *Clematis* in RANUNCULACEAE
        5. Plants trees or shrubs mostly over 3dm high; sepals not blue, less than 15mm long
          6. Leaflets mostly 3 or 5; bundle scars 3; fruit a double samara ACERACEAE
          6. Leaflets mostly 5 or 7; bundle scars more than 3; fruit a single samara OLEACEAE
      4. Leaves simple
        7. Leaves and branches opposite or rarely subopposite
          8. Leaves 3 or 5 lobed ACERACEAE
          8. Leaves not lobed
            9. Ovary inferior; stamens 5 *Lonicera* in CAPRIFOLIACEAE
            9. Ovary superior, or inferior with usually 8 stamens

10. Ovary sometimes appearing inferior; perianth 4 lobed; stamens usually 8 ELAEAGNACEAE

10. Ovary superior; perianth lobes and stamens mostly 5 *Atriplex* in CHENOPODIACEAE

7. Leaves and branches alternate

11. Leaves pinnately lobed; plants monoecious, staminate flowers in catkins, pistillate solitary or few in a cluster; fruit an acorn FAGACEAE

11. Leaves not pinnately lobed; flowers various; fruit not an acorn

12. Flowers in catkins or hard cone-like structures

13. Branches spine-tipped; leaves less than 4mm wide, fleshy *Sarcobatus* in CHENOPODIACEAE

13. Branches not spine-tipped; leaves mostly over 4mm wide, not fleshy

14. Ovary becoming a many-seeded capsule; seeds bearing long hairs SALICACEAE

14. Ovary becoming a 1 seeded nut, nutlet, or samara; seeds without hairs BETULACEAE

12. Flowers not in catkins or hard cone-like structures

15. Leaf blades, or some of them, very asymmetrical at base, 1 side extending lower than the other, toothed; fruit a drupe or a samara which is winged all around ULMACEAE

15. Leaf blades symmetrical at base or nearly so, toothed or not; fruit various

16. Fruit a nut enclosed by a beaked involucre; leaves cordate or subcordate at base, serrate; Black Hills <u>Corylus</u> in BETULACEAE

16. Fruit and leaves not as above; widespread

17. Sepals and stamens 4; ovary sometimes appearing inferior; leaves silvery-scurfy at least beneath; flowers mostly bisexual ELAEAGNACEAE

17. Sepals and stamens only rarely 4; ovary superior; leaves various; flowers sometimes unisexual

18. Fruit a samara which is winged all around; some leaves doubly toothed, sometimes obscurely so <u>Ulmus</u> in ULMACEAE

18. Fruit not a samara; leaves not doubly toothed

19. Stamens many; style in fruit elongate, twisted, and plumose; leaves either toothed at least toward tip or entire with rolled margins <u>Cercocarpus</u> in ROSACEAE

19. Stamens usually about 5 or fewer; style not as above; leaves various

20. Leaves toothed, green <u>Rhamnus</u> in RHAMNACEAE

20. Leaves entire, often silvery or gray

21. Flowers in involucres; sepals petal-like; leaves pubescent, the hairs not stellate
Eriogonum in POLYGONACEAE

21. Flowers not in involucres (pistillate sometimes each subtended by 2 at least partly united bracts); sepals not petal-like or lacking; leaves usually scurfy or glabrous or else pubescent often with some stellate hairs CHENOPODIACEAE

3. Plants herbs, or semi-shrubs woody only at base, or vines

22. Plants woody vines with alternate, either palmately compound or cordate simple leaves and tendrils VITACEAE

22. Plants not as above (herbaceous vines run here)

23. Plants succulent annuals of alkaline areas with jointed stems and opposite or whorled branches; leaves scale-like, opposite, and connate; flowers sunken in depressions of the spikes Salicornia in CHENOPODIACEAE

23. Plants not as above

24. Plants usually with milky juice; inflorescence appearing like a flower, consisting of a cup-shaped involucre with usually 4 glands on its margin alternating with 4 teeth or lobes, each involucre subtending a stalked pistil and several to many stamens; involucres solitary, or clustered and axillary, or in cymes EUPHORBIACEAE

24. Plants not as above

25. Plants growing in water or mud; leaves all whorled or opposite, linear or narrowly lanceolate or some terminal ones club-shaped or obovate; flowers and fruits sessile or nearly so in leaf axils; stamens and styles 1 or 2

26. Leaves opposite, the plants prostrate CALLITRICHACEAE

26. Leaves whorled, the plants usually erect HIPPURIDACEAE

25. Plants without the above combination of characters

27. Plants with spiny ovaries and fruits; leaves spiny and thistle-like (except a rare alpine species); sepals deciduous, the petals usually 4 or 6 and 1-5cm long, white or yellow PAPAVERACEAE

27. Plants not as above

28. Plants dioecious annuals, prominently stellate-pubescent, with simple, entire, alternate leaves; sepals 5; stamens 8-12 *Croton* in EUPHORBIACEAE

28. Plants not as above

29. Pistils 2 to many or rarely solitary; stamens usually more than 10 (rarely deciduous) RANUNCULACEAE

29. Pistils solitary (carpels rarely partly distinct); stamens 10 or fewer

30. Plants dioecious, either vines with opposite leaves which are 3-5 lobed or annual herbs with palmately compound leaves CANNABACEAE

30. Plants not as above

31. Plants annual, monoecious vines with tendrils and alternate, simple, lobed leaves; most flowers staminate with 3 somewhat connate stamens; ovary inferior, weakly spiny CUCURBITACEAE

31. Plants not as above

32.(Next lead at left margin)

32. Ovary inferior, at least partly so; flowers mostly bisexual (rarely all pistillate)

33. Leaves all basal or nearly so, the blades ovate, cordate, or reniform SAXIFRAGACEAE

33. Leaves not as above

34. Stamens 10; styles 2 <u>Saxifraga</u> in SAXIFRAGACEAE

34. Stamens 3-5 (rarely lacking); styles solitary or occasionally 2

35. Leaves whorled at least in part RUBIACEAE

35. Leaves opposite or alternate (rarely opposite and deeply irregularly divided and superficially appearing whorled)

36. Leaves alternate SANTALACEAE

36. Leaves opposite

37. Flower clusters subtended by 4-6 separate or united involucral bracts; rhizomes lacking; perianth usually 5mm or more long; stamens usually 3-5 NYCTAGINACEAE

37. Flower clusters usually subtended by no more than 2 bracts; rhizomes present, or if lacking, the perianth 4mm or less long; stamens 3 or lacking VALERIANACEAE

32. Ovary superior; flowers sometimes unisexual

38. Middle and lower leaves opposite or whorled

39. Leaves whorled at least in part; flowers all staminate, the perianth 3-5 lobed RUBIACEAE

39. Leaves usually not whorled (rarely opposite and deeply irregularly divided and superficially appearing whorled); flowers various

40. Leaf margins conspicuously toothed; plants with sharp stinging hairs _Urtica_ in URTICACEAE

40. Leaf margins mostly entire or wavy, rarely lobed or divided; plants lacking stinging hairs (leaf tips sometimes pungent)

41. Plants either alpine annuals to 4cm high or desert annuals with flowers subtended by often spiny involucres; calyx 3 or 6 parted POLYGONACEAE

41. Plants not as above

42. Plants perennial with gray or silvery leaves from minute scales, the flowers unisexual _Atriplex_ in CHENOPODIACEAE

42. Plants not as above

43. Sepals separate at least to near base, mostly green or at least herbaceous, sometimes spinulose-tipped; flowers bisexual CARYOPHYLLACEAE

43. Sepals united (or lacking), green or not, not spinulose-tipped; flowers sometimes unisexual

44. Plants usually with rhizomes; flowers axillary *Glaux* in PRIMULACEAE

44. Plants without rhizomes; flowers mostly terminal

45. Flowers about 4mm or less long, each subtended by 3 scarious bracts and covered with dense woolly hairs; plants annual *Froelichia* in AMARANTHACEAE

45. Flowers not as above; plants annual or perennial

46. Flowers all staminate VALERIANACEAE

46. Flowers not all staminate

47. Styles solitary; perianth 5mm or more long NYCTAGINACEAE

47. Styles usually 2 or 3; perianth less than 3mm long (pistillate flower often subtended by 2 at least partly connate bracts) *Atriplex* in CHENOPODIACEAE

38. Middle and lower leaves alternate or leaves all basal or nearly so

48. Plants decumbent to ascending annuals with 2 sepals, 1 stamen, and 1 entire style *Calyptridium* in PORTULACACEAE

48. Plants not as above (flowers sometimes all pistillate and subtended by 2 partly or wholly connate bracts)

49. Stamens 2 with usually purple filaments mostly 5mm or more long; calyx 2 lobed, 1 lobe sometimes 2 toothed or again lobed *Besseya* in SCROPHULARIACEAE

49. Stamens and calyx not as above

50. Flowers perigynous; ovary and fruit 2 lobed at tip; ovules and seeds several per ovary SAXIFRAGACEAE

50. Flowers not perigynous (sometimes unisexual); ovary and fruit usually not lobed (sometimes slightly notched); ovules and seeds sometimes 1 or 2 per ovary

51. Plants with irregular flowers and much dissected leaves; anthers 6, 3 per filament FUMARIACEAE

51. Plants not as above

52. Stipules of some leaves completely sheathing the stem at the nodes (rarely deciduous but then some nodes of inflorescence usually with sheathing stipules) POLYGONACEAE

52. Stipules usually lacking

53. Styles and stigmas solitary; calyx of 4 sepals or 4 lobed or rarely 3 or 5 but then the flowers cleistogamous and 1.5-2.5mm long

54. Flowers cleistogamous, with 3 or 5 sepals; plants perennial; ovary 1 celled CISTACEAE

54. Flowers not cleistogamous, the sepals 4; plants annual or biennial; ovary 1 or 2 celled

55. Flowers, or at least some of them, unisexual; ovary 1 celled with 1 ovule _Parietaria_ in URTICACEAE

55. Flowers bisexual; ovary 2 celled with 1 ovule per cell _Lepidium_ in CRUCIFERAE

53. Styles or stigmas 2 or more; calyx only rarely in 4's

56. Flowers subtended by small involucres of united bracts; calyx often petal-like or with its parts in 2 whorls POLYGONACEAE

56. Flowers not subtended by involucres, often subtended by separate bracts (or each flower subtended by 2 united bracts); calyx usually not as above

57. Plants taprooted perennials with 5 white or pinkish petals 3-7mm long, the sepals 2 and deciduous; leaves linear, subterete, and clustered near base of plant _Talinum_ in PORTULACACEAE

57. Plants not as above

58. Perianth and usually the bracts scarious at least on margins, often with spinulose tips; annuals; leaves not gray-mealy nor succulent, some usually over 5mm wide AMARANTHACEAE

58. Perianth (sometimes lacking) and bracts often herbaceous, usually without spinulose tips; annuals or perennials; leaves often gray-mealy or succulent or all narrower than 5mm CHENOPODIACEAE

SERIES C. Dicots with calyx present and petals separate to base.

1. Plants woody trees, shrubs, or vines, woody throughout
  2. Leaves compound
    3. Leaves opposite
      4. Plants vines (rarely shrubby and less than 3dm high) — *Clematis* in RANUNCULACEAE
      4. Plants trees or shrubs over 5dm high — ACERACEAE
    3. Leaves alternate (ignore tendrils)
      5. Plants vines with tendrils — VITACEAE
      5. Plants not vines with tendrils
        6. Leaflets with spine-tipped teeth, the spines mostly over 1mm long — BERBERIDACEAE
        6. Leaflets without spine-tipped teeth
          7. Stamens more than 10 — ROSACEAE
          7. Stamens 10 or less
            8. Petals solitary, blue or purple — *Amorpha* in LEGUMINOSAE
            8. Petals usually 5, not blue or purple — ANACARDIACEAE
  2. Leaves simple
    9. Plants shrubby with scale-like leaves mostly about 1mm long, appearing like a juniper — TAMARICACEAE
    9. Plants not as above
      10. Plants vines with alternate, cordate leaves and tendrils — VITACEAE

10. Plants not as above

11. Leaves opposite or subopposite

12. Leaves 3-5 lobed ACERACEAE

12. Leaves not lobed

13. Stamens 10 or more HYDRANGEACEAE

13. Stamens less than 10

14. Year-old twigs red; flowers in a terminal inflorescence CORNACEAE

14. Year-old twigs not red; flowers axillary CELASTRACEAE

11. Leaves alternate (rarely fascicled but the branches alternate)

15. Leaves entire, pale beneath with yellowish resinous dots; stamens usually 8-12, the filaments pubescent near base *Ledum* in ERICACEAE

15. Leaves not as above; stamen number and filaments various

16. Stamens more than twice as many as sepals or calyx lobes ROSACEAE

16. Stamens not more than twice as many as sepals

17. Stamens as many as petals and opposite them RHAMNACEAE

17. Stamens not the same number as the petals, or if so, then alternate with them

18. Stems with thorns, spines, or prickles

19. Leaves 3-5 palmately lobed about halfway or more to base GROSSULARIACEAE

19. Leaves not as above *Crataegus* in ROSACEAE

18. Stems without thorns, spines, or prickles

20. Leaf blades mostly 3 or 5 lobed or else cordate-orbicular; fruit a berry GROSSULARIACEAE

20. Leaf blades not lobed, usually longer than wide so not cordate-orbicular; fruit not a berry

21. Stamens 10 or more ROSACEAE

21. Stamens 4, 5, or 8, or lacking

22. Stamens 4 or 5 or lacking CELASTRACEAE

22. Stamens 8 *Calylophus* in ONAGRACEAE

1. Plants herbaceous or sometimes woody only at base

23. Stems thick, green, succulent, and spiny; leaves minute or lacking; cactuses CACTACEAE

23. Stems not as above; leaves usually well developed

24. Corolla irregular (rarely with a single white to ochroleucous or purple to blue petal)

25. Ovary inferior, sometimes only partly so

26. Petals 4; stamens 8 ONAGRACEAE

26. Petals 5; stamens 5 *Heuchera* in SAXIFRAGACEAE

25. Ovary superior

27. Stamens more than 10; pistils 3-5 RANUNCULACEAE

27. Stamens 10 or less; pistil 1

28. Plants annual, to 10cm high, with opposite simple leaves and 2 or 3 sepals, the flowers solitary and axillary and less than 2mm long ELATINACEAE

28. Plants not as above

29. Flowers solitary on each peduncle; sepals 5; stamens with broad connectives exceeding the anthers in length; leaves simple and entire or merely toothed (1 species dissected), elliptic to cordate or reniform VIOLACEAE

29. Flowers in racemes, spikes, or panicles, or if solitary, the leaves dissected or compound (rarely with linear or narrowly oblanceolate simple leaves) and the stamens not as above; sepals 2 or 5

30. Sepals 2; leaves finely dissected (rarely lacking) FUMARIACEAE

30. Sepals 5; leaves not finely dissected although sometimes compound

31. Stamens 5; leaves simple and basal *Heuchera* in SAXIFRAGACEAE

31. Stamens 9 or 10, or if 5, the leaves compound

32. Leaves compound, or if (as rarely) simple, the blades linear to narrowly oblanceolate LEGUMINOSAE

32. Leaves simple, broader than narrowly oblanceolate ERICACEAE

24. Corolla regular or nearly so

33. Sepals 3; flowers sometimes unisexual; stipules of at least some leaves completely sheathing the stem, rarely deciduous; plants perennial or rarely annual with simple, alternate leaves, never spiny all over POLYGONACEAE

33. Sepals not 3 or the plants without the other characters

34. Stamens more than 10 (rarely deciduous but then with more than 10 pistils)

35. Plants dioecious, stellate-pubescent annuals with simple, entire leaves   Croton in EUPHORBIACEAE

35. Plants not as above

36. Pistils more than 1, simple with 1 carpel

37. Flowers hypogynous or nearly so, the sepals usually separate; leaves lacking stipules

38. Pistils usually 5, each with several ovules, becoming follicles; petals reddish-purple, never spurred; NW Wyo.   PAEONIACEAE

38. Pistils 2 to many, if with more than 1 ovule, then either the petals not reddish-purple or the petals spurred; widespread   RANUNCULACEAE

37. Flowers usually obviously perigynous or at least the sepals united well up from base; leaves often with stipules   ROSACEAE

36. Pistils solitary, usually of 2 or more united carpels which may rarely separate in fruit

39. Flowers hypogynous

40. Plants with spiny ovaries and fruits; leaves spiny and thistle-like (except a rare alpine species); sepals 2 or 3; petals 4 or 6; juice often milky or yellow   PAPAVERACEAE

40. Plants not as above

41. Plants with simple basal leaves; sepals 2, or if more, then with mostly 12-18 petals — _Lewisia_ in PORTULACACEAE

41. Plants not as above

42. Stamens all united by the filaments into a tube which surrounds the style or styles — MALVACEAE

42. Stamens not all united as above

43. Leaves simple, sometimes with secondary leaves in their axils

44. Leaves alternate — CISTACEAE

44. Leaves opposite — HYPERICACEAE

43. Leaves compound

45. Leaflets toothed; plants glabrous or sparsely pubescent — _Actaea_ in RANUNCULACEAE

45. Leaflets entire; plants glandular-pubescent — CAPPARACEAE

39. Flowers epigynous or perigynous

46. Filaments united into a tube which surrounds the style or styles — MALVACEAE

46. Filaments not as above

47. Ovary superior

48. Plants with simple leaves which are entire or nearly so; pistils solitary; petals purple — LYTHRACEAE

48. Plants without the above combination of characters — ROSACEAE

47. Ovary inferior — LOASACEAE

34. Stamens 10 or less or the flowers all pistillate

49. Plants dioecious, stellate-pubescent annuals with simple, entire, alternate leaves — Croton in EUPHORBIACEAE

49. Plants not as above

50. Flowers hypogynous or perigynous, the ovary superior or rarely lacking

51. Flowers hypogynous

52. Pistils more than 5 — RANUNCULACEAE

52. Pistils 1-5 or the flowers all staminate

53. Leaves succulent; pistils (carpels) mostly 4 or 5, separate or sometimes united at base — CRASSULACEAE

53. Leaves often not succulent; pistil 1, of 2 or more carpels that are united at least halfway above the base (if free or united only at base, the carpels 2 or rarely 3), or the flowers all staminate

54. Sepals usually 2 or 3

55. Leaves pinnately compound or divided, not fleshy — LIMNANTHACEAE

55. Leaves simple and entire, often fleshy

56. Petals more than the sepals, or if the same number, the flowers in a definite inflorescence — PORTULACACEAE

56. Petals as many as sepals, the flowers usually solitary in leaf axils — ELATINACEAE

54. Sepals 4 or 5

57. Plants in swamps and bogs, the leaves all basal, the blades mostly oblong-oblanceolate and covered with long, reddish, gland-tipped hairs DROSERACEAE

57. Plants not as above

58. Plants saprophytic, red or pink to yellowish, not green, the flowers mostly 4-merous with all parts pubescent *Hypopitys* in ERICACEAE

58. Plants not as above

59. Placentation free-central (at least above) or basal, or the flowers all staminate; fruit a capsule

60. Plants with fleshy alternate leaves, the flowers staminate and less than 5mm long; sepals and petals usually 5 each CRASSULACEAE

60. Plants not as above

61. Leaves usually opposite; ovary not notched at tip, the styles 1-5 or rarely obsolete CARYOPHYLLACEAE

61. Leaves alternate or basal; ovary notched at tip, the styles usually 2 SAXIFRAGACEAE

59. Placentation not free-central or basal (except sometimes with the fruit an achene); flowers not all staminate; fruit various

62. Flowers 4-merous

63. Ovary and fruit 2 celled with a membranous partition; stamens usually 6 (rarely 2 or 4), 4 longer than the other 2, usually included; leaves not palmately compound CRUCIFERAE

63. Ovary and fruit 1 celled; stamens 6 or more, about the same length, often long-exserted; at least the lower leaves palmately compound with 3-7 leaflets CAPPARACEAE

62. Flowers 5-merous

64. Leaves simple

65. Leaves palmately lobed or divided; styles never 2 GERANIACEAE

65. Leaves mostly entire or toothed, if palmately lobed or divided, then with 2 styles

66. Styles usually 3, or if 4 or 5, then with 10 stamens CARYOPHYLLACEAE

66. Styles 1 or 2 or lacking, or if 4 or 5, then with 5 stamens

67. Stigmas 4 and mostly sessile; stamens alternating with staminodia; leaves all basal but sometimes with a bract along the scape
Parnassia in SAXIFRAGACEAE

67. Stigmas, stamens, and leaves not as above

68. Styles 2 (carpels partially separate)
Saxifraga in SAXIFRAGACEAE

68. Styles 1 or 5

69. Filaments usually united at least at base, sometimes very slightly so; leaves linear or nearly so (rarely narrowly elliptic or lanceolate)
LINACEAE

69. Filaments free; leaves not linear
ERICACEAE

64. Leaves compound

70. Leaflets 3, obcordate and entire OXALIDACEAE

70. Leaflets mostly 5 or more, not obcordate, entire or not

71. Leaves pinnately compound, the leaflets entire; fruits bearing stout spines ZYGOPHYLLACEAE

71. Leaves palmately compound, or if pinnately compound, the leaflets toothed or lobed; fruits not bearing stout spines GERANIACEAE

51. Flowers perigynous

72. Leaves pinnately compound, or ternately compound or dissected and with 5 stamens (simple in a densely matted species with 10 stamens and pink or purple petals 2-4mm long in limestone crevices of the Big Horn Mts.) ROSACEAE

72. Leaves simple, or if ternately compound, then with 10 stamens

73. Styles solitary or lacking

74. Plants scapose or sometimes with 1 to several leaves along scape; flowers solitary and terminal or many in a panicle, the petals white SAXIFRAGACEAE

74. Plants with leafy stems; flowers solitary and axillary or in a bracteate terminal spike, the petals purple LYTHRACEAE

73. Styles 2-4 (carpels sometimes partially separate) SAXIFRAGACEAE

50. Flowers epigynous, the ovary partly or completely inferior

75. Flowers 2- or 4-merous

76. Flowers not subtended by white, petaloid bracts; fruit a capsule or nutlet ONAGRACEAE

76. Flowers subtended by 4 white, petaloid bracts mostly 1-2cm long; fruit a red drupe CORNACEAE

75. Flowers 5-merous

77. Inflorescence an umbel or sometimes capitate; styles 2 or 5

78. Styles and carpels 5 ARALIACEAE

78. Styles and carpels 2 UMBELLIFERAE

77. Inflorescence not umbellate or capitate; styles 0-4

79. Plants perennial, the leaves often all or mostly basal, the hairs usually not multicellular nor pustulate at base SAXIFRAGACEAE

79. Plants annual, the leaves mostly along the stem, the hairs multicellular and some usually pustulate at base LOASACEAE

SERIES D. Dicots with calyx present and petals united at least at base.

1. Plants parasitic or saprophytic, white, yellow, brown, pink, red, or purple, not green
  2. Plants parasitic, attached to above ground parts of host plant, the stems twining ... Cuscuta in CONVOLVULACEAE
  2. Plants saprophytic, or if parasitic, attached to underground parts of host plant, the stems not twining
    3. Corolla irregular, not urn shaped, well over 8mm long ... OROBANCHACEAE
    3. Corolla regular, urn shaped, 5-8mm long ... Pterospora in ERICACEAE
1. Plants not parasitic (rarely so but definitely green) nor saprophytic, mostly green
  4. Stems thick, green, succulent, and spiny; leaves minute or lacking; cactuses ... CACTACEAE
  4. Stems not as above; leaves usually well developed
    5. Flowers in dense heads subtended by an involucre, the head usually appearing like a single flower (rarely 1 flower per head); ovary inferior; stamens united by their anthers (rarely free); flowers sometimes unisexual or neutral ... COMPOSITAE
    5. Flowers not in a head, or if so, lacking the other characters
      6. Plants woody trees, shrubs, or vines

7. Plants vines with tendrils VITACEAE

7. Plants not vines with tendrils

8. Leaves palmatifid with mostly 3-5 linear, spinulose-tipped segments, also with axillary fascicles of often simple linear leaves *Leptodactylon* in POLEMONIACEAE

8. Leaves not as above

9. Leaves compound

10. Leaves alternate *Amorpha* in LEGUMINOSAE

10. Leaves opposite *Sambucus* in CAPRIFOLIACEAE

9. Leaves simple (rarely with a pair of nearly distinct basal lobes)

11. Leaves opposite; ovary inferior; stamens 4 or 5 CAPRIFOLIACEAE

11. Leaves alternate or rarely opposite; ovary superior, or if inferior, stamens more than 5

12. Stamens 8-10 ERICACEAE

12. Stamens 5 SOLANACEAE

6. Plants herbs, rarely woody at very base, sometimes vine-like but then not woody

13. Perianth subtended by 3 scarious bracts (mistaken for sepals) and covered with dense woolly hairs; plants annual *Froelichia* in AMARANTHACEAE

13. Perianth not as above; plants annual to perennial

14. Plants with all basal simple leaves (rarely with only 2-3 opposite or whorled linear leaves) and regular flowers with 2 sepals PORTULACACEAE

14. Plants not as above

15. Plants with milky juice; ovaries and styles 2 but sharing a common stigma to which the stamens are adnate; pollen of each anther chamber coalescent in a sac-like mass, the sacs in pairs joined by a slender connective; hood-like structures borne from base of each stamen which often bear a slender horn-like appendage within ASCLEPIADACEAE

15. Plants not as above

16. Flowers all unisexual; ovary superior; stamens 10; leaves opposite and entire CARYOPHYLLACEAE

16. Flowers mostly bisexual, if unisexual, either with whorled or palmately lobed leaves or with 3 stamens and an inferior ovary

17. Stamens (or anthers) more numerous than corolla lobes (or calyx lobes if corolla lobes obscure)

18. Flowers regular or nearly so

19. Leaves compound

20. Leaflets 3, obcordate, entire OXALIDACEAE

20. Leaflets 3 or more, the shape various, toothed or lobed

21. Leaves basal except for an opposite pair on flowering stem ADOXACEAE

21. Leaves alternate MALVACEAE

19. Leaves simple

22. Stamens more than 10, all united by the filaments into a tube which surrounds the style or styles MALVACEAE

22. Stamens 10 or fewer (rarely 12), free or rarely united at very base

23. Plants annual; sepals 2; ovary half inferior <u>Portulaca</u> in PORTULACACEAE

23. Plants not as above

24. Leaves opposite; placentation free-central CARYOPHYLLACEAE

24. Leaves usually not opposite; placentation not free-central

25. Pistils (carpels) mostly 5, separate or united at base CRASSULACEAE

25. Pistils solitary with fully united carpels ERICACEAE

18. Flowers irregular

26. Stamens more than 10 RANUNCULACEAE

26. Stamens 10 or fewer

27. Anthers 10 or rarely 9 LEGUMINOSAE

27. Anthers 4-8

28. Anthers 6; locule 1; leaves dissected FUMARIACEAE

28. Anthers 4 or 8; locules 2; leaves various

29. Anthers 8; filaments united POLYGALACEAE

29. Anthers 4 (or apparently 8); filaments free SCROPHULARIACEAE

17. Stamens not more numerous than corolla lobes

30. Stamens usually as many as corolla lobes (1 rarely vestigial) and opposite them; placentation free-central or basal with 1 locule; ovary not 4 lobed; corolla regular PRIMULACEAE

30. Stamens alternate with corolla lobes (or opposite calyx lobes) or fewer; placentation various; ovary 4 lobed or not; corolla regular or irregular

31. (Next lead at left margin)

31. Ovary superior (rarely lacking)

32. Corolla of 1 basal larger petal and 4 usually smaller petals alternating with stamens at tip of filament tube *Petalostemon* in LEGUMINOSAE

32. Corolla not as above

33. Corolla regular or nearly so

34. Anther bearing stamens either as many as corolla lobes or else at least 5

35. Ovary lacking; leaves whorled at least in part RUBIACEAE

35. Ovary present; leaves various

36. Ovaries 2 but with only a single enlarged stigma; plants with milky juice; leaves opposite APOCYNACEAE

36. Ovaries solitary (sometimes deeply 4 lobed); plants without milky juice; leaves opposite or not

37. Ovary (or at least the fruits except when only 1 or 2 nutlets develop) 4 lobed or prominently 4 grooved (rarely capped by an umbrella-like stigma)

38. Leaves usually alternate, at least in part, rarely opposite; stamens 5 BORAGINACEAE

38. Leaves opposite; stamens 4 *Mentha* in LABIATAE

37. Ovary not 4 lobed or 4 grooved (sometimes 4 nerved)

39. Ovary 1 celled; placentation parietal; leaves either compound with 3 broad leaflets or else simple and opposite or whorled and entire or nearly so

40. Leaves simple and opposite or whorled

41. Plants desert annuals with mostly prostrate or decumbent stems to 10cm long, pubescent all over with usually stiff, coarse hairs HYDROPHYLLACEAE

41. Plants not as above GENTIANACEAE

40. Leaves compound with 3 leaflets MENYANTHACEAE

39. Ovary 2-10 celled, the placentation various, or if 1 celled, the leaves not as above

42. Stamens 4

43. Leaves mostly palmately divided into 3(1-5) linear, spinulose-tipped segments, the plants mostly mat-forming perennials
Leptodactylon in POLEMONIACEAE

43. Leaves not as above; plants not mat-forming
PLANTAGINACEAE

42. Stamens 5 or rarely more

44. Branches of style 3 or stigma 3 lobed, sometimes obscurely so, the locules usually 3
POLEMONIACEAE

44. Branches of style 2 or stigma 2 lobed (rarely with 2 styles), or unbranched and unlobed, the locules 1-3

45. Leaves mostly palmately divided into 3(1-5) linear, spinulose-tipped segments, the plants mostly mat-forming perennials
Leptodactylon in POLEMONIACEAE

45. Leaves not as above; plants not mat-forming perennials

46. Plants annual; leaves pinnatifid to bipinnatifid, the segments all acicular; calyx lobes acicular and somewhat unequal; corolla usually 1cm or less long
Navarretia in POLEMONIACEAE

46. Plants not as above

47. Ovary 1 celled with 2 parietal placentae which sometimes intrude and meet but do not join (rarely with 1 basal ovule); fruit a capsule

HYDROPHYLLACEAE

47. Ovary 2 celled (rarely incompletely 2) with axile placentation; fruit a capsule or berry

48. Ovules 4 or fewer; plants either with twining stems and sagittate or hastate leaf blades and the corolla 1.5-7cm long, or with 2 distinct styles which are each deeply 2 cleft, or with simple and entire leaf blades mostly 6 or more times longer than wide and the corolla 3.5-10cm long

CONVOLVULACEAE

48. Ovules usually more than 4; plants not fitting any of the above 3 categories

49. Plants coarse, taprooted biennials with woolly, stellate-pubescent leaves, not spiny; corolla yellow, less than 2cm long, the lobes over 2mm long *Verbascum* in SCROPHULARIACEAE

49. Plants without the above combination of characters SOLANACEAE

34. Anther bearing stamens 2-4 or rarely lacking, fewer than corolla lobes

50. Anther bearing stamens 4

51. Flowers either solitary in leaf axils or primarily in a basal rosette on long pedicels, or if in a terminal inflorescence, then with a bearded sterile filament SCROPHULARIACEAE

51. Flowers either densely clustered in leaf axils or in a terminal inflorescence, usually sessile or nearly so, never with a bearded sterile filament

52. Filaments about as long as anthers or shorter; style terminal on ovary or nearly so VERBENACEAE

52. Filaments much longer than anthers; style usually from near base between the 4 ovary lobes LABIATAE

50. Anther bearing stamens 2 or 3 (rarely lacking)

53. Leaves all basal; corolla scarious PLANTAGINACEAE

53. Leaves on stems; corolla not scarious

54. Plants annual vines with tendrils and alternate, simple, palmately lobed leaves CUCURBITACEAE

54. Plants not as above

55. Stamens 3; flowers all staminate VALERIANACEAE

55. Stamens 2; flowers not all staminate

56. Leaves lobed or very coarsely toothed; flowers in dense axillary clusters
Lycopus in LABIATAE

56. Leaves entire or toothed; flowers in loose terminal or axillary racemes or solitary in axils Veronica in SCROPHULARIACEAE

33. Corolla irregular

57. Anther bearing stamens 5

58. Corolla yellow Verbascum in SCROPHULARIACEAE

58. Corolla not yellow SOLANACEAE

57. Anther bearing stamens 2-4

59. Stamens 3; flowers all staminate VALERIANACEAE

59. Stamens 2 or 4; flowers not all staminate

60. Ovary 1 celled with 1 ovule, not 4 lobed; corolla 5-10mm long; leaves opposite, petioled; Black Hills
PHRYMACEAE

60. Ovary, if 1 celled, with 2 or more ovules, sometimes 4 lobed; corolla and leaves various; widespread

61. Ovules usually 1 or 2 per cell, the cells 2 or 4; leaves opposite; ovary usually 4 lobed

62. Anther bearing stamens 2 LABIATAE

62. Anther bearing stamens 4

63. Filaments about as long as anthers or shorter; calyx teeth usually 5 or fewer; corolla merely irregularly 4 or 5 lobed; style terminal on ovary or nearly so VERBENACEAE

63. Filaments obviously longer than the anthers, or if rarely not, the calyx with 10 hooked teeth; corolla often very irregular and 2 lipped; style often from near base between the 4 ovary lobes LABIATAE

61. Ovules 2 or more per cell, the cells 1 or 2; leaves opposite or alternate or all basal; ovary not 4 lobed

64. Leaves suborbicular to reniform or cordate, mostly 5cm or more long; flowers 3-5cm long, in racemes MARTYNIACEAE

64. Leaves usually not as above; flowers various SCROPHULARIACEAE

31. Ovary inferior

65. Leaves opposite or whorled

66. Leaves whorled at least in part RUBIACEAE

66. Leaves opposite

67. Ovary and fruit bearing hooked hairs RUBIACEAE

67. Ovary and fruit lacking hooked hairs

68. Stamens 3 VALERIANACEAE

68. Stamens 4 _Linnaea_ in CAPRIFOLIACEAE

65. Leaves alternate or all basal

69. Leaves compound MENYANTHACEAE

69. Leaves simple

70. Plants bearing tendrils and climbing CUCURBITACEAE

70. Plants without tendrils, not climbing CAMPANULACEAE

FAMILIES

Arranged Alphabetically Within the Major Groups of

PTERIDOPHYTES

GYMNOSPERMS

ANGIOSPERMS

A
B
C
D
E
F
G
J. Dern

Stems jointed, with longitudinal grooves and ridges, usually hollow (with a central cavity) except at the nodes; central cavity usually surrounded by vallecular cavities opposite each groove; leaves teeth-like, united below to from a sheath which surrounds the stem; regularly whorled branches present or not; sporangia borne in a terminal cylindrical strobilus; spores all similar.

Equisetum L. Horsetail; Scouring Rush

1. Strobili sharp-pointed at tip; stems evergreen, not dying back over winter, without regularly whorled branches; stems all alike
   2. Stems 3-12 ridged; central cavity less than half the diameter of stem; teeth mostly persistent *E. variegatum*
   2. Stems 14-40 ridged; central cavity over half the diameter of stem; teeth often deciduous
      3. Middle sheaths of mature stems with a medial or basal black band and an apical black band; sheaths not much longer than wide *E. hyemale*

Figure 1. Pteridophytes. A. *Equisetum arvense*, fertile stem (X 0.7) at left, vegetative stem (X 0.3) at right. B. *Isoetes*, whole plant below (X 0.7), base of leaf bearing megaspores above (X 4). C. *Botrychium lunaria*, whole plant below (X 0.6), sporangium above (X 10). D. *Marsilea vestita* (X 0.6). E. *Lycopodium annotinum*, portion of branch (X 0.7). F. *Selaginella densa* (X 1.2). G. *Cystopteris fragilis*, leaf at right (X 0.6), sorus covered by indusium at center (X 10), sporangium at left (X 80).

3. Middle sheaths of mature stems with only an apical black band; sheaths about twice as long as wide — _E. laevigatum_

1. Strobili rounded at tip; stems withering each year; some stems with regular whorls of branches; fertile stems flesh-colored or brownish in 1 species

4. Stems flesh-colored or brownish (with little or no chlorophyll), fertile — _E. arvense_

4. Stems green, fertile or sterile

5. Stems without regularly whorled branches

6. Teeth mostly deciduous, united; vallecular cavities present — _E. laevigatum_

6. Teeth persistent and separate; vallecular cavities lacking except sometimes below — _E. fluviatile_

5. Stems with regularly whorled branches

7. Central cavity about 4/5 the diameter of main stem; vallecular cavities lacking except sometimes below; NW Wyo. — _E. fluviatile_

7. Central cavity mostly less than 1/2 the diameter of main stem; vallecular cavities present; statewide — _E. arvense_

Equisetum arvense L., Sp. Pl. 1061. 1753. Stems of 2 kinds, the fertile white to brown, appearing early in spring and soon withering, without whorls of branches; strobili rounded at tip; sterile stems green, to 6dm high, with regular whorls of branches; teeth persistent; central cavity 1/4-1/2 the diameter of stem, the vallecular cavities smaller. Mostly moist areas. Statewide.

<u>Equisetum</u> <u>fluviatile</u> L., Sp. Pl. 1062. 1753. Stems all green, branched or not, to 15dm high; teeth persistent, not united; central cavity over 3/4 the diameter of stem; vallecular cavities lacking, at least above; strobili rounded at tip. Moist areas, often in water. Yellowstone and Teton Parks.

<u>Equisetum</u> <u>hyemale</u> L., Sp. Pl. 1062. 1753. Stems all green, without whorls of branches, to 15dm high, 18-46 ridged; sheaths barely longer than wide, often ashy-gray in color, usually with both apical and medial black bands; teeth often deciduous, usually united; central cavity over 3/4 the diameter of stem; vallecular cavities very small; strobili pointed at tip. Moist areas. Statewide. <u>E</u>. <u>robustum</u> A. Br., <u>E</u>. <u>prealtum</u> Raf.

<u>Equisetum</u> <u>laevigatum</u> A. Br., Am. Journ. Sci. 46:87. 1844. Stems all green, unbranched or rarely sporadically branched, to 10dm high, 14-30 ridged; sheaths about twice as long as wide, with only an apical black band, at least on the upper sheaths; teeth deciduous; central cavity 2/3-4/5 the diameter of stem; vallecular cavities very small; strobili mostly rounded at tip, rarely pointed. Moist or dry areas. Statewide. <u>E</u>. <u>kansanum</u> Schaffn.

<u>Equisetum</u> <u>variegatum</u> Schleich. ex Weber & Mohr, Bot. Taschenb. 1807: 60,447. 1807. Stems all green, lacking whorls of branches, to 5dm high, 3-12 ridged; teeth mostly persistent; central cavity 1/4-1/2 the diameter of stem; vallecular cavities about half as large as central cavity; strobili pointed at tip. Moist areas. Statewide.

ISOETACEAE Quillwort Family

Plants aquatic or amphibious; stem minute; leaves elongate and narrow, widened at base, clustered on a lobed stock, ligulate, with 4 longitudinal air cavities around a central vascular bundle, rarely with peripheral strands of supporting tissue also, the air cavities interrupted by transverse partitions; sporangium near base of sporophyll; spores of 2 kinds in separate sporangia.

Isoetes L. Quillwort

Megaspores mostly 0.5-0.8mm in diameter, the surface with jagged crests or high ridges *I. lacustris*

Megaspores mostly less than 0.5mm in diameter, the surface with tubercles or wrinkles *I. bolanderi*

Isoetes bolanderi Engelm. in Parry, Am. Nat. 8:214. 1874. Leaves mostly less than 15cm long; megaspores 0.3-0.5mm in diameter, the surface with low tubercles or wrinkles. Submerged or in mud. NW, SE.

Isoetes lacustris L., Sp. Pl. 1100. 1753. Leaves mostly less than 20cm long; megaspores 0.5-0.8mm in diameter, the surface with crests or high ridges. Submerged. Yellowstone Lake. *I. occidentalis* Henderson, *I. paupercula* (Engelm.) A. A. Eaton.

LYCOPODIACEAE Club Moss Family

Stems mostly creeping with upright branches, evergreen; leaves simple, sessile, and scale-like, without ligules; sporophylls forming terminal, cylindrical strobili, or in zones on the branch alternating with zones of sterile leaves; spores all alike.

Lycopodium L. Club Moss

Sporophylls like vegetative leaves, in zones on the branch alternating with sterile zones; creeping stem very short or none ..... *L. selago*

Sporophylls in terminal strobili, different in shape from vegetative leaves; creeping stems long ..... *L. annotinum*

Lycopodium annotinum L., Sp. Pl. 1103. 1753. Stems long-creeping; branches to 25cm high; strobili 1 per branch, sessile, 1-4cm long. Moist woods. NW.

Lycopodium selago L., Sp. Pl. 1102. 1753. Stems very short, not long-creeping; branches to 30cm high; sporophylls like vegetative leaves, in zones on the branch alternating with sterile zones. Moist rocks and woods. Teton Park.

MARSILEACEAE Pepperwort Family

Plants aquatic or amphibious; stem a creeping rhizome; leaves resembling a four-leaf clover, the leaflets on a long petiole; spores of two kinds; sori in a sporocarp which arises from the stem or is adnate to a petiole, each sorus bearing megasporangia and microsporangia.

Marsilea L. Pepperwort

Marsilea vestita Hook. & Grev., Icon. Fil. 2:pl. 159. 1830. Stems creeping; leaves with 4 leaflets at tip of petioles to 15cm long, resembling a four-leaf clover; sporocarps on short peduncles from the stem or adnate to a petiole. In water or on shores. SW, NW, NE. *M. mucronata* A. Br., *M. oligospora* Goodding.

OPHIOGLOSSACEAE Grape Fern Family

Stems underground, short, erect or ascending; roots fleshy and radially spreading from the stem; above ground portion of plant consisting of a leaf divided into a sterile foliaceous part and a fertile non-foliaceous part on a common petiole; sometimes more than 1 leaf is present; sporangia 2 valved, without an annulus; spores all alike.

<u>Botrychium</u> Sw. Grape Fern

1. Sterile part of leaf 2-4 times compound, the blade broadly triangular and mostly over 5cm long
   2. Sterile part of leaf stalked, attached to common petiole near ground level *B. multifidum*
   2. Sterile part of leaf sessile or nearly so, attached to common petiole well above ground level *B. virginianum*
1. Sterile part of leaf simple to twice compound, the blade triangular or not and mostly less than 7cm long
   3. Sterile part of leaf sessile and about as long as wide, triangular, sharply dissected, the ultimate segments mostly acute *B. lanceolatum*
   3. Sterile part of leaf stalked, or if sessile, mostly longer than wide, less sharply dissected, the ultimate segments blunt or rounded
      4. Sterile part of leaf with 3 or more pair of similar, fan-shaped segments or leaflets; sterile part of leaf usually arising from well above ground level *B. lunaria*
      4. Sterile part of leaf with 1-4 pair of mostly obovate, dissimilar segments or leaflets, occasionally twice compound; sterile part of leaf often arising from near ground level *B. simplex*

Botrychium lanceolatum (Gmel.) Angstr., Bot. Notiser 1854:68. 1854. Plants to 35cm high; sterile leaf blade deltoid, mostly once compound, 1-6cm long, 1-9cm wide, attached above middle of plant, sessile or nearly so. Moist areas in the mountains. Fremont Co.

Botrychium lunaria (L.) Sw., Schrader's Journ. Bot. 1800(2):110. 1802. Moonwort. Plants to 25cm high; sterile leaf blade ovate-oblong to oblong, simple or once compound, 1.5-10cm long, 0.7-3.5cm wide, attached at or below middle of plant, sessile or with a stalk to 8mm long. Moist areas in the mountains. NW, NE, SE. *B. minganense* Vict.

Botrychium multifidum (Gmel.) Trevisan, Atti Soc. Sci. Nat. Ital. 17:241. 1874. Plants to 40cm high; sterile leaf blades deltoid, mostly 3 times compound, 3-25cm long and about as wide, attached near ground level, stalked. Moist areas in the mountains. NW. *B. salaifolium* Presl, *B. coulteri* Underw.

Botrychium simplex E. Hitchc., Am. Journ. Sci. 6:103. 1823. Plants to 20cm high; sterile leaf blade ovate to oblong, simple to twice compound, to 6cm long and 4cm wide, stalked, often attached near ground level, especially if twice compound. Moist areas in the mountains. Yellowstone Park.

Botrychium virginianum (L.) Sw., Schrader's Journ. Bot. 1800(2):111. 1802. Plants to 50cm high; sterile leaf blades deltoid, about 3 times compound, 5-20cm long, 5-30cm wide, attached at middle of plant or above, sessile or with a stalk to 8mm long. Moist woods and thickets. Park and Teton Cos.

POLYPODIACEAE Fern Family

Stem a rhizome or caudex, often scaly or hairy; sporangia usually in groups (sori) which are on the lower surface of a vegetative-like or rarely a modified leaf, with an annulus, each sorus often covered by a thin, membranous outgrowth of the leaf (indusium); spores all alike.

1. Leaflets white-mealy beneath; leaf branching zig-zagged; S Wyo. <u>Notholaena</u>
1. Leaflets and branching not as above; more widespread
   2. Leaves with only 1-5 linear leaflets at tip; plants appearing grass-like <u>Asplenium</u>
   2. Leaves not as above; plants not appearing grass-like
      3. Leaves woolly-hairy beneath, the hairs completely obscuring the leaf surface or nearly so <u>Cheilanthes</u>
      3. Leaves glabrous or sparsely hairy beneath, the hairs, if present, not obscuring the leaf surface
         4. Sori near the leaflet margins, appearing elongate, usually at least partly covered by the rolled or reflexed leaflet margins
            5. Ultimate fertile leaflets with a mucronate tip <u>Aspidotis</u>
            5. Ultimate fertile leaflets lacking a mucronate tip
               6. Leaves of 2 kinds, the fertile usually longer than the sterile and with rolled margins <u>Cryptogramma</u>
               6. Leaves mostly similar

7. Sori discontinuous and covered by reflexed tips of leaflet lobes *Adiantum*

7. Sori appearing continuous around the leaflet margins, covered by the rolled leaflet margins

8. Leaves scattered along a horizontal rhizome; plants mostly in damp soil and over 25cm high; often in large colonies *Pteridium*

8. Leaves densely clustered; plants mainly in rock crevices and mostly less than 25cm high; not in large colonies

9. Ultimate segments of leaves mostly over 5mm long; leaves glabrous or sparsely hairy beneath *Pellaea*

9. Ultimate segments of leaves much less than 5mm long; leaves woolly-hairy beneath *Cheilanthes*

4. Sori on or along the veins between the leaflet or segment margin and its midrib, elongate or round in outline; leaflet margins usually flat

10. Sori elongate in outline

11. Leaves once compound below *Asplenium*

11. Leaves at least twice compound below *Athyrium*

10. Sori round in outline

12. Indusia lacking

13. Leaves simple or once compound below *Polypodium*

13. Leaves at least nearly twice compound below

14. Leaf blades triangular; leaves scattered, without persisting petiole bases <u>Gymnocarpium</u>

14. Leaf blades elliptic to lanceolate; leaves usually crowded, with clustered, persistent petiole bases

15. Leaf blades 2-7dm long, 3 times compound below or nearly so; persistent petiole bases usually over 3mm wide <u>Athyrium</u>

15. Leaf blades mostly less than 3.5dm long, 2 times compound below or nearly so; persistent petiole bases, if present, usually less than 2mm wide (indusia present but inconspicuous)

See separation 17 below

12. Indusia present

16. Indusia mostly under the sori and divided into hair-like segments, or else covering sori like hoods from below

17. Indusia attached at the center of sori, split into narrow, spreading segments; leaves clustered with persistent, old petiole bases; veins of lowest primary leaflet usually not prominent to margin <u>Woodsia</u>

17. Indusia attached toward the side of sori, covering sori like hoods which bend back at maturity; leaves scattered or in small clusters without persisting petiole bases; veins of lowest primary leaflet usually prominent to margin <u>Cystopteris</u>

16. Indusia mostly above the sori, round in outline or nearly so, sometimes with a cleft on one side
    18. Indusia without a cleft; leaves once or twice compound below — Polystichum
    18. Indusia with a cleft on one side; leaves at least nearly twice compound below — Dryopteris

Adiantum L. Maidenhair

Adiantum pedatum L., Sp. Pl. 1095. 1753. Plants to 7dm high; petioles reddish-brown or purplish, 1-5dm long, few together or solitary; leaf blades reniform-orbicular, deciduous, glabrous, 2-4 times compound; sori elongate, covered by reflexed tips of leaflet lobes. Moist woods and rocks. Teton Park.

Aspidotis (Nutt. ex Baker) Copel. Indian's Dream

Aspidotis densa (Brack.) Lellinger, Am. Fern Journ. 58:141. 1968. Plants to 25cm high, evergreen; petioles reddish-brown, to 20cm long, clustered with a few old petiole bases; sterile leaves rare or lacking; fertile leaf blades ovate to deltoid, glabrous, 3 times compound; leaflets mucronate-tipped; sori appearing elongate, partly or wholly covered by rolled leaflet margins. Rocks and slopes in the mountains. Teton Park. Cryptogramma densa (Brack.) Diels, Cheilanthes siliquosa Maxon, Pellaea densa (Brack.) Hook.

Asplenium L. Spleenwort

Leaves with only 1-5 linear leaflets at tip *A. septentrionale*
Leaves with more than 5 non-linear leaflets
Rachis reddish-brown *A. trichomanes*
Rachis green or yellowish *A. viride*

Asplenium septentrionale (L.) Hoffm., Deutschl. Fl. (Crypt.) 12. 1796. Grass Fern. Plants to 20cm high; petioles reddish-brown below, green above, to 18cm long, densely clustered; leaves evergreen with 1-5 linear leaflets at tip; sori elongate. Rocks. SE and reported from Teton Co.

Asplenium trichomanes L., Sp. Pl. 1080. 1753. Plants to 25cm high; petioles reddish-brown, to 10cm long, clustered; rachis reddish-brown; leaf blades elongate, evergreen, glabrous, once compound; sori elongate; indusia attached at side of sori. Rocks. Albany and Natrona Cos.

Asplenium viride Huds., Fl. Angl. 385. 1762. Plants to 15cm high; petioles brown below, green above, to 5cm long, clustered; rachis green; leaf blades elongate, deciduous, glabrous, once compound; sori and indusia similar to above species. Rocks, especially limestone. Teton and Carbon Cos.

Athyrium Roth Lady Fern

Leaves clustered, 2-4 times compound, deciduous; sori either round and lacking indusia, or elongate and curved with indusia.

Sori round; indusia lacking *A. distentifolium*
Sori elongate and curved or horseshoe-shaped; indusia present *A. filix-femina*

Athyrium distentifolium Tausch ex Opiz, Tent. Fl. Crypt. Boem. 1:14. 1820. Plants to 10dm high; petioles yellowish, 0.5-3dm long, clustered with persistent petiole bases; leaf blades elliptic to lanceolate, deciduous, glabrous, 2-4 times compound; sori round; indusia lacking. Moist alpine or subalpine areas. Teton and Albany Cos. *A. americanum* (Butters) Maxon, *A. alpestre* (Hoppe) Rylands ex Moore.

Athyrium filix-femina (L.) Roth, Tent. Fl. Germ. 3(1):65. 1799. Plants to 20dm high; petioles yellow to brown, 0.5-3dm long, clustered with persistent petiole bases; leaf blades elliptic, deciduous, mostly glabrous, 2-3 times compound; sori lunate to horseshoe-shaped; indusia attached at side of sori. NW, NE, SE. *Asplenium filix-femina* (L.) Bernh.

Cheilanthes Sw. Lip Fern

Cheilanthes feei Moore, Ind. Fil. xxxviii. 1857. Plants to 25cm high; petioles reddish-brown, to 10cm long, clustered; leaf blades ovate-lanceolate to deltoid, evergreen, densely woolly-hairy beneath, 3 times compound; sori on leaflet margins, the margins somewhat rolled. Dry rocks. Statewide.

Cryptogramma R. Br. Rock Brake

Leaves dimorphic, the fertile longer than the sterile and with rolled or reflexed leaflet margins; sori appearing elongate, partly or wholly covered by rolled or reflexed leaflet margins.

Reference: Löve, A. et al. 1971. Arctic & Alp. Res. 3:139-165.

Leaves densely clustered, the blades firm *C. acrostichoides*
Leaves mostly scattered, the blades thin *C. stelleri*

Cryptogramma acrostichoides R. Br. in Richards. in Frankl., Narr. 1st Journ. 754, 767. 1823. Plants to 35cm high; petioles green to yellowish, to 20cm long, densely clustered with persistent, old petiole bases; fertile leaves deciduous; sterile leaves evergreen, the blades ovate to ovate-lanceolate, glabrous, mostly twice compound. Rock crevices in the mountains. NW, SE. *C. crispa* (L.) R. Br. ex Hook. of authors.

Cryptogramma stelleri (Gmel.) Prantl, Bot. Jahrb. 3:413. 1882. Plants to 20cm high, not evergreen; petioles reddish-brown at least below, often greenish above, to 15cm long, not clustered; sterile leaf blades elliptic to deltoid, glabrous, once or twice compound. Rocks, usually wet limestone. Fremont and Teton Cos.

Cystopteris Bernh.

Cystopteris fragilis (L.) Bernh., Schrader's Neues Journ. Bot. 1(2): 26, 27. 1805. Brittle Fern. Plants to 35cm high; petioles green to brown, to 20cm long, loosely clustered, without persistent petiole bases; leaf blades elliptic to lanceolate, deciduous, glabrous, mostly twice compound; sori round; indusia attached below and toward side of sori, bending back like hoods at maturity, frequently inconspicuous. Rocks and woods. Statewide. *Filix fragilis* (L.) Underw.

Dryopteris Adans. Shield Fern

Leaves clustered, semi-evergreen; sori round; indusia attached by a notch or cleft at one side.

Reference: Britton, D. M. 1972. Canad. Field-Nat. 86:241-247.

Leaf blades somewhat triangular, usually widest near base, 3 times compound below or nearly so **D. assimilis**

Leaf blades elliptic, widest near middle, twice compound below or nearly so **D. filix-mas**

Dryopteris assimilis Walker, Am. Journ. Bot. 48:607. 1961. Plants to 15dm high; petioles green to brown, scaly, 15-45cm long, clustered; leaf blades deltoid to ovate, glabrous, mostly 3 times compound. Moist woods. NW. D. austriaca (Jacq.) Woynar ex Schinz & Thell. of authors.

Dryopteris filix-mas (L.) Schott, Gen. Fil. sub Polystichum. 1834. Male Fern. Plants to 10dm high; petioles green to brown, scaly, 2-20cm long, clustered; leaf blades mostly elliptic, glabrous, scaly, mostly twice compound. Moist, usually rocky and wooded areas. NE, SE, NW.

Gymnocarpium Newm. Oak Fern

Gymnocarpium dryopteris (L.) Newm., Phytologist 4:371. 1851. Plants to 50cm high; petioles greenish, 5-35cm long, not clustered; leaf blades deltoid, deciduous, glabrous or sparsely hairy, 3 times compound; sori round; indusia lacking. Moist, usually wooded areas in the mountains. Teton and Carbon Cos. Phegopteris dryopteris (L.) Fee, Dryopteris disjuncta (Rupr.) Morton, Carpogymnia dryopteris (L.) Löve & Löve.

<u>Notholaena</u> R. Br.

Reference: Tryon, R. 1956. Contr. Gray Herb. 179:1-106.

<u>Notholaena</u> <u>fendleri</u> Kunze, Farrnkr. 2:87. 1851. Zig-zag Cloak Fern. Plants to 40cm high; petioles reddish-brown, to 25cm long, clustered; leaf blades deltoid-ovate, evergreen, much branched in a zig-zag manner; leaflets white-mealy beneath; sori submarginal. Dry rocks. Laramie Co.

<u>Pellaea</u> Link Cliff Brake

Leaves clustered; sori marginal and appearing elongate, partly or wholly covered by rolled leaflet margins.

Persistent petiole bases usually more numerous than leafed petioles; lower part of petioles with a series of horizontal grooves; middle and lower leaflets usually 2 lobed or segmented and sessile <u>P</u>. <u>breweri</u>

Persistent petiole bases mostly fewer than leafed petioles; lower part of petioles usually lacking horizontal grooves; middle and lower primary leaflets often compound with short petiolules

Spores 64 per sporangium; lowest primary leaflet mostly divided into 3 or fewer secondary leaflets on a very short petiolule *P. occidentalis*

Spores 32 per sporangium; lowest primary leaflet mostly divided into more than 3 secondary leaflets and/or with a prominent petiolule *P. glabella*

Pellaea breweri D. C. Eaton in Gray, Proc. Am. Acad. 6:555. 1865. Plants to 20cm high; petioles reddish-brown, 0.5-9cm long, densely clustered with persistent petiole bases, lower part of petiole with a series of horizontal grooves; leaf blades oblong to oblong-lanceolate, deciduous, glabrous, mostly once compound; sori marginal and appearing elongate, partly or wholly covered by rolled leaflet margins. Rocks, usually limestone. NW, SW, SE.

Pellaea glabella Mett. ex Kuhn, Linnaea 36:87. 1869. Plants to 20cm high; petioles reddish-brown, 1-8cm long, clustered with few persistent petiole bases, horizontal grooves on lower part of petiole usually lacking; leaf blades oblong to elliptic, evergreen, glabrous, once to twice compound; sori similar to above species. Rocks. Teton and Crook Cos. *P. suksdorfiana* Butters.

Pellaea occidentalis (E. Nels.) Rydb., Mem. N. Y. Bot. Gard. 1:466. 1900. Similar to *P. glabella* but averaging smaller with the lowest primary leaflet mostly divided into 3 or fewer secondary leaflets on a very short petiolule and with 64 spores per sporangium rather than 32. Rocks, usually limestone. NW, NE, SE.

Polypodium L.

Polypodium vulgare L., Sp. Pl. 1085. 1753. Plants to 25cm high; petioles green or yellow, 1-12cm long, not clustered; leaf blades oblong to lanceolate, evergreen, glabrous, simple or once compound; sori round; indusia lacking. Rocks. Albany Co. *P. hesperium* Maxon.

Polystichum Roth

Leaves clustered, evergreen, the leaflets somewhat spiny on margins; sori round; indusia attached above and at center of sori.

Leaves once compound below	***P. lonchitis***
Leaves twice compound below or nearly so	***P. scopulinum***

__Polystichum__ __lonchitis__ (L.) Roth, Tent. Fl. Germ. 3(1):71. 1799. Holly Fern. Plants to 6dm high; petioles green to brown, scaly, 1-8cm long, clustered; leaf blades linear to narrowly elliptic, once compound, the leaflets with spines about 1mm long on their margins. Moist, usually shaded areas in the mountains. NW, SE.

__Polystichum__ __scopulinum__ (D. C. Eaton) Maxon, Fern Bull. 8:29. 1900. Rock Shield Fern. Plants to 4dm high; petioles yellow or green above, reddish-brown at base, 2-20cm long, scaly, clustered; leaf blades oblong to narrowly elliptic, twice compound or nearly so, the leaflets with short spines on the margins. Rocks. Yellowstone Park. The type locality is Upper Teton Canyon, Teton Co.

__Pteridium__ Scop. Bracken Fern

__Pteridium__ __aquilinum__ (L.) Kuhn in Decken, Reis. Ost-Afr. Bot. 3(3):11. 1879. Plants to 20dm high; rhizomes hairy, not scaly; petioles green or yellow, 4-50cm long, not clustered; leaf blades deltoid, deciduous, hairy or not, 2-3 times compound; sori marginal and elongate, often inconspicuous, sometimes covered by rolled leaflet margins. Moist to dry soil or talus. NW, NE, SE.

<u>Woodsia</u> R. Br.

Leaves clustered with usually persistent, old petiole bases, deciduous; sori round; indusia attached beneath sori, at maturity splitting into narrow segments and spreading in all directions, often inconspicuous.

Leaf blades and petioles with white hairs and glandular ... *W. scopulina*

Leaf blades and petioles glabrous and somewhat glandular ... *W. oregana*

Woodsia oregana D. C. Eaton, Can. Nat. & Geol. n. s. 2:89. 1865. Plants to 30cm high; petioles usually reddish-brown below and yellowish above, 1-10cm long, clustered; leaf blades linear to narrowly elliptic, glabrous, somewhat glandular, mostly twice compound. Rocks and moist areas. Statewide.

Woodsia scopulina D. C. Eaton, Can. Nat. & Geol. n. s. 2:91. 1865. Plants differing from above species in being slightly larger (to 40cm high) with hairy petioles and leaf blades. Rocks, especially talus. Statewide.

Stems creeping or climbing, branched, evergreen; leaves simple, sessile, and scale-like, ligulate at least when developing; spores of 2 kinds, usually 1-4 megaspores and numerous microspores in separate sporangia; sporangia in terminal strobili which usually are 4 angled.

<u>Selaginella</u> Beauv. Spike Moss

1. Strobili not 4 angled; leaves without a dorsal groove, bristly-margined, not bristle-tipped — <u>S. selaginoides</u>
1. Strobili 4 angled; leaves with a dorsal groove, ciliate-margined and bristle-tipped (bristle tip lacking in <u>S. mutica</u>)
   2. Leaves not bristle-tipped — <u>S. mutica</u>
   2. Leaves bristle-tipped
      3. Vegetative leaves on lower or convex side of branch longer than the others at the same level; widespread — <u>S. densa</u>
      3. Vegetative leaves about equal in length at the same level on the branch; Black Hills and SE Wyo.
         4. Broadest sporophylls about 4 times as broad as leaves; Black Hills area — <u>S. rupestris</u>
         4. Broadest sporophylls about 2 times as broad as leaves; SE Wyo. — <u>S. underwoodii</u>

Selaginella densa Rydb., Mem. N. Y. Bot. Gard. 1:7. 1900. Stems creeping; branches compact, curved-ascending, often forming cushion mats; leaves to 3mm long; terminal bristles 0.3-2mm long; strobili 4 angled; megasporangia usually below microsporangia in strobilus, rarely 1 or 2 megaspores rather than 4 per sporangium. Rocky areas, often amongst sagebrush. Statewide. *S. scopulorum* Maxon, *S. standleyi* Maxon.

Selaginella mutica D. C. Eaton ex Underw., Bull. Torrey Club 25:128. 1898. Stems creeping; branching distant; leaves 1-2mm long, terminal bristles lacking; strobili 4 angled; megasporangia usually below microsporangia in strobilus. Rocky areas. Carbon Co.

Selaginella rupestris (L.) Spring in Mart., Fl. Bras. 1(2):118. 1840. Stems creeping; branching more distant than in *S. densa*; leaves to 2.5mm long, only 1/4 as wide as sporophylls; terminal bristles 0.5-1.5mm long, conspicuously ciliate; strobili 4 angled; microspores usually lacking, megaspores usually 1 or 2 per sporangium. Rocky and sandy areas. Crook Co.

Selaginella selaginoides (L.) Link, Fil. Sp. Hort. Berol. 158. 1841. Stems creeping; branching distant; leaves 0.5-4mm long, without terminal bristles; strobili not 4 angled; megasporangia below microsporangia in strobilus. Moist, usually mossy areas. Sublette and Teton Cos.

Selaginella underwoodii Hieron. in Engl. & Prantl, Nat. Pflanz. 1(4):714. 1900. Stems creeping; branching distant; leaves 1.5-2.5mm long, terminal bristles 0.2-1mm long; strobili 4 angled, somewhat inconspicuous; megasporangia usually in 2 vertical rows beside 2 vertical rows of microsporangia. Moist, rocky areas, usually with moss. Albany and Laramie Cos.

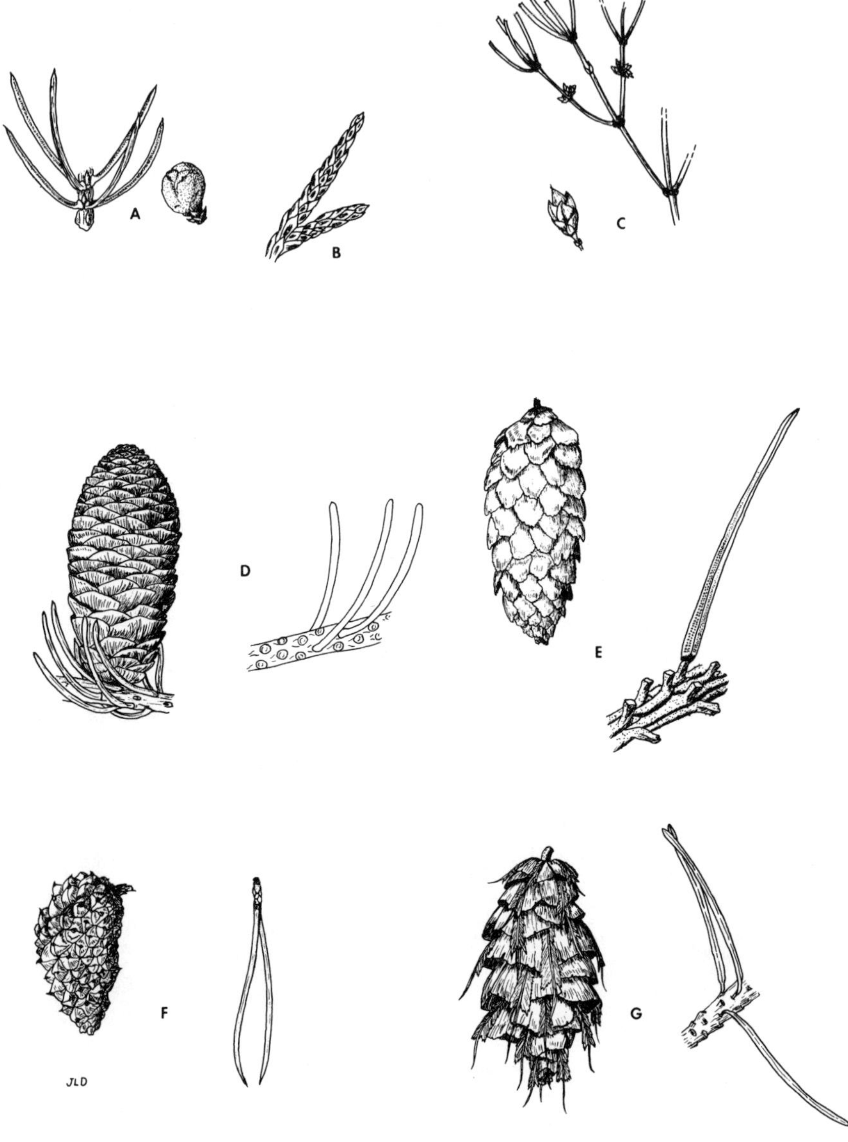
A
B
C
D
E
F
G
JLD

Trees or shrubs with scale-like, opposite, overlapping leaves, or whorled, needle-like leaves; plants monoecious or dioecious; staminate cones ovoid or globose with 12-16 stamens; ovulate cones globose and berry-like by coalescence of the fleshy scales, usually blue or brownish.

Juniperus L. Juniper

Reference: Vasek, F. C. 1966. Brittonia 18:350-372.

1. Leaves in whorls of 3, needle-like, whitish on upper side; plants shrubby J. communis
1. Leaves mostly opposite and scale-like, usually not whitened above; plants shrubby or tree-like
 2. Plants shrubby, rarely over 3dm high; leaves strongly apiculate J. horizontalis
 2. Plants usually tree-like, mostly over 3dm high; leaves usually not apiculate

---

Figure 2. Gymnosperms. A. Juniperus communis, portion of stem with leaves at left (X 2), berry-like cone at right (X 1.3). B. Juniperus scopulorum, tip of branch with scale-like leaves (X 2). C. Ephedra viridis, cone at left (X 1.3), portion of plant at right (X 0.3). D. Abies lasiocarpa, cone at left (X 0.7), portion of branch with leaves at right (X 1). E. Picea engelmannii, cone at left (X 1), portion of branch with leaf at right (X 2). F. Pinus contorta, cone at left (X 0.7), fascicled leaves at right (X 0.7). G. Pseudotsuga menziesii, cone at left (X 0.7), portion of branch with leaves at right (X 1).

3. Leaves mostly longer than wide, their margins entire *J. scopulorum*

3. Leaves mostly about as long as wide, their margins minutely ciliate or denticulate *J. osteosperma*

Juniperus communis L., Sp. Pl. 1040. 1753. Common Juniper. Shrub to 3m high; leaves needle-like, 5-15mm long, whorled, the upper side whitened; ovulate cones berry-like, mostly 6-9mm long, 1-3 seeded. Rocky hills, mountain slopes, and woods. Statewide.

Juniperus horizontalis Moench, Meth. 699. 1794. Creeping Juniper. Shrub to 3dm high; leaves scale-like, 0.5-5mm long, opposite, the tip apiculate, margins entire; ovulate cones berry-like, mostly 5-8mm long, 1-6 seeded. Foothill slopes to plains. NW, NE, SE. *J. sabina* L. of authors.

Juniperus osteosperma (Torr.) Little, Leafl. West. Bot. 5:125. 1948. Utah Juniper. Tree to 7m high; leaves scale-like, 1-3mm long, mostly opposite, acute or acuminate, margins minutely ciliate or denticulate; ovulate cones berry-like, mostly 7-10mm long, 1-2 seeded. Desert hills and canyons. NW, SW, SE. *J. utahensis* (Engelm.) Lem., *J. knightii* A. Nels.

Juniperus scopulorum Sarg., Silva N. Am. 14:93. 1902. Rocky Mountain Juniper. Tree to 10m high, rarely shrubby; leaves scale-like, 1-7mm long, mostly opposite and acute, margins entire; ovulate cones berry-like, mostly 5-6mm long, 1-4 seeded. Canyons, hills, and river bottoms. Statewide.

EPHEDRACEAE Jointfir Family

Shrubs with jointed, opposite branches, the younger branches green, resembling a horsetail or scouring rush; leaves scale-like, 2 per node, usually united below; plants dioecious, the cones axillary.

Ephedra L. Jointfir

Reference: Cutler, H. C. 1939. Ann. Mo. Bot. Gard. 26:373-428.

Ephedra viridis Coville, Contr. U. S. Nat. Herb. 4:220. 1893. Shrub to 1m high; leaves 2 per node, 1.5-9mm long, connate at least below; staminate cones 5-7mm long, anthers 5-8, sessile; pistillate cones 6-10mm long; seeds 5-8mm long. Dry hills. Sweetwater Co.

PINACEAE Pine Family

Trees with needle-like or linear evergreen leaves which are spirally arranged on the stems either singly or in fascicles of 2-5; plants monoecious; ovulate cones most conspicuous, with woody or subwoody scales on each of which are borne 2 ovules or seeds.

1. Leaves borne in clusters of 2-5 *Pinus*
1. Leaves borne singly
   2. Leaves sharp-pointed; twigs rough with short pegs where the leaves have fallen off *Picea*
   2. Leaves blunt at tip; twigs relatively smooth where the leaves have fallen off
      3. Leaves narrowed to a very short petiole at base; cones hanging downward, with 3 lobed bracts protruding from between the scales *Pseudotsuga*
      3. Leaves about the same width all the way to base; cones erect, without protruding bracts, falling apart at maturity *Abies*

Abies Mill. Fir

Leaves borne singly, blunt at tip, without petioles or nearly so, the twigs with smooth round scars where the leaves have fallen off; ovulate cones borne upright on tree, the scales thin and falling individually.

Resin ducts in cross-section of leaf at lateral margins just beneath epidermis; some leaves usually over 3cm long *A. concolor*

Resin ducts in cross-section of leaf usually between midvein and lateral margins, definitely away from epidermis; leaves rarely over 3cm long *A. lasiocarpa*

Abies concolor (Gord.) Lindl. ex Hildebr., Verbr. Conif. 261. 1861. White Fir. Tree to 40m high; leaves linear, 1.5-7cm long; ovulate cones yellow or brown to greenish-purple, mostly 7-10cm long. Foothills and lower mountains. Reported from SW Wyo. (Cary, N. Am. Fauna No. 42. 1917).

Abies lasiocarpa (Hook.) Nutt., N. Am. Sylva 3:138. 1849. Subalpine Fir. Tree to 30m high; leaves linear, 1-3cm long; ovulate cones purplish, 6-10cm long. Subalpine slopes and moist forests. NW, SW, SE.

Picea A. Dietr. Spruce

Leaves borne singly, sharp-pointed and mostly 4 angled, the twigs rough with short pegs where the leaves have fallen off; ovulate cones pendant, falling intact, the scales thin.

Cone scales entire on the ends, broadest near tip; leaves mostly less than 2.5cm long; mostly in the Black Hills *P. glauca*

Cone scales erose on the ends, broadest toward the middle or base; leaves often over 2.5cm long; widespread

Younger twigs usually finely pubescent; cones mostly less than 6cm long *P. engelmannii*

Younger twigs glabrous; cones mostly over 6cm long *P. pungens*

Picea engelmannii Parry ex Engelm., Trans. Acad. Sci. St. Louis 2:212. 1863. Engelmann Spruce. Tree to 40m high; leaves needle-like, 1.5-3cm long; twigs usually pubescent when young; ovulate cones mostly 4-5cm long, rarely to 6cm or slightly more; cone scales widest below middle, erose at tip. Subalpine slopes and moist forests. Statewide.

Picea glauca (Moench) Voss, Mitt. Deutsch. Dendrol. Ges. 16:93. 1907. White Spruce. Tree to 25m high; leaves needle-like, 1.2-2.5cm long; twigs glabrous; ovulate cones mostly 2.5-3.5cm long, rarely to 6cm long; cone scales widest near tip, entire. Swamps and forests. NE, NW.

Picea pungens Engelm., Gard. Chron. n. s. 11:334. 1879. Blue Spruce. Tree to 40m high; leaves needle-like, 1.2-3.5cm long; twigs glabrous; ovulate cones mostly 5.5-10cm long; cone scales widest at or below middle, erose at tip. Swamps and forests. NW, SW, SE.

Pinus L. Pine

Leaves borne in clusters of 2-5, pointed at tip; ovulate cones with mostly thick woody scales.

1. Leaves mostly 5 in a cluster
  2. Cones mostly over 8cm long, falling intact from the tree; cone scales thinning toward tip; lower elevations to timberline *P. flexilis*
  2. Cones mostly less than 8cm long, seldom falling intact from the tree; cone scales thickened toward tip; mostly near timberline in NW Wyo. *P. albicaulis*
1. Leaves mostly 2 or 3 in a cluster
  3. Leaves mostly over 7cm long, in clusters of 2 and 3; cones mostly over 6cm long *P. ponderosa*
  3. Leaves mostly less than 6cm long, mostly in clusters of 2; cones mostly less than 6cm long
    4. Cone scales prickly near tip; seeds winged; widespread in montane forests *P. contorta*
    4. Cone scales not prickly near tip; seeds not winged; in juniper woodlands of SW Wyo. *P. edulis*

Pinus albicaulis Engelm., Trans. Acad. Sci. St. Louis 2:209. 1863. Whitebark Pine. Tree to 20m high; leaves needle-like, 5 per fascicle, mostly 4-7cm long; ovulate cones mostly 3.5-8cm long, rarely falling from tree intact, the scales thickened toward tip. Mostly subalpine. NW.

Pinus contorta Dougl. ex Loud., Arbor. Fruticet. Brit. 4:2292. 1838. Lodgepole Pine. Tree to 30m high; leaves needle-like, 2 per fascicle, mostly 3-6cm long; ovulate cones mostly 3-6cm long, tending to remain on tree for several years; seeds winged, 2-5mm long. Drier or well drained forests. Statewide. *P. murrayana* Balf.

Pinus edulis Engelm. in Wisliz., Mem. North. Mex. 88. 1848. Pinyon Pine. Tree to 6m high; leaves needle-like, mostly 2 per fascicle, rarely 3, mostly 3-5cm long; ovulate cones 2-5cm long; seeds large and wingless. Dry juniper forests. Sweetwater Co.

Pinus flexilis James, Rep. Long's Exp. Rocky Mts. 2:27, 34. 1823. Limber Pine. Tree to 15m high; leaves needle-like, 5 per fascicle, mostly 4-7cm long; ovulate cones mostly 8-15cm long, rarely only 5cm, falling from the tree entire, the scales thinning toward tip. Wind swept foothills to subalpine. Statewide.

Pinus ponderosa Laws. & Laws., Agr. Man. 354. 1836. Ponderosa Pine. Tree to 60m high; leaves needle-like, 2 or 3 per fascicle, rarely more, mostly 7-25cm long; ovulate cones mostly 6-14cm long. Foothills and lower forests. NE, SE. *P. scopulorum* (Engelm.) Lemmon.

Pseudotsuga Carr. Douglas-fir

Pseudotsuga menziesii (Mirb.) Franco, Bol. Soc. Brot., ser. 2, 24:74. 1950. Tree to 60m high; leaves linear, mostly 2-3cm long, very short petioled, rounded at tip; ovulate cones 4-10cm long, pendant, with prominently exserted 3 lobed bracts. Foothills to forests well up in the mountains. NW, SW, SE. *P. mucronata* (Raf.) Sudw., *P. taxifolia* (Poir.) Britt. ex Sudw.

ACERACEAE Maple Family

Trees or shrubs with opposite leaves which are simple and palmately lobed or compound; plants polygamous, monoecious, or dioecious; flowers regular, in umbels, corymbs, or panicles; sepals 4 or 5, separate or not, sometimes deciduous; petals 4 or 5 and separate or lacking; stamens 4-10 but usually 8; ovary 1, superior; carpels 2; styles 1 or 2; locules 2; placentation axile; ovules 1 or 2 per carpel; fruit a samara.

Acer L. Maple

Reference: Desmarais, Y. 1952. Brittonia 7:347-387.

Leaves compound, the terminal leaflet stalked A. negundo
Leaves simple or rarely compound but the terminal leaflet sessile
  Sinuses of leaves narrowly acute, the margins bidentate or serrate; petals usually present A. glabrum
  Sinuses of leaves rounded and broad, the margins coarsely sinuate-toothed or lobed; petals lacking A. grandidentatum

Acer glabrum Torr., Ann. Lyc. N. Y. 2:172. 1828. Rocky Mountain Maple. Shrub or tree to 10m high; leaf blades mostly cordate-orbicular, palmately 3-5 lobed or rarely divided into 3 leaflets, 1-10cm long, serrate or bidentate; flowers axillary, corymbose; sepals 3-5mm long; petals as long as or shorter than sepals or rarely lacking. Woods and slopes. Statewide.

Acer grandidentatum Nutt. in T. & G., Fl. N. Am. 1:247. 1838. Big Tooth Maple. Shrub or tree to 8m high; leaf blades cordate-orbicular, 3-5 palmately lobed, coarsely toothed or lobed, 3-9cm long; flowers umbellate or corymbose; calyx about 5mm long; petals lacking. Slopes and canyons. Lincoln and Uinta Cos.

Acer negundo L., Sp. Pl. 1056. 1753. Boxelder. Tree, or rarely shrubby, to 20m high; leaves mostly ternately compound, the leaflets mostly lanceolate or ovate, coarsely toothed or lobed, 2-15cm long; flowers unisexual, the staminate in axillary clusters, the pistillate in axillary racemes; sepals 0.5-3mm long; petals lacking. Streambanks and canyons or disturbed areas. NE, SE.

ADOXACEAE Moschatel Family

Perennial herbs; main stem a short rhizome; leaves 3-foliolate or ternately compound, mostly basal but with a single pair of opposite leaves on flower stems; flowers bisexual, regular or nearly so, in head-like cymes of about 5; sepals usually 2 or 3; petals 4-6, united; stamens 4-6 but appearing twice as many from the filaments divided with each segment bearing half an anther, paired at corolla sinuses; ovary 1, about half inferior; carpels 3-5; style 3-5(6) cleft; locules 3-5; ovule 1 per locule, pendulous; fruit a berry with 3-5 nutlets.

Adoxa L. **Moschatel**

Adoxa moschatellina L., Sp. Pl. 367. 1753. Plants to 2dm high; leaves petioled, the blades 0.5-8cm long; lateral flowers mostly with 3 sepals and 5 corolla lobes, the terminal mostly with 2 sepals and 4 corolla lobes; corolla rotate, 5-8mm wide. Moist woods and rock crevices in the mountains. Albany Co. and Yellowstone Park.

ALISMATACEAE Water Plantain Family

Aquatic or marshland herbs with sheathing basal leaves with long petioles; inflorescence a raceme or panicle; flowers bisexual or unisexual, regular; sepals and petals 3 each, all separate, the petals early deciduous; stamens 6 to many; pistils many; ovary superior; style 1; locule 1; ovules 1 to several, usually basal; fruit an achene.

Flowers bisexual; ovaries in a circle on a flattened receptacle; leaves not sagittate Alisma
Flowers often unisexual; ovaries in a spherical head on a rounded receptacle; leaves usually sagittate Sagittaria

Alisma L. Water Plantain

References: Bjorkqvist, I. 1967. Opera Botanica 17, 128pp.
1968. Opera Botanica 19, 138pp.

Leaf blades usually less than 25mm wide; achenes with 3 ridges and 2 grooves on their edge; inflorescence barely exceeding the leaves or shorter — *A. gramineum*

Leaf blades mostly over 25mm wide; achenes often with only 2 ridges and 1 groove on their edge; inflorescence much exceeding the leaves — *A. triviale*

Alisma gramineum Lej., Fl. Spa 1:175. 1811. Plants to 5dm high; leaves petioled, the blades linear and about as wide as petioles to narrowly elliptic, mostly 5-15cm long, 3-30mm wide, sometimes emersed or floating; panicle often not exceeding the leaves; sepals about 2.5mm long, the petals slightly larger and white with often reddish or purplish tinging; achenes about 2.5mm long, 2 grooved on the edge. In water or wet areas. NW, SE. *A. geyeri* Torr.

Alisma triviale Pursh, Fl. Am. Sept. 252. 1814. Plants to 12dm high; leaves petioled, the blades ovate to oblong-lanceolate, 6-15cm long, 2-15cm wide; panicle exceeding the leaves; sepals 2.5-4mm long; petals mostly 3-6mm long and white to pink or purplish; achenes about 2.5-3mm long, grooved once or occasionally twice on the edge. In water or wet areas. Statewide. *A. plantago-aquatica* L. of authors.

Sagittaria L. Arrowhead

Beak of achene, or mature style, less than 0.5mm long and pointing forward *S. cuneata*

Beak of achene, or mature style, usually over 0.5mm long and pointing at a right angle to the achene or pistil axis *S. latifolia*

Sagittaria cuneata Sheld., Bull. Torrey Club 20:283. 1893. Plants emersed or submersed, to 5dm high; leaf blades linear, ovate, cordate, or sagittate, 2-18cm long, 0.2-10cm wide; sepals 4-8mm long; petals 7-12mm long, white; achenes mostly 2-2.5mm long, the beak 0.2-0.4mm long and directed longitudinally. In water or wet areas. Statewide. *S. arifolia* Nutt. ex Smith, *S. hebetiloba* A. Nels.

Sagittaria latifolia Willd., Sp. Pl. 4:409. 1805. Plants usually emersed, to 9dm high; leaf blades mostly sagittate, rarely lacking when submersed, 5-25cm long and about as wide; sepals 5-10mm long; petals 10-20mm long, white; achenes 2.5-4mm long, the beak 0.5-1.5mm long and laterally directed. In water or wet areas. Goshen Co.

AMARANTHACEAE Pigweed Family

Annual herbs with simple, usually entire, alternate or opposite leaves; flowers bisexual or unisexual, regular or slightly irregular, usually subtended by scarious bracts, in racemes or spikes or axillary clusters; sepals 3-5, usually dry and membranous, separate or not; petals lacking; stamens 2-5, the filaments often united into a tube; ovary 1, superior; carpels 2-3; styles 1-3; locule 1; ovule 1, basal; fruit usually a utricle or nutlet.

Leaves alternate; plants not woolly-hairy — Amaranthus

Leaves opposite; plants somewhat woolly-hairy — Froelichia

Amaranthus L. Pigweed

Reference: Fernald, M. L. 1945. Rhodora 47:139-140.

1. Inflorescences both terminal and axillary
   2. Plants dioecious; bracts of inflorescence mostly ovate and 3mm long or less; flower clusters somewhat loose in a narrow, elongate inflorescence — A. arenicola
   2. Plants monoecious but the staminate flowers often few; bracts of inflorescence mostly narrowly lanceolate and 4mm or more long; flowers often densely clustered in a somewhat short and stout inflorescence
      3. Sepals of pistillate flowers mostly rounded or truncate below the spinulose tip; stamens 5; plants often conspicuously pubescent in the inflorescence and on underside of leaves — A. retroflexus

3. Sepals of pistillate flowers mostly gradually tapered to the spinulose tip; stamens usually 3; plants glabrous to very sparsely pubescent ... *A. powellii*

1. Inflorescences all axillary clusters

4. Sepals 3; seeds 0.6-1mm long ... *A. albus*

4. Sepals 4 or 5; seeds 1.3-2mm long ... **A. blitoides**

Amaranthus albus L., Sp. Pl. 2:1404. 1763. Plants to 1m high, often becoming tumbleweeds; leaf blades ovate-rhombic to obovate, mostly 1-3cm long, the petioles to 4.5cm long; flowers in axillary clusters; bracts spinulose at tip; sepals 3, about 1.5mm long; stamens 2 or 3. Disturbed areas. SW, SE, NE. *A. graecizans* of authors, not of L.

Amaranthus arenicola Johnston, Journ. Arnold Arb. 29:193. 1948. Plants dioecious, erect, to 2m high; leaf blades oblong or elliptic, 1-7cm long, the petioles to 6cm long; flowers primarily in a terminal, elongate inflorescence; bracts often spinulose at tip; sepals 5, 1.5-3mm long; stamens 5. Disturbed areas. SE, NE.

Amaranthus blitoides Wats., Proc. Am. Acad. 12:273. 1877. Plants prostrate with stems to 8dm long; leaf blades elliptic, obovate, or ovate, 0.5-4cm long; flowers axillary, densely clustered; bracts spinulose at tip; sepals 4 or 5, unequal, 1-4mm long; stamens usually 4. Disturbed areas and sagebrush. NW, NE, SE.

Amaranthus powellii Wats., Proc. Am. Acad. 10:347. 1875. Plants erect, to 2m high; leaf blades lanceolate, ovate, or elliptic, 1-8cm long, the petioles about as long as blades; flowers in spike-like terminal and axillary clusters; bracts spinulose at tip; sepals 4 or 5, mostly 2-4mm long; stamens 3 or rarely 4. Disturbed areas. Laramie and Albany Cos.

Amaranthus retroflexus L., Sp. Pl. 991. 1753. Plants erect, to 1m high; leaf blades lanceolate or ovate to obovate, 1-10cm long, the petioles as long as blade or shorter; flowers in terminal and axillary spikes; bracts spinulose at tip; sepals 4 or 5, 2.5-4mm long; stamens mostly 5. Disturbed areas. SW, SE, NE.

Froelichia Moench Snake Cotton

Froelichia gracilis (Hook.) Moq. in DC., Prodr. 13(2):420. 1849. Usually basally branched annual to 5dm high; leaves opposite, short petioled or sessile, the blades linear to lanceolate or elliptic, 2-5cm long, 2-15mm wide, pubescent; flowers in opposite and terminal spikes; calyx 5 cleft, 2-4mm long, subtended by 3 scarious bracts and covered with dense woolly hairs; stamens 5, the filaments united. Mostly sandy soil of hills and plains. Weston Co.

ANACARDIACEAE Sumac Family

Shrubs with compound, alternate leaves; flowers usually unisexual, regular, in a panicle, raceme, or thyrse; sepals usually 5, united at base; petals usually 5, mostly distinct; fertile stamens usually 5, sometimes with 10 vestigial stamens; ovary 1, superior; carpels usually 3; styles 1-3; locule 1; ovule 1 per locule; placentation parietal or basal; fruit a drupe.

Leaflets 3 or rarely 5, the lateral ones mostly over 3.5cm long and 2.5cm wide; mature fruits whitish, glabrous; poison ivy *Toxicodendron*

Leaflets more than 5, or if 3 or 5, the lateral ones mostly less than 3cm long and 2.5cm wide; mature fruits usually red and pubescent *Rhus*

Rhus L.

Leaflets more than 5 *R. glabra*

Leaflets 3 or 5 *R. trilobata*

Rhus glabra L., Sp. Pl. 265. 1753. Sumac. Shrub to 3m high; leaflets 7-29, lanceolate to elliptic, 2-12cm long, serrate; flowers in panicles; sepals about 2mm long; petals yellowish, 1-2mm long; fruits reddish, usually stellate pubescent. Hills and slopes. NE, SE. *R. cismontana* Greene, *R. asplenifolia* Greene.

Rhus trilobata Nutt. in T. & G., Fl. N. Am. 1:219. 1838. Skunkbush. Shrub to 4m high; leaflets usually 3, ovate, elliptic, obovate, or cuneate, 0.5-5cm long, lobed or crenate; flowers in short racemes; sepals about 1.5mm long; petals yellow-green, about 2-3mm long; fruit red-orange, puberulent. Hills and slopes. Statewide.

Toxicodendron Miller

Reference: Gillis, W. T. 1971. Rhodora 73:72-237, 370-443, 465-540.

Toxicodendron rydbergii (Small ex Rydb.) Greene, Leafl. Bot. Observ. Crit. 1:117. 1905. Poison Ivy. Shrubs mostly less than 1m high; leaflets usually 3, mostly ovate, 3-13cm long, usually crenulate to shallowly lobed; flowers in panicles, 2-4mm long; fruit whitish and glabrous. Woods and rocky areas, usually below 7000 feet. Statewide. *Rhus rydbergii* Small ex Rydb. There is still some question as to whether or not our plants are a species distinct from *T. radicans* (L.) Kuntze.

APOCYNACEAE Dogbane Family

Rhizomatous, perennial herbs with milky juice; leaves simple, opposite, entire; flowers perfect, regular, in cymes; sepals 5, united; petals 5, united; stamens 5, alternate with the corolla lobes and attached to the tube; ovaries 2, superior or nearly so; carpels 1 per ovary; style 1, very short and enlarged; locule 1 per ovary; ovules many; placentation parietal; fruit a follicle.

References: Woodson, R. E., Jr. 1938. N. Am. Fl. 29:103-192.
Boivin, B. 1966. Nat. Canad. 93:107-128.

Apocynum L. Dogbane

Flowers usually (4) 5mm long or more; corolla pink, often more than twice as long as the calyx; corolla lobes mostly spreading or reflexed; leaves usually spreading or drooping — _A. androsaemifolium_

Flowers usually less than 5mm long; corolla white or greenish, usually less than twice as long as the calyx; corolla lobes mostly erect; leaves ascending

- Follicles 11cm or less long; seed coma 1-2cm long; lower leaves of main stem usually sessile or nearly so and cordate at base — _A. sibiricum_
- Follicles usually 12cm long or more at maturity; seed coma 2-3cm long; leaves of main stem usually petioled and often not cordate at base — _A. cannabinum_

Apocynum _androsaemifolium_ L., Sp. Pl. 2:311. 1762. Plants to 6dm high; leaf blades elliptic, ovate, or subcordate, 1-8cm long; calyx 1.5-4mm long; corolla pinkish, 4-9mm long. Hills, slopes, and open woods. NW, NE, SE. Hybridizes with the next species. _A_. _pumilum_ (Gray) Greene.

Apocynum _cannabinum_ L., Sp. Pl. 213. 1753. Plants to 1m high; leaf blades oblong-ovate to lanceolate, 2-11cm long; calyx 2-4mm long; corolla white or greenish-white, 2-4.5mm long; follicles somewhat sickle-shaped, mostly 12-18cm long when mature. Plains, hills, slopes, and disturbed areas. NW, NE, SE. Hybridizes with the above species (_A_. x _medium_ Greene).

Apocynum _sibiricum_ Jacq., Hort. Vindob. 3:37, pl. 66. 1776. Similar to _A_. _cannabinum_ and intergrading with it but tending to have most lower leaves sessile or subsessile and cordate or clasping at base and the follicles nearly straight and 4-11cm long. Plains, hills, slopes, and disturbed areas. NW, NE, SE. _A_. _hypericifolium_ Ait.

ARACEAE Arum Family

Perennial herbs with short, fleshy, underground stems and large, somewhat net-veined leaves; inflorescence a fleshy spike subtended by a spathe; flowers crowded, bisexual, regular; perianth 4 lobed; stamens 4; pistils solitary; ovary superior; style usually lacking; locules 2; ovules 2-4; placentation various; fruit berry-like.

Lysichiton Schott Skunk Cabbage

Reference: St. John, H. & E. Hultén. 1956. Bull. Torrey Club 83:151-153.

Lysichiton americanum Hultén & St. John, Svensk Bot. Tidskr. 25(4):455. 1931. Plants subacaulescent, to 9dm high; leaves mostly petioled, the blades to 7dm long, lanceolate or elliptic to oblong-ovate or oval; spathe yellow or white; flowers greenish-yellow; perianth lobes 1.5-4mm long. Swamps. Yellowstone Park.

ARALIACEAE Ginseng Family

Herbs with compound, mostly basal leaves, the leaflets toothed; flowers bisexual, regular, in umbels; calyx 5 lobed; petals 5, separate; stamens 5; ovary 1, inferior; carpels 5; styles 5; locules 5; ovules 1 per locule; placentation axile or apical; fruit a berry.

Aralia L. Wild Sarsaparilla

Reference: Smith, A. C. 1944. N. Am. Fl. 28B:3-41.

Aralia *nudicaulis* L., Sp. Pl. 274. 1753. Perennial to 7dm high; leaves basal or nearly so, mostly twice pinnately compound, the leaflets ovate to obovate, 4-13cm long, 2.5-8cm wide, serrate; flowers in compound umbels on a stalk from the base; calyx lobes minute; petals greenish-white, about 2mm long; berries purple, 6-8mm long. Moist woods and thickets. Crook Co.

ASCLEPIADACEAE Milkweed Family

Perennial herbs with milky juice; leaves simple, usually entire, opposite, alternate, or whorled; flowers regular and perfect, in axillary or terminal umbels; sepals 5, united at base or separate; petals 5, united at base, usually reflexed; stamens 5, borne on the corolla tube and alternate with the lobes, all joined together and usually adnate to the stigma, with a hood-like appendage at the base of each stamen and an inward curving appendage or horn (rarely lacking) included in the hood or exserted from it; pollen of each chamber coalescent in a mass forming a sac-like structure (pollinium), these in pairs (1 each from adjacent anthers), each with a slender connective (translator arm) which are joined by a dark gland; ovaries 2, superior or nearly so; carpel 1 per ovary; style 1 per ovary, sharing a single stigma; locule 1 per ovary; ovules many; placentation parietal; fruit a follicle; seeds bearing long silky hairs.

Asclepias L. Milkweed

1. Leaves linear, 4mm wide or less
   2. Leaves mostly 4cm or less long, spirally arranged and very crowded; plants mostly less than 20cm high — _A. pumila_
   2. Leaves mostly over 4cm long, mostly whorled; plants mostly over 20cm high — _A. subverticillata_
1. Leaves not linear, over 4mm wide

3. Leaves oval or nearly so, barely longer than wide, mostly 7cm or less long; lower half of stem glabrous *A. cryptoceras*

3. Leaves only rarely oval, often over 7cm long; lower half of stem often hairy

4. Hoods lacking horns within; corolla greenish *A. viridiflora*

4. Hoods with horns within; corolla pink, purple, or rose

5. Leaves broadly rounded to subcordate at base, at least some over 5cm wide *A. speciosa*

5. Leaves not with combined characters as above *A. hallii*

Asclepias cryptoceras Wats., Bot. King Exp. 283. 1871. Plants to 3dm high; leaves mostly opposite, the blades oval or ovate to cordate, 2-7cm long; sepals 5-9mm long; corolla greenish-yellow, 8-15mm long; hoods pinkish or brownish-purple, 5-8mm long, mostly enclosing the horns. Plains, hills, and slopes. NW, SW, SE.

Asclepias hallii Gray, Proc. Am. Acad. 12:69. 1876. Plants to 6dm high; leaves opposite or alternate, the blades lanceolate, 4-15cm long; sepals about 3mm long; corolla rose or purple, 6-9mm long; hoods rose to cream, 4-7mm long; horns shorter than the hoods, sharply incurved. Hills and banks. Albany Co. *A. curvipes* A. Nels.

Asclepias pumila (Gray) Vail in Britt. & Brown, Ill. Fl. 3:12. 1898. Plants to 2(3)dm high; leaves mostly spirally alternate and very crowded, the blades usually filiform, 1-4cm long, about 1mm or less wide, revolute; sepals about 2mm long; corolla white or tinged with rose or yellow-green, 2-4mm long; hoods greenish-white, about 2mm long; horns as long as to about twice as long as hoods. Plains and hills. NE, SE.

Asclepias speciosa Torr., Ann. Lyc. N. Y. 2:218. 1828. Plants to 12dm high; leaves mostly opposite, the blades oblong-lanceolate to ovate-oblong, 3-20cm long; sepals 4-6mm long; corolla pinkish to reddish-purple, 9-15mm long; hoods pink to yellowish, 10-16mm long; horns much shorter than the hoods. Disturbed areas and streambanks. Statewide.

Asclepias subverticillata (Gray) Vail, Bull. Torrey Club 25:178. 1898. Plants to 12dm high; leaves mostly whorled, the blades linear, 2-13cm long; sepals 1.5-3mm long; corolla white or greenish-purple tinged, 3-5mm long; hoods whitish, 1.5-2.5mm long; horns slightly longer than the hoods. Plains, hills, and roadsides. Carbon Co.

<u>Asclepias</u> <u>viridiflora</u> Raf., Med. Repos. II, 5:360. 1808. Plants to 9dm high; leaves opposite or alternate, the blades oval or ovate to lanceolate or linear-lanceolate, 2-15cm long; sepals 2-4mm long; corolla greenish, 5-7mm long; hoods greenish or purplish tinged, 4-5mm long; horns lacking. Plains, hills, and roadsides. NE, SE. <u>Acerates</u> <u>viridiflora</u> (Raf.) Pursh ex Eaton.

BERBERIDACEAE Barberry Family

Shrubs with alternate, compound leaves, the leaflets spiny on the margins; flowers bisexual, regular, usually in a raceme; sepals 6, separate, in 2 whorls; petals 6, separate, in 2 whorls; stamens 6; ovary 1, superior; carpel 1; style 1 or lacking; locule 1; ovules 2 to several; placentation usually basal; fruit a berry.

<u>Mahonia</u> Nutt. Oregon Grape

<u>Mahonia</u> <u>repens</u> (Lindl.) G. Don, Gen. Syst. Gard. 1:118. 1831. Shrub to 3dm high; leaflets ovate-lanceolate or elliptic, mostly 3-8cm long and 2-5cm wide, spiny on the margins; outer sepals greenish, 2-3mm long; inner sepals yellow, 4-7mm long; petals yellow, 4-7mm long; berry blue. Woods and slopes. Statewide. <u>Berberis</u> <u>repens</u> Lindl., <u>B</u>. <u>aquifolium</u> Pursh of authors, <u>M</u>. <u>nana</u> (Greene) Fedde.

BETULACEAE Birch Family

Trees or shrubs with alternate, simple, toothed leaves; flowers unisexual, the plants monoecious; staminate flowers in catkins, pistillate in catkins, strobiloid inflorescences, or merely clustered; calyx 2-4 parted or lacking; petals lacking; stamens 1-10; pistil 1; ovary inferior or apparently superior; carpels 2; styles 1 or 2; locules 1 or 2; ovules 1 or 2 per locule; placentation axile; fruit a small samara, nutlet, or nut.

1. Plants trees with smooth, whitish bark; Black Hills area — Betula
1. Plants not as above
   2. Leaves usually cordate or subcordate at base; fruit a nut or nutlet enclosed by an involucre; staminate flowers without a calyx; stamens 3 or more with each bract or scale; twigs without blister-like roughenings
      3. Plants usually tree-like; fruit a nutlet 6mm or less long, enclosed in a bladdery involucre; pistillate inflorescence elongate, bearing several to many nutlets — Ostrya
      3. Plants usually shrub-like; fruit a nut usually over 6mm long, enclosed in a tightly appressed, beaked involucre; pistillate inflorescence bearing usually only 1 or 2 nuts — Corylus
   2. Leaves usually not cordate at base, rarely subcordate; fruit a nutlet or samara, not enclosed by an involucre; staminate flowers with a 2-4 parted calyx; stamens 2 or 4 with each perianth; twigs often with blister-like, resinous roughenings

4. Bracts of pistillate inflorescence woody, forming persistent cone-like structures; buds usually stalked; twigs usually without blister-like roughenings *Alnus*
4. Bracts of pistillate inflorescence not woody, deciduous with the mature nutlets; buds not stalked; twigs often with blister-like roughenings *Betula*

Alnus Ehrh. Alder

Alnus incana (L.) Moench, Meth. 424. 1794. Shrub, often tree-like, to 10m high; buds often stalked; twigs usually smooth, often pubescent; leaves petioled, the blades elliptic to ovate, doubly serrate, 3-10cm long, 2-8cm wide; stamens mostly 4; pistillate inflorescence a hard cone-like structure 9-15mm long; fruit a nutlet. Streambanks, lakeshores, and moist woods. NW, SW, SE. *A. tenuifolia* Nutt.

Betula L. Birch

Trees or shrubs, the twigs often with blister-like, resinous roughenings; flowers in catkins, the staminate with 2-3 perianth segments and usually 2(4) stamens; fruit a winged nutlet.

Reference: Dugle, J. R. 1966. Canad. Journ. Bot. 44:929-1007.

Plants trees with relatively smooth, whitish bark (coppery when young); tufts of white hairs usually present in vein axils on underside of leaves; Black Hills area — *B. papyrifera*

Plants mostly shrubs without white bark; tufts of white hairs usually lacking in vein axils on underside of leaves; widespread

- Leaves orbicular to oval or obovate, with mostly rounded teeth; plants mostly less than 2m high; mostly subalpine or montane; wings much narrower than nutlet — *B. glandulosa*
- Leaves subcordate to ovate with pointed teeth; plants often over 2m high; mostly at lower and middle elevations; wings often as wide as or wider than nutlet — *B. occidentalis*

Betula glandulosa Michx., Fl. Bor. Am. 2:180. 1803. Bog Birch. Shrub to 3m high; twigs resinous and warty; leaves petioled, the blades mostly oval to obovate, 0.5-4cm long, 0.5-3cm wide, bluntly serrate, minutely glandular; pistillate catkins 0.5-2.5cm long. Bogs, streambanks, and lakeshores. Statewide.

Betula occidentalis Hook., Fl. Bor. Am. 2:155. 1838. Water Birch. Shrub or small tree to 10m high; twigs resinous and somewhat warty; leaves petioled, the blades ovate to subcordate, 1-6cm long, 1-5cm wide, once or twice serrate, usually minutely glandular; pistillate catkins 1-4cm long. Streambanks and lakeshores. Statewide. *B. fontinalis* Sarg.

Betula papyrifera Marsh., Arb. Am. 19. 1785. Paper Birch. Tree to 20m high, usually with white, papery bark; twigs sometimes resinous, slightly warty; leaves petioled, the blades ovate to subcordate, acuminate, 2-8cm long, 1-6cm wide, once or twice serrate, glandular or not; pistillate catkins 1-5cm long. Woods. Crook Co.

Corylus L. Hazelnut

Corylus cornuta Marsh., Arb. Am. 37. 1785. Beaked Hazelnut. Shrub to 3m high; leaves petioled, the blades oval-oblong to ovate, usually cordate at base, 3-10cm long, 1-8cm wide, doubly serrate; stamens usually 8; pistillate flowers in clusters at ends of short branches; fruit a nut enclosed by a jug-like involucre. Woods and thickets. Crook Co. *C. rostrata* Ait.

Ostrya Scop. Ironwood

Ostrya virginiana (Mill.) Koch, Dendr. II, 2:6. 1873. Tree to 20m high; leaves petioled, the blades mostly ovate, usually cordate at base, 2-10cm long, 1-6cm wide, once or twice serrate; stamens 3 or more; pistillate catkins mostly 3-5cm long; fruit a flattened nutlet enclosed by a bladdery involucre. Woods. Crook Co.

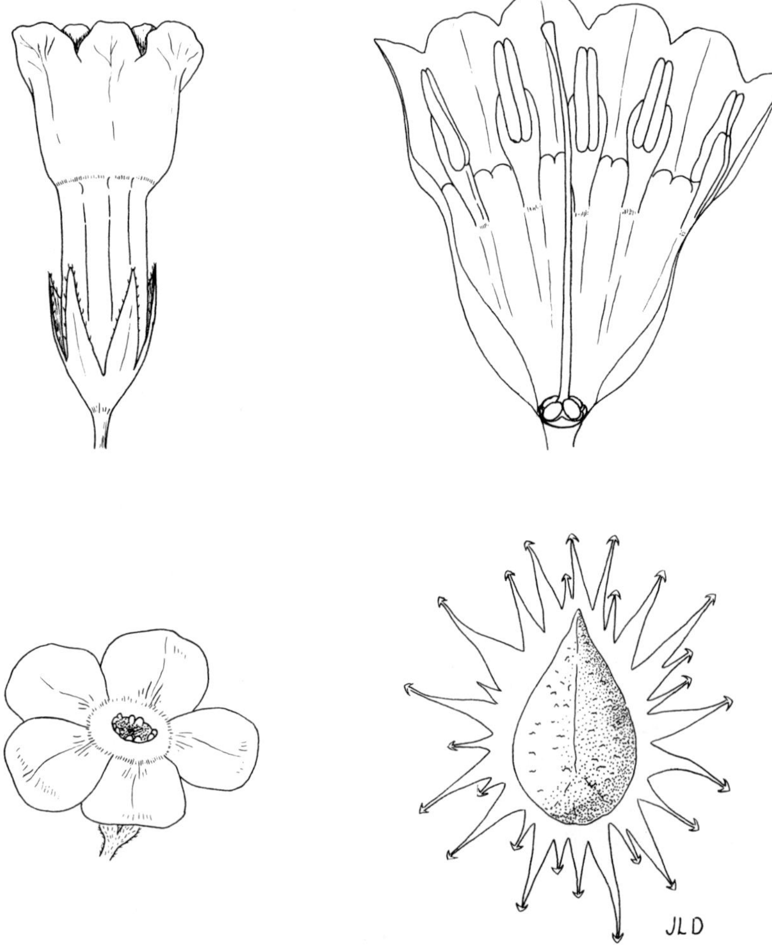

Herbs with simple, usually entire and alternate (rarely some opposite) leaves; flowers perfect, regular, mostly in helicoid or scorpioid cymes, occasionally in false racemes or panicles, rarely solitary or in axillary clusters; sepals 5, united or not; petals 5, united; stamens 5, attached to corolla tube and alternate with its lobes; ovary 1 but usually deeply 4 divided or 4 grooved, superior; carpels 2; style 1; locules apparently 4; ovules 1 per locule; placentae axile or apparently basal; fruit 1-4 nutlets.

1. Ovary merely 4 grooved; style terminal or nearly so, or lacking; plants either glabrous and succulent or with the style 2 cleft
   2. Style entire or lacking; plants succulent — *Heliotropium*
   2. Style 2 cleft; plants not succulent — *Coldenia*
1. Ovary usually deeply 4 parted; style arising from between the nearly distinct lobes; plants not succulent, often pubescent; style usually not cleft
   3. Corolla reddish-purple; fruits covered with hooked (glochidiate) prickles — *Cynoglossum*
   3. Corolla not reddish-purple; fruits either without prickles, with simple prickles, or with hooked prickles only on the margins (rarely with a few reduced intermarginal prickles)

Figure 3. Boraginaceae. Upper left: flower of *Mertensia* (X 4). Upper right: same flower opened along one side (note 4 lobed ovary and fornices between stamens). Lower left: flower of *Hackelia* (X 7). Lower right: nutlet of *Hackelia* (X 7).

4. Plants with greenish-white, white, or rarely yellowish corollas 10-16mm long, hairy on outside; corolla lobes mostly erect; style long exserted from corolla; nutlets broadly attached at base to a flat gynobase <u>Onosmodium</u>

4. Plants not with the above combination of characters

5. Corolla blue or occasionally pinkish, tubular to funnelform, 5mm long or more, the tube often much exceeding the calyx <u>Mertensia</u>

5. Corolla yellow, orange, or white, or if blue, then usually salverform or rotate and usually less than 5mm long, the tube often equal to or little exceeding the calyx

6. Nutlets and ovary segments broadly attached at their base to a flat gynobase; corolla yellow; perennials <u>Lithospermum</u>

6. Nutlets and ovary segments usually apically, medially, or basilaterally attached to a broad or narrow pyramidal gynobase; corolla yellow or not; annual to perennial

7. Plants annual, with procumbent stems which are angled and bear stiff, retrorse hairs mostly on the angles; calyx in fruit much enlarged and veiny, usually with 2 teeth between adjacent lobes; corolla blue or pink-purple, 2-3mm long <u>Asperugo</u>

7\. Plants not with the above combination of characters

8\. Plants less than 10cm high, densely caespitose, usually forming cushion mats; leaves mostly 2cm or less long and 3mm or less wide; corolla limb usually blue, rarely white or cream; plants often alpine or subalpine, rarely down to the foothills where the corolla definitely blue Eritrichium

8\. Plants not with the above combination of characters

9\. Nutlets bearing hooked or barbed prickles along the margins

10\. Nutlets widely spreading in fruit, attached at the end; plants prostrate to weakly ascending annuals with white corollas 1-1.5mm long Pectocarya

10\. Nutlets usually not widely spreading, attached near or below middle; plants mostly upright, annual to perennial with blue or white corollas 1-9mm long

11\. Pedicels erect or ascending in fruit; inflorescence bracteate; styles often surpassing the mature nutlets Lappula

11\. Pedicels recurved or deflexed in fruit; inflorescence often naked or nearly so; styles usually shorter than the mature nutlets Hackelia

9. Nutlets without hooked or barbed prickles, sometimes with distally branched, minute bristles but these scattered over the nutlet

12. Corolla blue *Myosotis*

12. Corolla white to yellow or orange

13. Corolla yellow or orange; nutlets keeled on ventral side; annuals *Amsinckia*

13. Corolla white or cream colored, or if yellow, the nutlets with a longitudinal groove-scar or slit on ventral side (rarely closed by meeting of edges), not keeled; annuals to perennials

14. Nutlets with a groove-scar or slit running most of their length on ventral side (occasionally closed by meeting of edges); plants annual to perennial *Cryptantha*

14. Nutlets smooth or keeled on ventral side; plants mostly annual

15. Nutlets smooth, winged or keeled around the tip and lateral edges *Myosotis*

15. Nutlets keeled on ventral side and wrinkled or tuberculate, not winged or keeled around tip and lateral edges *Plagiobothrys*

Amsinckia Lehm. Tarweed

Annuals with coarse, glassy hairs; corolla salverform, yellow to orange, little exceeding calyx; nutlets tuberculate or rugose.

Reference: Ray, P. M. & H. F. Chisaki. 1957. Am. Journ. Bot. 44:529-536.

Corolla throat obstructed by hairy fornices; stamens attached below middle of corolla tube _A. lycopsoides_

Corolla throat open, fornices not well developed; stamens attached above middle of corolla tube _A. menziesii_

Amsinckia lycopsoides Lehm., Del. Sem. Hort. Hamb. 1831:1, 7. 1831. Plants to 7dm high; leaf blades linear or oblong to lanceolate or oblanceolate or rarely ovate, 1-13cm long; calyx 5-10mm long; corolla yellow to orange, the tube 5-8mm long, the lobes 1-3mm long, hairy in the throat; nutlets tuberculate and often rugose. Plains, hills, and disturbed areas. Albany Co.

Amsinckia menziesii (Lehm.) Nels. & Macbr., Bot. Gaz. 61:36. 1916. Plants to 7dm high; leaf blades linear or lanceolate to elliptic or ovate, 1-12cm long; calyx 5-10mm long; corolla yellow, the tube 3-7mm long, the lobes 0.5-2mm long, glabrous in the throat; nutlets tuberculate and often rugose. Plains, hills, and disturbed areas. Sheridan Co.

Asperugo L. Catchweed

Asperugo procumbens L., Sp. Pl. 138. 1753. Annual with weak, retrorsely hispid stems to 12dm long; leaves alternate or sometimes opposite, the blades oblanceolate to elliptic or rarely lanceolate, 1-10cm long, entire or nearly so; flowers on short recurved pedicels at or near the nodes; calyx 5 lobed, usually with 2 teeth between adjacent lobes, becoming enlarged (1-2cm wide) and veiny in fruit; corolla campanulate, blue or pink-purple, 2-3mm long; nutlets smooth or slightly tuberculate. Disturbed areas. SW, SE.

Coldenia L.

Coldenia nuttallii Hook., Journ. Bot. & Kew Misc. 3:296. 1851. Prostrate, dichotomously branched annual to 2cm high; leaves usually crowded at nodes or ends of branches, the blades elliptic, ovate, or suborbicular, 2-8mm long, entire; flowers in axillary clusters; calyx 2-4mm long; corolla tubular-campanulate, 3-5mm long, the tube white to greenish-yellow, the limb pink to lavender or sometimes white; style nearly terminal and 2 cleft; nutlets smooth. Plains and hills especially where sandy. NW, SW.

Cryptantha Lehm. Miner's Candle

Annual to perennial herbs, often pungently hairy; corolla rotate to salverform or campanulate; filaments very short, attached at or below middle of corolla tube or sometimes slightly above middle.

Reference: Higgins, L. C. 1971. Brigham Young Univ. Sci. Bull. (Biol.) 13(4):1-63.

1. Plants annual; corolla limb usually less than 2.5mm wide
   2. Calyx circumscissile at maturity, the persistent basal half scarious, different in texture from the more herbaceous deciduous upper half; plants usually less than 6cm high ... *C. circumscissa*
   2. Calyx not circumscissile; plants often over 6cm high
      3. Nutlets of 2 kinds, 1 longer than the other 3, the 3 smaller ones tuberculate or papillate, the longer often smooth or granular
         4. Inflorescence with bracts subtending most of the flowers ... *C. minima*
         4. Inflorescence without bracts or with a few only at the base ... *C. kelseyana*
      3. Nutlets all alike or nearly so (occasionally 1 or more fail to develop), or else the nutlets (or the 3 similar ones) all smooth
         5. Nutlets tuberculate (often minutely so) or spiculate at least above
            6. Nutlets tuberculate, ovate, 0.8-1.5mm wide *C. ambigua*
            6. Nutlets spiculate, lanceolate, 0.5-0.7mm wide ... *C. scoparia*
         5. Nutlets smooth
            7. Ventral groove of nutlet strongly offset from center ... *C. affinis*
            7. Ventral groove of nutlet in the center of nutlet

8. Margins of nutlets sharply angled, especially above *C. watsonii*

8. Margins of nutlets rounded or blunt

9. Nutlets ovate, averaging 0.8-1.2mm wide; plants often few branched or unbranched *C. torreyana*

9. Nutlets lanceolate, averaging 0.5-0.7mm wide; plants usually rather diffusely branched *C. fendleri*

1. Plants biennial or perennial; corolla limb usually well over 3mm wide

10. Corolla tube 6mm or more long, usually evidently longer than the calyx

11. Nutlets smooth; corolla yellow *C. flava*

11. Nutlets papillate or wrinkled; corolla white or yellow *C. flavoculata*

10. Corolla tube 5mm long or less, about equal to or shorter than the calyx

12. Dorsal surface of nutlets smooth or nearly so; nutlet margins in fruit separated or else the fruits strongly curved dorsally and appearing depressed from top *C. jamesii*

12. Dorsal surface of nutlets wrinkled, tuberculate, or spiny; nutlet margins in fruit touching each other or only slightly separated, the nutlets usually not appearing depressed

13. Ventral surface of nutlets smooth or nearly so
  14. Inflorescence spicate; bracts greatly exceeding the flowers *C. virgata*
  14. Inflorescence sometimes narrow but not spicate; bracts not greatly exceeding the flowers except sometimes below
    15. Inflorescence usually much branched, often over 5cm wide; not known west of Albany Co. *C. thyrsiflora*
    15. Inflorescence usually little branched, rarely over 4cm wide; Carbon Co. *C. stricta*
13. Ventral surface of nutlets wrinkled, tuberculate, or muricate
  16. Plants densely caespitose from a branched caudex, usually 15cm or less high; style not exceeding nutlets by more than 0.5mm; leaf blades usually 6mm or less wide
    17. Nutlets usually exclusively muricate; hairs of leaves closely appressed to blade and tending to follow the margins giving a smooth-looking appearance *C. cana*
    17. Nutlets not exclusively muricate; hairs of leaves slightly ascending from blade and spreading beyond margins giving a ragged-looking appearance *C. caespitosa*

16. Plants caespitose or not, mostly over 15cm high; style often exceeding nutlets by over 1mm; leaf blades often over 6mm wide

 18. Leaves with pustulate hairs on lower surface along with some non-pustulate but with only non-pustulate hairs on upper surface — C. sericea

 18. Leaves with pustulate and smaller non-pustulate hairs on both surfaces

  19. Inflorescence usually long and narrow; statewide — C. celosioides

  19. Inflorescence usually somewhat oval or diffusely branched; eastern plains — C. thyrsiflora

Cryptantha affinis (Gray) Greene, Pittonia 1:119. 1887.
Annual to 3dm high; leaf blades linear or oblong to oblanceolate, 1-4cm long; calyx 2-4mm long; corolla white, 1-3mm long; nutlets smooth or very finely granular. Plains, hills, woods, and slopes. Teton Co. and Yellowstone Park.

Cryptantha ambigua (Gray) Greene, Pittonia 1:113. 1887.
Annual to 3dm high; leaf blades linear to oblanceolate, 0.5-3cm long; calyx 2-6mm long; corolla white, 1-3mm long; nutlets somewhat granular or with scattered tubercles above. Plains, hills, and slopes. NW, SW, SE. C. multicaulis A. Nels.

Cryptantha _caespitosa_ (A. Nels.) Payson, Ann. Mo. Bot. Gard. 14:281. 1927. Caespitose perennial, often mat forming, to 15cm high; leaf blades linear to oblanceolate or obovate, 0.5-3cm long; calyx 3-8mm long; corolla white or drying yellowish, the tube 3-5mm long, the lobes 1-3mm long; nutlets rugulose and tuberculate. Plains and hills. NW, SW, SE, possibly endemic to Wyoming. _Oreocarya_ _caespitosa_ A. Nels.

Cryptantha _cana_ (A. Nels.) Payson, Ann. Mo. Bot. Gard. 14:316. 1927. Perennial to 15(20)cm high; leaf blades oblanceolate to rarely linear, 1-6cm long; calyx 2-7mm long; corolla white, the tube 2-5mm long, the lobes 1-5mm long; nutlets usually muricate. Plains and hills. NW, NE, SE. _Oreocarya_ _cana_ A. Nels.

Cryptantha _celosioides_ (Eastw.) Payson, Ann. Mo. Bot. Gard. 14:299. 1927. Biennial or perennial to 5dm high; leaf blades mostly oblanceolate or spatulate, 0.5-7cm long; calyx 3-10mm long; corolla white, the tube 3-5mm long, the lobes 1-5mm long; nutlets roughened on both sides. Plains, hills, and slopes. Statewide. _C_. _bradburyana_ Payson, _C_. _macounii_ (Eastw.) Payson, _Oreocarya_ _affinis_ Greene, _O_. _glomerata_ (Pursh) Greene.

<u>Cryptantha</u> <u>circumscissa</u> (H. & A.) Johnst., Contr. Gray Herb. n. s. 68:55. 1923. Annual to 6cm high; leaf blades linear, 3-13mm long; flowers solitary in upper axils; calyx 2-4mm long, circumscissile slightly below the middle, the persistent basal part scarious and obviously different in texture from the herbaceous deciduous part; corolla white, 1-2.5mm long; nutlets smooth or nearly so. Plains and hills, often where sandy. SW, SE.

<u>Cryptantha</u> <u>fendleri</u> (Gray) Greene, Pittonia 1:120. 1887. Annual to 5dm high; leaf blades linear to oblanceolate, 0.5-5cm long; calyx 2-6mm long; corolla white, 1-2.5mm long; nutlets mostly smooth. Plains, hills, and slopes. Statewide. <u>C</u>. <u>ramulosissima</u> A. Nels., <u>C</u>. <u>wyomingensis</u> Gand., <u>C</u>. <u>pattersonii</u> (Gray) Greene. These are all completely intergradient with <u>C</u>. <u>fendleri</u>.

<u>Cryptantha</u> <u>flava</u> (A. Nels.) Payson, Ann. Mo. Bot. Gard. 14:259. 1927. Perennial to 4dm high; leaf blades oblanceolate or linear, 1-9cm long; calyx 5-10mm long; corolla yellow, the tube 7-12mm long, the lobes 2-4mm long; nutlets smooth. Plains and hills. SW, SE. <u>Oreocarya</u> <u>flava</u> A. Nels.

Cryptantha flavoculata (A. Nels.) Payson, Ann. Mo. Bot. Gard. 14:334. 1927. Perennial to 4dm high; leaf blades linear or oblanceolate, 0.5-7cm long; calyx 5-10mm long; corolla white or pale yellow, the tube 6-12mm long, the lobes 1.5-5mm long; nutlets papillate to wrinkled. Plains and hills. Statewide. *Oreocarya flavoculata* A. Nels.

Cryptantha jamesii (Torr.) Payson, Ann. Mo. Bot. Gard. 14:242. 1927. Perennial to 6dm high; leaf blades linear or oblanceolate, 1-8cm long; calyx 3-7mm long; corolla white, the tube 2.5-4mm long, the lobes 1-3mm long; nutlets smooth or nearly so. Plains and hills. NE, SE. *Oreocarya suffruticosa* (Torr.) Greene.

Cryptantha kelseyana Greene, Pittonia 2:232. 1892. Annual to 35cm high; leaf blades linear to oblanceolate, 5-45mm long; calyx 2-7mm long; corolla white, 1-3mm long; nutlets tuberculate or the one larger one often smooth or finely granular. Plains and hills. Statewide.

Cryptantha minima Rydb., Bull. Torrey Club 28:31. 1901. Annual to 2dm high; leaf blades oblanceolate or linear, 0.5-3cm long; calyx 2-7mm long; corolla white, 1-3mm long; nutlets papillate or tuberculate or the larger one smooth or granular. Plains and hills. NW, NE, SE.

Cryptantha scoparia A. Nels., Bot. Gaz. 54:144. 1912. Annual to 4dm high; leaf blades mostly linear, 0.5-4cm long; calyx 2-6mm long; corolla white, 1-2.5mm long; nutlets with antrorse spiculate papillae. Plains and hills. Sweetwater Co. *C. muriculata* var. *montana* A. Nels.

Cryptantha sericea (Gray) Payson, Ann. Mo. Bot. Gard. 14:286. 1927. Perennial to 5dm high; leaf blades mostly oblanceolate, 1-7cm long; calyx 2.5-8mm long; corolla white, the tube 2.5-5mm long, the lobes 1-5mm long; nutlets tuberculate and somewhat rugulose. Plains and hills. SW, SE. *Oreocarya sericea* (Gray) Greene.

Cryptantha stricta (Osterh.) Payson, Ann. Mo. Bot. Gard. 14:264. 1927. Perennial to 4dm high; leaf blades oblanceolate to linear, 1-7cm long; calyx 4-9mm long; corolla white, the tube 3-5mm long, the lobes 1-4mm long; nutlets tuberculate or rugulose dorsally. Plains and hills. Carbon Co.

Cryptantha thyrsiflora (Greene) Payson, Ann. Mo. Bot. Gard. 14:283. 1927. Perennial or sometimes biennial to 4dm high; leaf blades oblanceolate to oblong, 1-10cm long; calyx 2-8mm long; corolla white, the tube 2-4mm long, the lobes 1-5mm long; nutlets rugose dorsally or with a few tubercles. Plains and hills. NE, SE. *Oreocarya thyrsiflora* Greene.

Cryptantha torreyana (Gray) Greene, Pittonia 1:118. 1887. Annual to 4dm high; leaf blades linear to oblanceolate, 0.5-5cm long; calyx 2-6mm long; corolla white, 1-3mm long; nutlets smooth or rarely granular. Plains, hills, woods, and slopes. Statewide. *C. flexuosa* (A. Nels.) A. Nels.

Cryptantha virgata (Porter) Payson, Ann. Mo. Bot. Gard. 14:270. 1927. Biennial or short lived perennial to 7dm high; leaf blades mostly oblanceolate, 1-12cm long; calyx 3-11mm long; corolla white, the tube 3-5mm long, the lobes 1-5mm long; nutlets usually rugose dorsally and often tuberculate. Plains and hills. Albany Co. *Oreocarya virgata* (Porter) Greene.

Cryptantha watsonii (Gray) Greene, Pittonia 1:120. 1887. Annual to 3dm high; leaf blades linear to oblanceolate, 0.5-4cm long; calyx 1.5-5mm long; corolla white, 1-3mm long; nutlets smooth. Plains and hills. NW, SW, SE.

Cynoglossum L. Hound's Tongue

Cynoglossum _officinale_ L., Sp. Pl. 134. 1753. Biennial to 12dm high; leaf blades mostly oblanceolate to elliptic or lanceolate, 4-30cm long; flowers in somewhat scorpioid cymes; calyx 3-8mm long; corolla rotate-salverform, reddish-purple, the tube 3-5mm long, the lobes 1-3mm long; nutlets covered with glochidiate prickles. Disturbed areas. NW, NE, SE.

Eritrichium Schrad. Alpine Forget-me-not

Mat forming perennials with crowded, mostly basal leaves; flowers in cymose clusters; corolla rotate-salverform, the limb blue or occasionally white or cream, the tube and throat often yellow.

Mature leaves densely appressed hairy, the leaf surface often obscured; hairs usually not forming an apical tuft or fringe at tip of leaf _E_. _howardii_

Mature leaves loosely long hairy, the leaf surface easily visible; hairs often more numerous toward leaf tip forming an apical tuft or fringe _E_. _nanum_

Eritrichium howardii (Gray) Rydb., Mem. N. Y. Bot. Gard. 1:327. 1900. Plants to 1dm high; leaf blades linear-oblanceolate, linear-elliptic, or linear, 3-20mm long, 0.5-2mm wide; calyx 3-5mm long; corolla tube 2-4mm long, the lobes 2-5mm long; nutlets smooth or hispidulous on back. Rocky slopes and ridges. Park and Sheridan Cos.

Eritrichium nanum (Vill.) Schrad., Asperif. 16. 1820. Plants to 1dm high; leaf blades oblong-elliptic, lanceolate, or ovate, 2-10mm long, 0.5-3mm wide; calyx 1.5-3mm long; corolla tube 1.5-4mm long, the lobes 1-3mm long; nutlets smooth. Ridges and slopes in the high mountains. Statewide. *E. argenteum* Wight, *E. elongatum* (Rydb.) Wight.

Hackelia Opiz Stickseed

Annual to perennial herbs with flowers in false racemes or panicles; inflorescence often not bracteate or sparsely so; corolla rotate or rotate-salverform, blue or white; nutlets with glochidiate prickles along the margins, sometimes also a few on the back.

1. Corolla limb 1.5-3mm wide; mature nutlets 2-3mm long on dorsal side (excluding prickles); plants annual or biennial — *H. deflexa*

1. Corolla limb mostly over 3mm wide; mature nutlets mostly 3-5mm long on dorsal side; plants biennial or perennial
   2. Corolla white, sometimes tinged with blue — *H. patens*
   2. Corolla dark blue, sometimes drying pinkish
      3. Plants biennial or rarely a short lived perennial, often with a single stem from a taproot and simple crown; corolla limb mostly 3-6mm wide — *H. floribunda*
      3. Plants perennial, usually with several stems from a taproot and branched caudex; corolla limb mostly 6-11mm wide — *H. micrantha*

Hackelia deflexa (Wahlenb.) Opiz in Bercht., Fl. Böhm. 2(2):147. 1839. Annual or biennial to 1m high; leaf blades oblanceolate to lanceolate, 2-15cm long; calyx 1-3mm long; corolla blue or sometimes white, the tube 1-2mm long, the lobes 1-2mm long. Open woods and thickets. Sheridan and Crook Cos. *Lappula americana* (Gray) Rydb.

Hackelia floribunda (Lehm.) Johnst., Contr. Gray Herb. n. s. 68:46. 1923. Biennial or short lived perennial to 1m high; leaf blades oblanceolate to lanceolate or oblong, 3-20cm long; calyx 1.5-3mm long; corolla blue with a yellow throat, the tube 1.5-3mm long, the lobes 1-3mm long. Meadows, woods, and disturbed areas. Statewide. *Lappula floribunda* (Lehm.) Greene.

Hackelia micrantha (Eastw.) Gentry, Madroño 21(7):490. 1972. Perennial to 1m high; leaf blades oblanceolate to lanceolate or oblong, 2-25cm long; calyx 1-3mm long; corolla blue, the throat yellow or white, the tube 1.5-4mm long, the lobes 2-4mm long. Meadows, slopes, and woods in the mountains. NW, SW. *H. jessicae* (McGregor) Brand.

Hackelia patens (Nutt.) Johnst., Journ. Arnold Arb. 16:194. 1935. Perennial to 8dm high; leaf blades oblanceolate to lanceolate or oblong, 1-20cm long; calyx 2-3mm long; corolla white with a yellow throat, sometimes tinged with blue, the tube 2-4mm long, the lobes 2-5mm long. Plains, hills, woods, and slopes. NW, SW. *Lappula subdecumbens* (Parry) A. Nels., *L. caerulescens* Rydb.

Heliotropium L.

Heliotropium curassavicum L., Sp. Pl. 130. 1753. Somewhat succulent, glabrous perennial with stems to 6dm long; leaf blades mostly oblanceolate to obovate, 1-6cm long; flowers in helicoid cymes; calyx 2-3mm long; corolla rotate-salverform, white or tinged with blue, the tube 2-4mm long, the lobes 2-5mm long; nutlets somewhat wrinkled or roughened. Shores and ditches, often where alkaline. NW, SW, SE. *H. spathulatum* Rydb.

Lappula Fabr. Stickseed

Annual or biennial herbs with flowers in false bracteate racemes; corolla funnelform-campanulate to salverform or rarely rotate, usually blue, occasionally white; nutlets with glochidiate prickles on the margins.

Marginal prickles on nutlets in at least 2 rows ..... *L. myosotis*
Marginal prickles on nutlets in 1 row ..... *L. redowskii*

Lappula myosotis Wolf, Gen. Pl. 17. 1776. Plants to 5dm high; leaf blades oblanceolate to lanceolate or linear, 0.5-5cm long; calyx 2-4mm long; corolla usually blue, the tube 1.5-2mm long, the lobes 0.5-2mm long; marginal prickles of nutlets in 2 or 3 rows. Hills, canyons, and disturbed areas. Statewide. *L. echinata* Gilib., *L. fremontii* (Torr.) Greene, *L. erecta* A. Nels., *L. cenchroides* (*cenchrusoides*) A. Nels., *L. lappula* (L.) Karst.

Lappula redowskii (Hornem.) Greene, Pittonia 2:182. 1891. Plants to 5dm high; leaf blades oblanceolate to lanceolate or linear, 0.5-5cm long; calyx 1.5-5mm long; corolla blue or white, the tube 1-2mm long, the lobes 0.5-2mm long; marginal prickles of nutlets in 1 row. Plains, hills, and disturbed areas. Statewide. *L. texana* (Scheele) Britt., *L. occidentalis* (Wats.) Greene, *L. foliosa* A. Nels., *L. cucullata* A. Nels.

Lithospermum L. Stoneseed; Puccoon

Perennial herbs with flowers in bracteate cymes or solitary in the upper axils; corolla salverform to campanulate, yellow or rarely orange-yellow; nutlets smooth or pitted, basally attached.

Corolla 10mm or less long, the tube about as long as the calyx, the lobes entire; nutlets often over 4mm long — *L. ruderale*

Corolla 10mm long or more, the tube usually about twice as long as the calyx or more, the lobes often erose or toothed; nutlets mostly 4mm long or less

- Corollas, or some of them, usually over 20mm long, the tube often 3-4 times the length of the calyx (later flowers smaller); stamens usually all borne near top of corolla tube; widespread — *L. incisum*
- Corollas not over 20mm long, the tube about twice the length of the calyx; stamens of some or all flowers borne near or below middle of corolla tube; SE Wyo. — *L. multiflorum*

Lithospermum incisum Lehm., Asperif. 2:303. 1818. Plants to 4dm high; leaf blades oblanceolate, lanceolate, linear, or oblong, 1-6cm long; calyx 5-10mm long; corolla tube (9)12-30mm long, the lobes 3-10mm long and erose or toothed; cleistogamous flowers often present later in season; nutlets somewhat pitted. Plains and hills. Statewide. *L. angustifolium* Michx., *L. asperum* A. Nels.

Lithospermum multiflorum Torr. ex Gray, Proc. Am. Acad. 10:51. 1874. Plants to 6dm high; leaf blades linear to lanceolate, 1-6cm long; calyx 4-7mm long; corolla tube 8-13mm long, the lobes 2-4mm long and entire or erose; cleistogamous flowers lacking; nutlets smooth or sparsely pitted. Hills, canyons, and slopes. Laramie Co.

Lithospermum ruderale Dougl. ex Lehm., Stirp. Pug. 2:28. 1830. Plants to 6dm high; leaf blades lanceolate or linear to elliptic, 2-10cm long; calyx 4-6mm long; corolla tube 4-7mm long, the lobes 2-5mm long and entire; nutlets smooth or very sparsely pitted. Plains, hills, woods, meadows, and slopes. Statewide. L. pilosum Nutt.

Mertensia Roth Bluebell

Perennial herbs with flowers in usually bractless cymes at ends of stems and branches; corolla tubular to funnelform, usually abruptly expanded at the throat, blue or occasionally pinkish; nutlets usually rugose or roughened. This genus is badly in need of revision. Species at the end of the key may not key out satisfactorily.

1. Plants usually over 4dm high, mostly in moist or shaded areas; cauline leaves with distinct lateral veins, the middle leaves often over 6cm long
   2. Calyx lobes 3-7mm long, acute at tip; corolla limb slightly longer than tube — *M. arizonica*
   2. Calyx lobes 1-3mm long, usually obtuse at tip; corolla limb usually about as long as tube or shorter — *M. ciliata*
1. Plants usually less than 4dm high, often in dry or open areas; cauline leaves usually lacking distinct lateral veins, the middle leaves only rarely over 6cm long
   3. Filaments attached down in the corolla tube, the anthers not projecting beyond the throat; filaments usually shorter and narrower than the anthers
      4. Leaves glabrous except on margins — *M. humilis*
      4. Leaves pubescent above and glabrous beneath
         5. Backs of calyx lobes glabrous; plants mostly alpine or subalpine — *M. alpina*
         5. Backs of calyx lobes pubescent; plants mostly not alpine or subalpine — *M. brevistyla*
   3. Filaments attached near throat of corolla, the anthers projecting beyond the throat; filaments often longer and wider than the anthers
      6. Limb of corolla shorter than the tube

7. Tube of mature corolla only slightly longer than the limb; some corolla tubes usually with a ring of hairs within below middle (the hairs rarely scattered) *M. viridis*

7. Tube of mature corolla usually 1.3-2 times as long as the limb; corolla tubes without a ring of hairs within *M. oblongifolia*

6. Limb of corolla longer than or equal to the tube

8. Leaves pubescent on both surfaces *M. lanceolata*

8. Leaves pubescent only above or glabrous on both surfaces

9. Calyx lobes pubescent on back *M. fusiformis*

9. Calyx lobes glabrous on back

10. Plants mostly of the plains and foothills, rarely to 8500 ft., in eastern Wyo. *M. lanceolata*

10. Plants mostly of the mountains, rarely down to 7000 ft., in western and SE Wyo. *M. viridis*

Mertensia alpina (Torr.) G. Don, Gen. Hist. Dichl. Pl. 4:372. 1838. Plants to 25(40)cm high; leaf blades lanceolate to oblanceolate, 1-6cm long; calyx 2-5mm long; corolla 5-12mm long, the tube 3-7mm long. Meadows and slopes in the high mountains. Statewide. *M. tweedyi* Rydb.

Mertensia arizonica Greene, Pittonia 3:197. 1897. Plants to 8dm high; leaf blades ovate to oblong-lanceolate or oblanceolate, 5-15cm long; calyx 4-8mm long; corolla 1-2cm long, the tube 4-9mm long. Woods and streambanks. Uinta and Sheridan Cos.

Mertensia brevistyla Wats., Bot. King Exp. 239. 1871. Plants to 4dm high; leaf blades lanceolate to oblanceolate, 1-10cm long; calyx 2-5mm long; corolla 6-10mm long, the tube 2-5mm long. Meadows, woods, and slopes. Carbon Co.

Mertensia ciliata (James ex Torr.) G. Don, Gen. Hist. Dichl. Pl. 4:372. 1838. Plants to 15dm high; leaf blades ovate or lanceolate to elliptic, or the lower sometimes deltoid or cordate, 1.5-20cm long; calyx 1-3.5mm long; corolla 8-17mm long, the tube 4-9mm long. Streambanks, wet meadows, and damp thickets or woods. Statewide.

Mertensia fusiformis Greene, Pittonia 4:89. 1899. Plants to 4dm high; leaf blades mostly oblong or oblanceolate to elliptic, 2-10cm long; calyx 3-6mm long; corolla 8-14mm long, the tube 3-7mm long. Plains, hills, and slopes. Sublette Co.

Mertensia humilis Rydb., Bull. Torrey Club 36:681. 1909. Plants to 2dm high; leaf blades ovate to oblanceolate, 1-6cm long; calyx 2.5-5mm long; corolla 6-13mm long, the tube 3-7mm long. Plains, hills, and slopes. NE, SE.

Mertensia lanceolata (Pursh) DC. ex A. DC. in DC., Prodr. 10:88. 1846. Plants to 45cm high; leaf blades mostly lanceolate to oblanceolate or oblong, 1-10cm long; calyx 2-7mm long; corolla 5-15mm long, the tube 2-8mm long. Plains, hills, and slopes. NE, SE. *M. linearis* Greene.

Mertensia oblongifolia (Nutt.) G. Don, Gen. Hist. Dichl. Pl. 4:372. 1838. Plants to 4dm high; leaf blades lanceolate to elliptic or oblanceolate, 2-13cm long; calyx 2-6mm long; corolla 1-2cm long, the tube 5-12mm long. Plains, hills, slopes, and meadows. SW, NW, NE. *M. oreophila* Williams, *M. foliosa* A. Nels.

Mertensia viridis (A. Nels.) A. Nels., Bull. Torrey Club 26:244. 1899. Plants to 4dm high; leaf blades oblanceolate to elliptic or lance-ovate, 1-10cm long; calyx 2-5mm long; corolla 6-15mm long, the tube 3-9mm long. Slopes and ridges in the mountains. NW, SW, SE. *M. coriacea* A. Nels., *M. amoena* A. Nels., *M. coronata* A. Nels., *M. perplexa* Rydb.

Myosotis L. Forget-me-not

Annual to perennial herbs with flowers in helicoid false racemes; corolla rotate to salverform, blue or white; nutlets smooth.

Calyx tube closely strigose, the hairs not spreading nor uncinate; stems often with a creeping or stoloniferous base _M. scorpioides_

Calyx tube with some loose or spreading somewhat uncinate hairs; stems not creeping or stoloniferous

Corolla limb 4-8mm wide, normally blue; plants perennial _M. alpestris_

Corolla limb 1-4mm wide, white; plants annual _M. verna_

Myosotis alpestris Schmidt, Fl. Boëm. 3:26. 1794. Fibrous rooted perennial to 4(6)dm high; leaf blades oblanceolate to lanceolate, 1-13cm long; calyx 1.5-4mm long; corolla blue, very rarely white, the tube 2-5mm long, the lobes 1-4mm long. Meadows and slopes in the mountains. SW, NW, NE.

Myosotis scorpioides L., Sp. Pl. 131. 1753. Fibrous rooted perennial to 6dm high, sometimes stoloniferous; leaf blades oblanceolate to lance-elliptic, 1-8cm long; calyx 1.5-5mm long; corolla blue, the tube 2-5mm long, the lobes 1-4mm long. Shallow water and wet places. Teton Co.

Myosotis verna Nutt., Gen. Pl. 2:Add. 1818. Fibrous rooted annual to 4dm high; leaf blades oblanceolate-obovate to elliptic or oblong, 1.5-5cm long; calyx 1.5-5mm long; corolla white, the tube 1.5-5mm long, the lobes 0.5-1.5mm long. Open areas. Crook Co.

Onosmodium Michx. Marbleseed

Onosmodium molle Michx., Fl. Bor. Am. 1:133. 1803. Perennial to 7dm high; leaf blades elliptic or lanceolate to ovate, 1-10cm long; flowers in helicoid cymes; calyx 5-9mm long; corolla tubular, white, greenish-white, or yellowish, 10-16mm long; nutlets smooth. Plains and hills. NE, SE. *O. occidentale* Mack.

Pectocarya DC. Combseed

Pectocarya linearis (R. & P.) A. DC., Prodr. 10:120. 1846. Prostrate or ascending annual with stems to 2dm long; leaf blades linear, 0.2-2cm long; flowers in false racemes; calyx 0.5-2mm long; corolla tubular, white, 1-1.5mm long; nutlets spreading, with hooked prickles on margins. Plains and hills, often where sandy. Carbon and Sweetwater Cos. *P. penicillata* (H. & A.) A. DC., *P. miser* A. Nels.

<u>Plagiobothrys</u> F. & M. Popcorn Flower

<u>Plagiobothrys</u> <u>scouleri</u> (H. & A.) Johnst., Contr. Gray Herb. n. s. 68:75. 1923. Annual with stems to 2dm long, often prostrate; leaf blades linear, opposite below, alternate above, 0.2-6cm long; flowers in a false raceme or spike; calyx 1.5-4mm long; corolla campanulate-salverform, white, 1.5-2mm long; nutlets rugose and sometimes tuberculate or with minute bristles. Moist places. Statewide. <u>Allocarya</u> <u>scopulorum</u> Greene, <u>A</u>. <u>nelsonii</u> Greene, <u>P</u>. <u>scopulorum</u> (Greene) Johnst., <u>P</u>. <u>hispidulus</u> (Greene) Johnst.

CACTACEAE Cactus Family

Plants with fleshy, usually leafless stems bearing spines in clusters; flowers perfect, regular, solitary or clustered; sepals and petals numerous, grading into one another, separate or united at base; stamens many; ovary 1, inferior; carpels 3 to many; style 1; locule 1; ovules numerous; placentation parietal; fruit usually berry-like.

Reference: Benson, L. 1970. Cactaceae. In D. S. Correll & M. C. Johnston, Manual of the Vascular Plants of Texas.

1. Stems flat or cylindrical and jointed; clusters of short, minutely barbed bristles usually present near base of the large spines — Opuntia
1. Stems globose or oval or rarely cylindrical, not jointed (clusters at ground level may appear jointed); clusters of short bristles lacking near base of the large spines
   2. Flowers arising from the sides of the stem; ovary and fruit spiny — Echinocereus
   2. Flowers arising from near the top of the stem; ovary and fruit usually not spiny
      3. Flowers arising from near base of the spine-bearing tubercles away from the base of spines — Coryphantha
      3. Flowers arising from near tip of the spine-bearing tubercles near base of the spines — Pediocactus

<u>Coryphantha</u> (Engelm.) Lem. Pincushion Cactus

Stems mostly globose, the spines at tips of mammillate tubercles; flowers near tip of stem, attached near base of tubercles; ovary and fruit lacking spines.

Petals greenish-white or yellowish; spines all white
<u>C</u>. <u>missouriensis</u>

Petals pink or pink-purple; some spines usually brown or reddish
<u>C</u>. <u>vivipara</u>

<u>Coryphantha</u> <u>missouriensis</u> (Sweet) Britt. & Rose in Britt. & Brown, Ill. Fl. 2nd ed. 2:570. 1913. Plants to 8cm high; areoles with 1 central spine 9-20mm long (or this lacking) and 10-20 smaller spines; flowers mostly 2-4cm long, the petals greenish-white or yellowish or sometimes reddish tinged. Plains and hills. Crook and Platte Cos. <u>Mammillaria</u> <u>missouriensis</u> Sweet, <u>Neobesseya</u> <u>missouriensis</u> (Sweet) Britt. & Rose.

<u>Coryphantha</u> <u>vivipara</u> (Nutt.) Britt. & Rose in Britt. & Brown, Ill. Fl. 2nd ed. 2:571. 1913. Plants to 12cm high; areoles with mostly 3-6 central spines 8-20mm long and 10-20 smaller spines; flowers 2-5cm long, the petals pink or pink-purple. Plains and hills. NW, NE, SE. <u>Mammillaria</u> <u>vivipara</u> (Nutt.) Haw.

Echinocereus Engelm. Hedgehog Cactus

Echinocereus viridiflorus Engelm. in Wisl., Mem. North. Mex. 91. 1848. Stems globose to cylindric, to 15cm high, with longitudinal ribs (or coalescent tubercles when young) on which the spines are borne; areoles with 0-3 central spines 7-20mm long and about 16 shorter radial spines; flowers 1.5-2.5cm long, borne on side of stem and from top of ribs, the petals greenish-yellow to reddish or purplish; ovary and fruit spiny. Plains and hills. SE.

Opuntia Mill. Prickly Pear Cactus

Stems flattened or terete, jointed; flowers in areoles; petals yellow to red in age.

Stem segments oval to cylindric, about as thick as wide, easily detached *O. fragilis*

Stem segments flattened, much wider than thick, not easily detached *O. polyacantha*

<u>Opuntia</u> <u>fragilis</u> (Nutt.) Haw., Suppl. Pl. Succ. 82. 1819. Plants somewhat spreading, to 2dm high; segments of stem slightly flattened to terete, the upper ones usually 2-5cm long; flowers 2.5-5cm long, the petals yellow. Plains and hills. NW, NE, SE.

<u>Opuntia</u> <u>polyacantha</u> Haw., Suppl. Pl. Succ. 82. 1819. Plants to 4dm high; segments of stem strongly flattened, the larger 5-15cm long; flowers 4-7cm long, the petals yellow or red in age. Plains and hills. Statewide. <u>O</u>. <u>rutila</u> Nutt.

<u>Pediocactus</u> Britt. & Rose Barrel Cactus

References: Benson, L. 1961. Cact. & Succ. Journ. 33:49-54.
1962. Cact. & Succ. Journ. 34:17-19, 57-61, 163-168.

<u>Pediocactus</u> <u>simpsonii</u> (Engelm.) Britt. & Rose in Britt. & Brown, Ill. Fl. 2nd ed. 2:570. 1913. Stems subglobose to depressed, to 15cm high; areoles on low tubercles which are in spiral rows, the central 4-12 spines 7-20mm long, with 10-30 smaller marginal spines; flowers 1.5-4cm long, borne near tip of stem, attached near tip of tubercles, the petals pink to white or sometimes purplish. Plains and hills. SW, SE. <u>Echinocactus</u> <u>simpsonii</u> Engelm.

CALLITRICHACEAE Water Starwort Family

Mostly aquatic herbs with opposite, entire, simple leaves; plants monoecious; flowers axillary, the 2 sexes often close together to appear like a single flower; perianth lacking but 2 bracts sometimes present; stamen 1; ovary 1; carpels 2; styles 2; locules 2 or 4; ovules 1 per locule; placentation axile; fruit a schizocarp, splitting into 2 or 4 segments with 1 seed each, the fruits 0.5-2mm long.

Callitriche L. Water Starwort

Leaf bases at each node not connected by a wing; floral bracts lacking — *C. hermaphroditica*

Leaf bases at each node usually connected by a narrow wing; floral bracts present but sometimes deciduous

- Height of fruit exceeding width by 0.2mm; carpels winged at summit; reticulation on mericarp in vertical lines — *C. palustris*
- Height of fruit exceeding width by 0.1mm or less; carpels wingless (sometimes with thickened margins); reticulation on mericarp not in vertical lines — *C. heterophylla*

Callitriche hermaphroditica L., Cent. Pl. 1:31. 1755. Stems to 2dm long; leaves linear, 3-15mm long, 1 nerved; floral bracts lacking. In ponds and streams. NW, SW, SE. *C. bifida* (L.) Morong.

Callitriche heterophylla Pursh, Fl. Am. Sept. 3. 1814. Stems to 4dm long; leaves linear and 1 nerved under water, floating and land leaves obovate and 3 nerved, 5-25mm long; floral bracts present. In ponds and streams and on mud. Sublette Co.

Callitriche palustris L., Sp. Pl. 969. 1753. Differing from above species by characters in the key. In ponds and streams and on mud. Statewide. *C. verna* L.

CAMPANULACEAE Bellflower Family

Herbs with simple, usually alternate leaves; flowers bisexual, regular or irregular, solitary and terminal or axillary, or in racemes, spikes, or cymes; calyx lobes usually 5, rarely 4; petals usually 5, united (lower flowers sometimes lacking corollas); stamens usually 5, free from the corolla or nearly so, sometimes united into a tube; ovary 1, at least half inferior; carpels 2-5; style 1; locules 1-5; ovules few to many; placentation usually axile, sometimes parietal or apparently basal; fruit a capsule.

Reference: McVaugh, R. 1943. N. Am. Fl. 32A:1-134.

1. Filaments of stamens separate (at least just below anthers); corolla regular
   2. Flowers usually distinctly pedicelled; perennials Campanula
   2. Flowers sessile or nearly so; annuals
      3. Flowers in axils of bracts Triodanis
      3. Flowers and bracts on opposite sides of stem Heterocodon
1. Filaments of stamens united into a tube; corolla irregular
   4. Flowers on pedicels; corolla tube nearly equaling the limb Porterella
   4. Flowers sessile (do not mistake hypanthium for a pedicel); corolla tube much shorter than the limb Downingia

Campanula L. Harebell; Bellflower

Perennial herbs with flowers solitary or in a racemiform inflorescence; corolla campanulate; filaments and anthers distinct except sometimes the very base of filaments.

Reference: Shetler, S. G. 1963. Rhodora 65:319-337.

Plants alpine or subalpine, mostly less than 15cm high; flowers mostly solitary on each stem; anthers 1-2.5mm long C. uniflora

Plants not with the above combination of characters

Plants usually with several flowers; lower leaves with mostly scabrous or entire margins at base; widespread
C. rotundifolia

Plants often with solitary flowers; lower leaves with ciliate margins at base; SE Wyo. *C. parryi*

Campanula parryi Gray, Syn. Fl. 2nd ed. 2(1):395. 1886. Plants to 35cm high, with rhizomes; leaf blades obovate, oblanceolate, elliptic, or linear, 1-5cm long; flowers mostly solitary but sometimes 2-4; calyx lobes 6-20mm long; corolla blue, 1-2cm long. Meadows, slopes, and streambanks. SE.

Campanula rotundifolia L., Sp. Pl. 163. 1753. Plants to 8dm high, rhizomatous or ultimately taprooted; leaf blades oblanceolate or linear (rarely lanceolate) or sometimes the lower ovate to suborbicular or cordate, 0.5-9cm long; flowers usually several or sometimes solitary; calyx lobes 3-12mm long; corolla blue, 1-2.5cm long. Plains, hills, slopes, and woods. Statewide.

Campanula uniflora L., Sp. Pl. 163. 1753. Plants to 15cm high, taprooted; leaf blades oblanceolate or elliptic to linear or the lower obovate, 0.3-3cm long; flowers usually solitary; calyx lobes 2-5mm long; corolla blue, 4-12mm long. High mountain slopes and meadows. Statewide.

Downingia Torr.

Downingia laeta (Greene) Greene, Leafl. Bot. Observ. Crit. 2:45. 1910. Fibrous rooted, glabrous, somewhat succulent annual to 2dm high; leaves sessile, linear to lanceolate or elliptic, 5-20mm long; flowers sessile and solitary in axils, with a long hypanthium; calyx lobes 2-8mm long; corolla bilabiate, white, blue, or purplish and tinged with yellow or white, 3-7mm long; filaments and anthers connate. Wet meadows, shores, and swamps. Uinta and Albany Cos.

Heterocodon Nutt.

Heterocodon rariflorum Nutt., Trans. Am. Phil. Soc. II, 8:255. 1842. Annual with lax stems to 3dm long; leaves sessile, ovate to cordate-orbicular, toothed or lobed, 2-8mm long; flowers sessile or subsessile, opposite the bracts, the lower cleistogamous without a corolla; calyx lobes 1.5-4mm long; corolla of upper flowers blue, 3-6mm long, campanulate; anthers and filaments free. Moist places. Teton Co.

Porterella Torr.

Reference: McVaugh, R. 1940. Bull. Torrey Club 67:796-797.

Porterella carnosula (H. & A.) Torr. in Porter in Hayden, Rep. U. S. Geol. Surv. Mont. 5:488. 1872. Fibrous rooted, glabrous, somewhat succulent annual to 3dm high; leaves sessile, linear to lanceolate, 0.3-5cm long; flowers pedicelled, solitary in axils; calyx lobes 3-8mm long; corolla bilabiate, blue with a yellow or white eye, 8-16mm long; filaments and anthers connate. Wet meadows and shores or in shallow water. Teton Co. P. eximia A. Nels., Laurentia eximia (A. Nels.) A. Nels.

Triodanis Raf. Venus' Looking Glass

Annuals with mostly sessile leaves; flowers 1 to several in axils, the lower cleistogamous with a reduced or no corolla; corolla of upper flowers funnelform or tubular-funnelform to campanulate; filaments and anthers free.

Reference: McVaugh, R. 1948. Rhodora 50:38-49.

Bracts of the flower lanceolate to linear, mostly over 5 times longer than wide T. leptocarpa
Bracts of the flower ovate to orbicular or cordate, mostly less than 4 times longer than wide

Flower bracts usually somewhat cordate at base and clasping; openings of capsule broadly elliptic to round, 0.6-1.5mm wide *T. perfoliata*

Flower bracts predominantly ovate to deltoid, usually not clasping; openings of capsule linear or oblong, 0.2-0.4mm wide *T. holzingeri*

Triodanis holzingeri McVaugh, Wrightia 1:45. 1945. Plants to 6dm high; leaves mostly ovate or deltoid or the lower oblanceolate or obovate, 5-25mm long, toothed; calyx lobes 2-8mm long; upper corollas blue or purple, 5-9mm long. Plains, hills, and canyons. Platte Co.

Triodanis leptocarpa (Nutt.) Nieuwl., Am. Midl. Nat. 3:192. 1914. Plants to 5dm high; leaves lanceolate to linear or the lower oblanceolate or rarely obovate, 1-3.5cm long, the margins crenate or entire, often scabrous; calyx lobes 4-15mm long; upper corollas blue-violet, 6-10mm long. Plains and hills. Sheridan Co. *Specularia leptocarpa* (Nutt.) Gray.

Triodanis perfoliata (L.) Nieuwl., Am. Midl. Nat. 3:192. 1914. Plants to 6dm high; leaves ovate to orbicular or cordate or some lower ones obovate, 4-25mm long, toothed; calyx lobes 2-8mm long; upper corollas purple or blue, 6-13mm long. In many habitats. NE, SE. *Specularia perfoliata* (L.) DC.

CANNABACEAE Hemp Family

Herbs or vines with simple or compound, alternate or opposite leaves; flowers mostly regular, unisexual, the plants dioecious; staminate flowers in racemes or panicles, the pistillate in dense clusters or spikes; sepals 5, all joined to form a cup in the pistillate flowers; petals lacking; stamens 5; pistil 1; ovary superior or inferior; carpels 2; styles 1 or 2; locule 1; ovule 1, pendulous; fruit an achene.

Leaves palmately compound; stems erect	Cannabis
Leaves simple, usually 3-5 lobed; stems twining	Humulus

Cannabis L.

Cannabis sativa L., Sp. Pl. 1027. 1753. Marijuana. Herbaceous annual to 2m high; leaves petioled, mostly alternate, sometimes opposite, palmately compound, the blades 5-15cm long, 5-15cm wide, the leaflets toothed; staminate flowers in panicles; pistillate flowers in small clusters, closely enveloped by the calyx. Disturbed areas. Albany Co.

Humulus L.

Humulus lupulus L., Sp. Pl. 1028. 1753. Hops. Herbaceous, perennial vines to 10m long; leaves petioled, opposite, the blades often 3-5 lobed, serrate, 4-12cm long, 3-10cm wide; staminate flowers in panicles; pistillate flowers in foliaceous-bracteate spikes 2-5cm long, yellow-glandular. Thickets and disturbed areas. Statewide.

CAPPARACEAE Caper Family

Herbs with simple or compound, alternate leaves; flowers perfect, regular or slightly irregular, in racemes; sepals 4, separate or united near base; petals 4, separate; stamens 6-16; ovary 1, superior; carpels 2 or 4; style 1; locule 1; ovules few to many; placentation parietal; fruit a capsule, often resembling a legume.

Stamens 6; plants glabrous or sparsely pilose — *Cleome*

Stamens 8 or more; plants glandular-pubescent — *Polanisia*

Cleome L. Bee Plant

Annual, glabrous to sparsely pilose herbs with palmately compound leaves at least below, the leaflets mostly 3-5; petals yellow, purple, or pink, or rarely white; stamens 6, conspicuously exserted; fruit a stipitate capsule resembling a legume.

Petals yellow; leaflets mostly 5 on lower leaves — *C. lutea*

Petals pink to purple or rarely white; leaflets mostly 3 on lower leaves — *C. serrulata*

Cleome lutea Hook., Fl. Bor. Am. 1:70. 1830. Plants to 15dm high; leaflets 3-7, oblanceolate or elliptic, 1-6cm long, 2-20mm wide; sepals 1.5-3mm long; petals yellow, 5-8mm long; capsules oblong-elliptic, mostly 15-40mm long, the stipe 10-15mm long. Plains and hills. SW, NW, NE.

Cleome serrulata Pursh, Fl. Am. Sept. 441. 1814. Plants to 15dm high; leaflets mostly 3, elliptic-lanceolate to oblanceolate or rarely linear, 1-7cm long, 1-18mm wide; sepals 2-3mm long; petals purple to pink or rarely white, 8-12mm long; capsules oblong-elliptic, 3-6cm long, the stipe 1-2cm long. Plains and hills or disturbed areas. Statewide.

Polanisia Raf. Clammy Weed

Reference: Iltis, H. H. 1958. Brittonia 10:33-58.

Polanisia trachysperma T. & G., Fl. N. Am. 1:669. 1840. Glandular-pubescent annual to 8dm high; leaves mostly 3 foliolate below, the leaflets ovate or elliptic to oblanceolate or obovate, 1-3.5cm long, 2-18mm wide; sepals 3-4mm long, deciduous; petals white or cream to purplish, 6-10mm long; stamens 8-16, exserted; capsule oblong-lanceolate, 2.5-5cm long, the stipe obsolete or to 3mm long. Plains and hills. Statewide. *P. dodecandra* (L.) DC. var. *trachysperma* (T. & G.) Iltis.

CAPRIFOLIACEAE Honeysuckle Family

Shrubs or rarely subshrubs, herbs, or vines with simple or compound, opposite leaves; flowers usually bisexual, rarely neutral, regular or irregular, in modified cymes, occasionally paired; sepals 5, mostly united (occasionally minute); petals 5 or rarely 4, united; stamens 5 or rarely 4, attached to corolla tube and alternate with the lobes; ovary 1, inferior; carpels 2-5; style 1 or obsolete; locules 2-5 or 1 by abortion; ovules 1 to several per locule; placentation usually axile or pendulous; fruit a berry or drupe, rarely dry.

1. Plants herbaceous or woody only at base, trailing along ground, 10cm or less high; flowers paired, nodding on a long peduncle — <u>Linnaea</u>
1. Plants woody, mostly upright or twining, usually over 10cm high; flowers not as above
   2. Leaves pinnately compound — <u>Sambucus</u>
   2. Leaves simple
      3. Leaf margins mostly with sharp pointed teeth; fruit 1 seeded — <u>Viburnum</u>
      3. Leaf margins entire or wavy or rarely sinuately lobed, lacking sharp pointed teeth; fruit more than 1 seeded
         4. Corolla regular or merely bulged on 1 side near middle; fruit with 2 seeds or stones — <u>Symphoricarpos</u>
         4. Corolla irregular, either 2 lipped or spurred or gibbous at base; fruit with several seeds — <u>Lonicera</u>

<u>Linnaea</u> L. Twinflower

<u>Linnaea</u> <u>borealis</u> L., Sp. Pl. 631. 1753. Creeping evergreen herb or subshrub to 10cm high; leaf blades mostly obovate to orbicular, 3-25mm long, entire or few toothed toward tip; flowers mostly in pairs at tips of long peduncles; calyx 2-4mm long; corolla funnelform to campanulate, pinkish, 6-15mm long; stamens 4; fruit dry and 1 seeded. Woods and thickets. NW, NE, SE. <u>L</u>. <u>americana</u> Forbes.

<u>Lonicera</u> L. Honeysuckle

Shrubs or vine-like with mostly entire, simple leaves; flowers paired on axillary peduncles or in terminal clusters; corolla often spurred or gibbous near base; ovary 2 or 3 locular; ovules 3-8 per locule; fruit a berry.

1. Terminal leaves usually perfoliate; flowers mostly in terminal clusters; climbing or scrambling shrubs or vines <u>L</u>. <u>dioica</u>
1. Terminal leaves distinct; flowers axillary, paired; upright shrubs

2. Bracts at summit of peduncle well over 5mm wide *L. involucrata*

2. Bracts at summit of peduncle about 1mm wide or less

3. Inner bracts fused and forming a cup which encloses the 2 ovaries or fruits *L. caerulea*

3. Inner bracts inconspicuous or lacking, not enclosing the 2 readily apparent ovaries or fruits *L. utahensis*

Lonicera caerulea L., Sp. Pl. 174. 1753. Shrub to 2m high; leaf blades elliptic or oblong to obovate, 1-6cm long; flowers paired on axillary peduncles; bracts subtending each flower pair lance-linear, 3-6(10)mm long; corolla yellow, 8-13mm long; fruit usually red. Moist places in the mountains. Yellowstone Park.

Lonicera dioica L., Syst. Nat. 12th ed. 2:165. 1767. Twining shrub or vine to 6m long; leaf blades elliptic or ovate, 2-10cm long, the upper usually perfoliate; flowers terminal in several verticils; corolla yellow, purplish, or reddish, 15-30mm long; fruit red. Woods. Crook Co. *L. glaucescens* (Rydb.) Rydb.

Lonicera involucrata (Richards.) Banks ex Spreng., Syst. Veg. 1:759. 1824. Shrub to 4m high; leaf blades elliptic, ovate, or elliptic-obovate, 2-14cm long; flowers paired on axillary peduncles, subtended by a double pair of broad bracts 8-17mm long which become reddish-purple in fruit; corolla yellow or tinged with red, 1-2cm long; fruit black or dark purple. Moist woods and thickets. Statewide.

Lonicera utahensis Wats., Bot. King Exp. 133. 1871. Shrub to 2m high; leaf blades elliptic, ovate, or oblong, 1.5-9cm long; flowers paired on axillary peduncles, subtended by linear or lance-linear bracts 1-3mm long; corolla white to yellow, 13-23mm long; fruit red or orange. Woods and slopes in the mountains. SW, NW, NE.

Sambucus L. Elderberry

Shrubs with pinnately compound leaves and a large amount of pith in the stems; flowers in compound umbelliform or paniculiform, cymose inflorescences; corolla rotate or nearly so; ovary 3-5 celled, 1 ovule per cell; fruit a berry.

Inflorescence usually as long as or longer than wide, pyramid-like or oval; pith of year old stems usually brownish — S. racemosa

Inflorescence wider than long, somewhat flat-topped or broadly rounded; pith of year old stems white

  Plants stoloniferous, forming large clumps; stems only slightly woody; fruit not glaucous — S. canadensis

  Plants not stoloniferous, usually in one small clump; stem definitely woody; fruit glaucous — S. cerulea

Sambucus canadensis L., Sp. Pl. 269. 1753. Stoloniferous shrub to 3m high; leaflets mostly 5-11, ovate, oval, or elliptic, 2-12cm long, serrate; inflorescence flat-topped or umbrella shaped; corolla white, 2-3mm long; fruit usually dark purple or black. Disturbed areas. Washakie Co.

Sambucus cerulea Raf., Alsogr. Am. 48. 1838. Shrub to 4m high; leaflets mostly 5-11, lanceolate, ovate, or elliptic, 4-15cm long, serrate; inflorescence flat-topped or umbrella shaped; corolla white or cream, 2-3mm long; fruit blue or blue-black. Woods, thickets, and slopes. Sheridan Co. S. glauca Nutt.

<u>Sambucus</u> <u>racemosa</u> L., Sp. Pl. 270. 1753. Shrub to 4m high; leaflets mostly 5-7, lanceolate, ovate, or elliptic, 2-18cm long, serrate or rarely lobed; inflorescence pyramidal or oval, paniculiform; corolla white or cream, 2-4mm long; fruit red or purple-black or rarely yellowish or white. Woods, thickets, and slopes. Statewide. <u>S</u>. <u>melanocarpa</u> Gray, <u>S</u>. <u>microbotrys</u> Rydb., <u>S</u>. <u>pubens</u> Michx.

<u>Symphoricarpos</u> Duhamel Snowberry; Buckbrush

Shrubs with simple, entire or sometimes sinuate or lobed leaves; flowers in short racemes or spikes or occasionally solitary or paired; corolla campanulate to funnelform, white or pink; ovary 4 locular, 2 with abortive ovules, 2 with 1 normal ovule each; berry white or sometimes orange or purplish.

Corolla evidently longer than wide, usually long tapering to base, not bulged on one side, the lobes mostly 1/4-1/2 as long as the tube — <u>S</u>. <u>oreophilus</u>

Corolla little if at all longer than wide, rather abruptly tapering to base, often bulged on one side, the lobes mostly 1/2 as long to exceeding the tube

Style over 3mm long, usually hairy near middle, projecting from the corolla; anthers mostly 1.5-2mm long *S. occidentalis*

Style 3mm or less long, glabrous, included; anthers mostly 1-1.5mm long *S. albus*

Symphoricarpos albus (L.) Blake, Rhodora 16:118. 1914. Shrub to 2m high; leaf blades elliptic to ovate or oval, 1-7cm long; calyx 1-2mm long; corolla 5-7mm long; style 2-3mm long, glabrous. Woods, thickets, and slopes. Statewide. *S. racemosus* Michx., *S. rivularis* Suksd.

Symphoricarpos occidentalis Hook., Fl. Bor. Am. 1:285. 1833. Shrub to 15dm high; leaf blades elliptic to ovate or oval, 1-8cm long; calyx 1-2mm long; corolla 5-8mm long; style (3)4-7mm long, hairy near middle or occasionally glabrous. Woods, hills, and shores. Statewide.

Symphoricarpos oreophilus Gray, Journ. Linn. Bot. Soc. 14:12. 1873. Shrub to 15dm high; leaf blades elliptic to ovate or oval, 0.5-4cm long; calyx 1-2.5mm long; corolla 6-10mm long; style 2-6mm long, usually glabrous. Woods, meadows, hills, and slopes. Statewide. *S. tetonensis* A. Nels., *S. rotundifolius* Gray of authors, *S. vaccinioides* Rydb., *S. utahensis* Rydb.

Viburnum L.

Shrubs with simple, toothed or lobed leaves; flowers in umbelliform or paniculiform cymes; corolla campanulate to rotate; ovary 3 locular but 2 reduced and sterile, the fruit a 1 seeded drupe.

Leaves not lobed — V. lentago

Leaves, or some of them, 3 lobed

- Leaves with linear stipules at base of petioles — V. opulus
- Leaves without stipules — V. edule

Viburnum edule (Michx.) Raf., Med. Repos. II, 5:354. 1808. Squashberry. Shrub to 2.5m high; leaf blades somewhat oval and 3 lobed and toothed or some lanceolate or ovate and toothed or subentire but not lobed, 2-10cm long; calyx lobes about 0.5mm long; corolla whitish, 3-6mm long; fruit red or orange. Moist woods and swamps. Park and Albany Cos. V. pauciflorum Pylaie ex T. & G.

Viburnum lentago L., Sp. Pl. 268. 1753. Nannyberry. Shrub to 6m high; leaf blades ovate to elliptic-lanceolate, serrate, acuminate, 2-10cm long; calyx lobes 0.5-1mm long; corolla white or cream, 2-4mm long; fruit blue or blue-black. Woods, thickets, and streambanks. Sheridan and Crook Cos.

Viburnum opulus L., Sp. Pl. 268. 1753. High-bush Cranberry. Shrub to 4m high; leaf blades somewhat cuneate-oval to suborbicular, 3 lobed, 3-12cm long, subentire to serrate-dentate; calyx about 1mm or less long; corolla white, those of the perfect flowers about 2mm long, the neutral to 13mm long; fruit red. Moist woods and thickets. Crook Co.

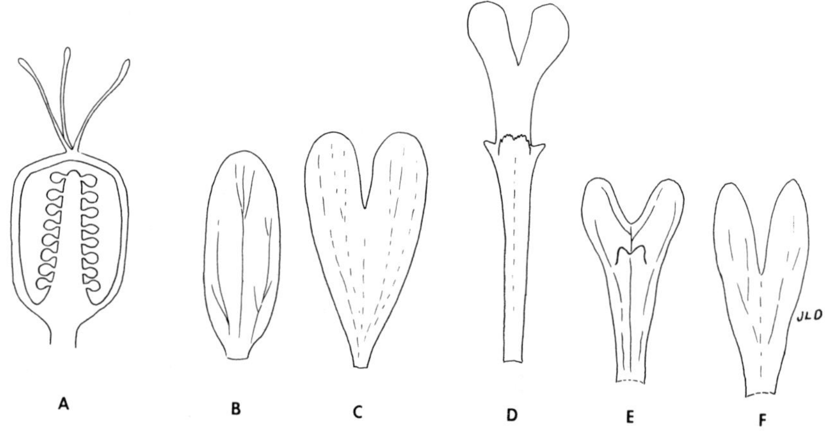

Figure 4. Caryophyllaceae. A. Longitudinal section of ovary showing free-central placentation (diagramatic). B-F. Petals: B. Arenaria hookeri (X 7); C. Cerastium arvense (X 4); D. Lychnis alba (X 2); E. Silene menziesii (X 4); F. Stellaria longipes (X 7).

Herbs with simple, usually opposite (rarely whorled) leaves; flowers bisexual or rarely unisexual, regular, solitary or in cymes; sepals 4 or 5, united or not; petals 3-5, usually separate, rarely lacking; stamens 3-10; ovary 1, superior; carpels 1-5; styles 1-5; locules 1-5; ovules usually numerous but sometimes solitary; placentation free-central, basal, or axile; fruit a capsule or utricle.

References: Maguire, B. 1950. Rhodora 52:233-245.
Hartman, R. L. 1971. The family Caryophyllaceae in Wyoming. M. S. Thesis. Univ. Wyo., Laramie.

1. Sepals united
   2. Flowers all staminate
      3. Plants erect biennials or perennials; calyx 12-25mm long, with acute, triangular calyx teeth mostly 2-5mm long <u>Lychnis</u>
      3. Plants either mat forming perennials, or annuals with mostly linear calyx teeth (rarely triangular with subulate tips and 2mm or less long), or perennials with the calyx 5-8mm long <u>Silene</u>
   2. Flowers pistillate or bisexual
      4. Styles solitary, cleft above; calyx lobes spine-tipped; petals lacking <u>Paronychia</u>
      4. Styles 2 or more; calyx lobes rarely spine-tipped; petals present or lacking

5. Calyx lobes 15-40mm long, usually longer than the tube — Agrostemma

5. Calyx lobes mostly less than 10mm long, usually shorter than the tube

6. Flowers immediately subtended by long-tapering bracts — Dianthus

6. Flowers not immediately subtended by bracts, the bracts lacking or usually at some distance below calyx

7. Calyx 1-3mm long — Gypsophila

7. Calyx usually 5mm or more long

8. Styles usually 2, rarely 3 (check several flowers)

9. Petals with 2 linear appendages at junction of claw and blade; calyx terete — Saponaria

9. Petals lacking appendages at junction of claw and blade; calyx 5 angled — Vaccaria

8. Styles 3-5

10. Styles usually 5; valves of capsule usually 5 or 10 — Lychnis

10. Styles usually 3; valves of capsule usually 3 or 6 — Silene

1. Sepals separate or nearly so

11. Styles solitary and cleft above, or lacking and with 3 stigmas; calyx lobes spine-tipped; petals lacking or minute

12. Plants perennial; stipules prominent and scarious; fruit a 1 seeded utricle <u>Paronychia</u>

12. Plants annual; stipules inconspicuous or lacking; fruit a several seeded capsule <u>Loeflingia</u>

11. Styles 2 or more; calyx lobes rarely spine-tipped; petals present or lacking

13. Leaves with distinct scarious stipules <u>Spergularia</u>

13. Leaves lacking stipules (sometimes with secondary leaves in their axils or the bases of opposite pairs connate)

14. Styles usually 5; capsule dehiscent by 10 teeth or valves; petals lobed or notched; basal rosette of leaves usually lacking <u>Cerastium</u>

14. Styles either 3 or 4, or if 5, the capsule dehiscent by less than 10 teeth or valves, the petals usually entire, and often with a basal rosette of leaves present

15. Stems mostly less than 5cm long; basal rosette of leaves present; styles usually 5 <u>Sagina</u>

15. Stems often over 5cm long; basal rosette of leaves usually lacking; styles usually 3, rarely more

16. Petals deeply lobed or sometimes very small or lacking; stamens and petals inserted under the ovary; capsules 6 valved <u>Stellaria</u>

16. Petals entire or rarely shallowly notched (sometimes lacking in <u>A</u>. <u>rossii</u>); stamens and petals often inserted at the edge of a prominent disc; capsules 3 or 6 valved <u>Arenaria</u>

Agrostemma L. Corn Cockle

Agrostemma githago L., Sp. Pl. 435. 1753. Taprooted, hirsute annual to 1m high; leaves linear-lanceolate, 3-12cm long, sessile; flowers appearing terminal on main and lateral stems; calyx tube 12-16mm long at anthesis, the lobes usually longer than the tube; petals red, often drying purplish, about twice as long as calyx tube; stamens 10; styles 5; capsule 5 valved. Disturbed areas. NW, NE, SE.

Arenaria L. Sandwort

Flowers in open to capitate cymes or occasionally solitary; sepals 5, free or connate only at base; petals 5 or lacking, entire or slightly emarginate, white; stamens usually 10; styles usually 3; capsule 3 or 6 valved.

1. Leaves oblong to elliptic or lanceolate, usually 3mm or more wide _A. lateriflora_
1. Leaves linear or awl-shaped, less than 2mm wide
  2. Plants mostly matted and less than 10cm high; primary leaves mostly less than 10mm long, often with secondary leaves in their axils; capsule 3 valved or 6 valved in _A. hookeri_

3. Capsule 6 valved; sepals 6-8mm long; flowers several to many in a congested inflorescence *A. hookeri*

3. Capsule 3 valved; sepals 6mm or less long; flowers 1 to several in an open inflorescence

 4. Plants glabrous

  5. Inflorescence 1 or 2 flowered, the flowering stems usually less than 3cm high; leaves 1 nerved; plants alpine or subalpine *A. rossii*

  5. Inflorescence 1 to several flowered, the flowering stems over 5cm high; leaves 3 nerved; plants below subalpine *A. stricta*

 4. Plants glandular-pubescent

  6. Leaves mostly pungent, 3 nerved, sometimes obscurely so; seeds about 1.5mm long; sepals acuminate to pungent; flowering stems usually brittle and shattering at nodes when mature *A. nuttallii*

  6. Leaves mostly obtuse or rarely slightly mucronate, 1 or 3 nerved; seeds mostly 1mm or less long; sepals obtuse to acute; flowering stems not brittle

   7. Sepals 3-5mm long, usually with a purplish, hooded tip; seeds 0.8-1mm long *A. obtusiloba*

   7. Sepals 2.5-4mm long, not hooded although sometimes purplish; seeds rarely as much as 0.8mm long *A. rubella*

2. Plants mostly over 10cm high and usually not mat forming; leaves often over 10mm long, rarely with secondary leaves in their axils; capsule 6 valved
   8. Inflorescence congested, most of the flowers with pedicels shorter than or equal to the flower — A. congesta
   8. Inflorescence relatively open, many of the flowers on pedicels much longer than the flower
      9. Sepals ovate, acute; petals usually exceeding sepals by 1mm or more — A. kingii
      9. Sepals linear-lanceolate to lanceolate, somewhat acuminate; petals barely exceeding or not exceeding sepals
         10. Inflorescence glandular-pubescent; basal leaves often over 3cm long — A. fendleri
         10. Inflorescence glabrous; basal leaves less than 3cm long — A. eastwoodiae

Arenaria congesta Nutt. in T. & G., Fl. N. Am. 1:178. 1838. Caespitose perennial to 4dm high; leaves linear, 0.5-8cm long, mostly less than 1mm wide; flowers mostly in head-like clusters; sepals 3-6mm long; petals 1.5-2 times as long as calyx; capsule 6 valved. Sagebrush flats to alpine. Statewide.

<u>Arenaria</u> <u>eastwoodiae</u> Rydb., Bull. Torrey Club 31:406. 1904. Caespitose perennial to 2dm high; leaves linear, 0.5-2cm long, less than 1mm wide, pungent; flowers in an open cyme; sepals 4-6mm long; petals about equaling sepals; capsule 6 valved. Dry hills. Carbon Co.

<u>Arenaria</u> <u>fendleri</u> Gray, Pl. Fendl. 13. 1849. Caespitose perennial to 3dm high; leaves linear, 1-8cm long, less than 1mm wide, somewhat pungent; flowers in an open cyme; sepals 4-6mm long; petals about as long as sepals; capsule 6 valved. Plains and hills. Albany and Laramie Cos.

<u>Arenaria</u> <u>hookeri</u> Nutt. in T. & G., Fl. N. Am. 1:178. 1838. Caespitose, usually matted perennial to 15cm high; leaves linear, mostly 3-10mm long and about 1mm wide; flowers in head-like cymes; sepals 6-8mm long; petals slightly longer than calyx; capsule 6 valved. Plains to lower mountain slopes. Statewide. <u>A</u>. <u>pinetorum</u> A. Nels.

<u>Arenaria</u> <u>kingii</u> (Wats.) Jones, Proc. Calif. Acad. Sci. II, 5:627. 1895. Caespitose perennial to 3dm high; leaves linear, 0.5-6cm long, 0.5-1mm wide; flowers in an open cyme; sepals 3-5mm long; petals 1.5-2 times as long as sepals; capsule 6 valved. Sagebrush flats to alpine. SW. <u>A</u>. <u>uintahensis</u> A. Nels.

Arenaria lateriflora L., Sp. Pl. 423. 1753. Rhizomatous perennial to 2dm high; leaves elliptic-oblong to lanceolate, 0.5-3cm long, 2-12mm wide; flowers 1-6 in terminal and lateral cymes; sepals 2-3mm long; petals 2-3 times as long as sepals; capsule 6 valved. Moist woods, meadows, and slopes. Statewide. Moehringia lateriflora (L.) Fenzl.

Arenaria nuttallii Pax, Bot. Jahrb. 18:30. 1893. Caespitose perennial with a branching caudex, to 10cm high; leaves linear to narrowly lanceolate, 4-10mm long, 0.5-1.3mm wide; flowers in a loose cyme; sepals 4-6mm long; petals 2-7mm long; capsule 3 valved. Sagebrush flats to alpine. Statewide.

Arenaria obtusiloba (Rydb.) Fern., Rhodora 21:14. 1919. Caespitose perennial, often mat forming, to 10cm high; stems often trailing; leaves linear, mostly 3-10mm long, less than 1mm wide; flowers mostly 1 or 2 per stalk; sepals mostly 3-5mm long; petals 3-9mm long; capsule 3 valved. Alpine or subalpine or lower talus slopes. Statewide. A. sajanensis Willd. ex Schlecht. of authors.

Arenaria rossii R. Br. ex Richards. in Frankl., Narr. 1st Journ. 738. 1823. Mat forming perennial to 5cm high; leaves linear, mostly 4-6mm long, less than 1mm wide; flowers mostly 1 or 2 per stalk; sepals 2.5-3.5mm long; petals about equaling sepals or more often vestigial or lacking; capsule 3 valved. Alpine and subalpine. Park and Sublette Cos.

Arenaria rubella (Wahlenb.) Smith, Engl. Fl. 4:267. 1828. Mat forming perennial to 10cm high; leaves mostly linear, 3-10mm long, less than 1mm wide; flowers 1-7 in an open cyme; sepals 2.5-4mm long; petals slightly shorter to slightly longer than sepals; capsule 3 valved. Moist or rocky areas in the high mountains. Statewide. A. aequicaulis (A. Nels.) A. Nels.

Arenaria stricta Michx., Fl. Bor. Am. 1:274. 1803. Plants annual to perennial, to 25cm high; leaves linear, 3-10mm long, less than 1mm wide; flowers in an open cyme; sepals 2-3.5mm long; petals 1.5-4mm long; capsule 3 valved. Moist areas. Teton Co.

Cerastium L. Mouse-ear Chickweed

Flowers usually in terminal open or congested cymes; sepals 5, distinct to base; petals 5, rarely lacking, white, deeply notched or bilobed; stamens mostly 10, rarely 5; styles 5 or rarely 4; capsule dehiscing by 10 usually revolute-margined teeth.

1. Plants annual, sometimes decumbent but not rooting at nodes *C. nutans*
1. Plants biennial or perennial, often rooting at nodes
   2. Petals subequal to sepals *C. fontanum*
   2. Petals 1.5 times as long as sepals or more
      3. Bracts of inflorescence scarious margined; leaves of flowering stems usually with secondary leaves or sterile branches in their axils; alpine or not *C. arvense*
      3. Bracts of inflorescence usually not scarious margined; leaves of flowering stems without secondary leaves or sterile branches in their axils; mostly alpine *C. beeringianum*

Cerastium arvense L., Sp. Pl. 438. 1753. Caespitose perennial with trailing stems, to 5dm high; leaves linear to oblong-lanceolate, 0.5-3cm long, 0.5-5mm wide, often with fascicled secondary leaves in the axils; sepals 4-7mm long, glandular; petals usually about twice as long as calyx. Plains to alpine. Statewide. *C. fuegianum* (Hook.) A. Nels., *C. elongatum* Pursh, *C. oreophilum* Greene.

Cerastium beeringianum Cham. & Schlecht., Linnaea 1:62. 1826. Caespitose perennial to 2dm high, glandular, often trailing; leaves oblong-lanceolate to spatulate, 4-30mm long, 1-9mm wide; sepals 4-6mm long; petals usually 1.5-2 times as long as sepals. Alpine and subalpine. NW, SW, NE. *C. buffumae* A. Nels.

Cerastium fontanum Baumg., Enum. Stirp. Transs. 1:425. 1816. Biennial or perennial with upright to prostrate stems, glandular, to 5dm high; leaves mostly oblanceolate, 0.5-3cm long, 1-7mm wide; sepals 4-8mm long; petals about as long as sepals. Disturbed areas. NW, SE. C. vulgatum L. of authors.

Cerastium nutans Raf., Precis. Decouv. 36. 1814. Annual with decumbent to erect stems, usually glandular, to 4dm high; leaves lanceolate to spatulate, 0.8-6cm long, 3-12mm wide; sepals 3-5mm long; petals slightly shorter to 1.5 times as long as calyx or sometimes lacking. Woods and meadows. NE, SE. C. brachypodum (Engelm.) Robins., C. longipedunculatum Muhl.

Dianthus L. Pink

Dianthus armeria L., Sp. Pl. 410. 1753. Annual or biennial to 6dm high; leaves mostly linear, sessile, 3-8cm long, 1-4mm wide; flowers in a congested cyme; sepals 5, united, subtended by several pair of imbricate bracts, the calyx about 2cm long; petals 5, pink or red, 20-25mm long; stamens 10; styles 2; capsule dehiscing by 4 valves. Disturbed areas. Teton Co.

<u>Gypsophila</u> L. Baby's Breath

<u>Gypsophila paniculata</u> L., Sp. Pl. 407. 1753. Glabrous perennial to 1m high; leaves lanceolate to linear, sessile, 1-10cm long, 1-10mm wide; flowers many in a much branched cyme; sepals 5, united, about 2mm long, white margined; petals 5, white, slightly longer than calyx; stamens 10; styles 2; capsule 4 valved. Disturbed areas. Albany and Platte Cos.

<u>Loeflingia</u> L.

Reference: Barneby, R. C. & E. C. Twisselmann. 1970. Madroño 20:398-408.

<u>Loeflingia squarrosa</u> Nutt. in T. & G., Fl. N. Am. 1:174. 1838. Spreading, glandular-pubescent annual, the stems to 15cm long; leaves linear or linear-lanceolate, 4-6mm long, less than 1mm wide, spine-tipped; stipules often present but inconspicuous; flowers axillary and sessile or nearly so; sepals 5, 4-6mm long; petals 3-5, minute; stamens 3-5; style 1 or lacking, the stigmas 3; capsule 3 valved. Dry plains and hills. Sweetwater Co.

Lychnis L. Campion

Biennial or perennial herbs with opposite, entire leaves; flowers bisexual or unisexual, in a cyme; sepals 5, united; petals 5, often variously lobed and often with auricles or appendages at junction of blade and claw; stamens 10; styles 5 or rarely 4; capsules mostly 5 or 10 valved, rarely 4. _L. kingii_ Wats. has been reported for Wyoming but its presence could not be confirmed. It might not be distinct from _L. apetala_ from which it reputedly differs in having erect, not drooping buds.

Flowers unisexual; blades of petals over 7mm long _L. alba_
Flowers mostly perfect; blades of petals less than 5mm long
  Plants mostly less than 20cm high; flowers usually 1 or 2 per stem; seeds usually wing-margined; alpine _L. apetala_
  Plants mostly 20cm or more high; flowers 1 to several; seeds not wing-margined; alpine or not _L. drummondii_

Lychnis alba Mill., Gard. Dict. 8th ed. Lychnis No. 4. 1768. Plants usually dioecious, to 11dm high; cauline leaves lanceolate to elliptic, 2-12cm long, 5-35mm wide; calyx 12-25mm long; petals white, 2-3cm long, the blade bilobed to about the middle. Disturbed areas. Statewide.

Lychnis apetala L., Sp. Pl. 437. 1753. Stems to 25cm high; leaves mostly basal, short-petioled, the blades oblanceolate, 0.5-4cm long, 0.5-4mm wide; calyx 8-16mm long; petals pink to purple, slightly longer to shorter than calyx, shallowly bilobed. Alpine and subalpine. NW, NE. *L. montana* Wats., *Silene hitchguirei* Bocq.

Lychnis drummondii (Hook.) Wats., Bot. King Exp. 37. 1871. Stems to 5dm high; leaf blades linear to oblanceolate or obovate, 2-12cm long, 2-15mm wide; calyx 1-2cm long; petals white to pink, 12-20mm long, shallowly lobed at most. Sagebrush slopes to alpine talus. Statewide. *L. striata* Rydb., *L. pudica* Boivin. *Silene parryi* sometimes has 5 styles and is confused with *L. drummondii*. The petals of *S. parryi* have lateral teeth and a prominently bilobed blade 3-7mm long. The petals of *L. drummondii* lack lateral teeth and have a blade at most shallowly bilobed and 1-3mm long.

Paronychia Adans. Whitlow-wort

Low perennials with opposite leaves and scarious stipules; flowers solitary or in few flowered cymes; sepals 5, united at base, spinulose-tipped; petals lacking; stamens 5, sometimes alternating with staminodia; style 1, cleft at tip; fruit a 1 seeded utricle.

1. Plants mostly over 10cm high, not mat forming; some leaves over 10mm long *P. jamesii*
1. Plants mostly 10cm or less high, usually mat forming or the stems prostrate; leaves mostly all less than 10mm long
   2. Plants mostly alpine or subalpine; leaves usually ovate or obovate to elliptic or oblong, rarely linear, not spinulose-tipped, the stipules usually broadly ovate *P. pulvinata*
   2. Plants mostly below subalpine; leaves mostly linear, sometimes spinulose-tipped, the stipules mostly linear to lanceolate
      3. Flowers usually clustered; some leaves often over 6mm long, often widest at or above middle *P. depressa*
      3. Flowers mostly solitary or paired; leaves mostly less than 6mm long, usually widest near base *P. sessiliflora*

Paronychia depressa (T. & G.) Nutt. ex A. Nels., Bull. Torrey Club 26:236. 1899. Plants usually mat forming, to 10cm high; leaves linear, 2-10mm long, often widest at or above middle; stipules 1-7mm long; flowers clustered; calyx 2-3.5mm long. Dry plains and hills. NE, SE. *P. diffusa* A. Nels.

Paronychia jamesii T. & G., Fl. N. Am. 1:170. 1838. Stems caespitose, not mat forming, to 3dm high; leaves linear, 3-20mm long; stipules 1-15mm long; flowers numerous, in a cyme; calyx 2-3mm long. Dry plains and hills. Albany Co.

Paronychia pulvinata Gray, Proc. Acad. Phila. 1863:58. 1863. Plants mat forming, to 5cm high; leaves mostly ovate or obovate to oblong or elliptic, 3-6mm long; stipules 2-5mm long; flowers mostly solitary, scattered amongst the leaves; calyx 2-3mm long. Alpine and subalpine. Albany Co.

Paronychia sessiliflora Nutt., Gen. Pl. 1:160. 1818. Plants mat forming, to 8cm high; leaves linear, crowded, 2-6mm long, usually widest near base; flowers terminal and solitary or sometimes several; calyx 2.5-4.5mm long. Dry plains and hills. Statewide. *P. brevicuspis* (A. Nels.) Rydb.

Sagina L. Pearlwort

Sagina saginoides (L.) Karsten, Deutsche. Fl. Pharm. Bot. 539. 1881. Glabrous, often matted, annual or perennial to 5cm high; leaves rosulate and cauline, linear, mostly 3-15mm long; flowers solitary and terminal or in leaf axils; sepals 4 or 5, separate, 1.5-2.5mm long; petals usually 4 or 5, 1-3mm long, white; styles 4 or 5; stamens usually 10; capsule 4 or 5 valved. Moist areas. Statewide.

<u>Saponaria</u> L. Soapwort

<u>Saponaria</u> <u>officinalis</u> L., Sp. Pl. 408. 1753. Rhizomatous perennial to 9dm high; leaves sessile or short petioled, the blades ovate or lanceolate to obovate, mostly 3-10cm long, 1-3.5cm wide; flowers in congested cymes; calyx 15-25mm long, 5 lobed; petals 5, white to pink, 25-40mm long, with 2 linear appendages at junction of claw and blade; styles 2; stamens 10; capsule usually 4 valved. Disturbed areas. Lincoln Co.

<u>Silene</u> L. Catchfly

Annual to perennial herbs with opposite or whorled, entire leaves; flowers bisexual or rarely unisexual, solitary or in cymes; sepals 5, united; petals 5, white, pink, or purple, often with auricles or appendages at junction of blade and claw; stamens 10; styles usually 3, occasionally 4 or 5; capsules dehiscing by mostly 6 or rarely 8-10 teeth.

1. Plants alpine, forming dense cushion mats less than 6cm high; leaves mostly linear, less than 1.5mm wide; flowers pink or purple, solitary on each stem <u>S</u>. <u>acaulis</u>
1. Plants not as above

2. Plants annual, weedy
   3. Plants glabrous above, often glandular in bands beneath the nodes; blade of petals 2-4mm long — *S. antirrhina*
   3. Plants glandular-pubescent throughout; blade of petals mostly 7-10mm long — *S. noctiflora*
2. Plants perennial, usually not weedy
   4. Calyx glabrous except sometimes the margins ciliate — *S. cucubalus*
   4. Calyx pubescent, usually with glandular hairs
      5. Calyx 8mm or less long, the nerves often obscure, not purple — *S. menziesii*
      5. Calyx usually over 9mm long, the nerves prominent and often purple
         6. Petals 4 lobed or the middle 2 lobes again divided, with 4 appendages — *S. oregana*
         6. Petals 2 lobed, or 4 lobed with 2 appendages
            7. Claw of petal often erose-margined above, the blade usually with a pair of lateral teeth; capsules mostly 1 celled — *S. parryi*
            7. Claw of petal usually entire above, the blade lacking lateral teeth; capsule 3 or 4 celled nearly to tip — *S. repens*

<u>Silene</u> <u>acaulis</u> (L.) Jacq., Enum. Stirp. Vindob. 78, 242. 1762. Moss Campion. Mat forming perennial to 6cm high; leaves mostly basal, sessile, linear to linear-lanceolate, 3-10mm long, to 1.5mm wide; flowers solitary, often unisexual; calyx 4-8mm long; petals 6-12mm long, pink to lavender; styles 3; capsule usually 3 celled. Mostly alpine. Statewide.

<u>Silene</u> <u>antirrhina</u> L., Sp. Pl. 419. 1753. Annual to 8dm high; cauline leaves oblanceolate to linear, 1-6cm long, 1-13mm wide; calyx 4-10mm long; petals white to pink, 5-12mm long; styles 3; capsule usually 3 celled. Disturbed areas. NE, SE.

<u>Silene</u> <u>cucubalus</u> Wibel, Prim. Fl. Werth. 241. 1799. Perennial to 1m high; leaf blades elliptic-lanceolate to oblanceolate, 3-8cm long, 0.5-2cm wide; calyx 1-2cm long; petals white, 14-18mm long; styles usually 3; capsule usually 3 celled. Disturbed areas. Carbon and Park Cos.

<u>Silene</u> <u>menziesii</u> Hook., Fl. Bor. Am. 1:90, pl. 30. 1830. Usually dioecious perennial with decumbent or ascending stems to 3dm long; leaves elliptic to oblanceolate or obovate, 2-5cm long, 3-25mm wide, sessile or nearly so; calyx 5-8mm long; petals white, 6-10mm long; styles 3 or rarely 4; capsule 1 celled. Woods, slopes, and stream banks. Statewide.

Silene _noctiflora_ L., Sp. Pl. 419. 1753. Annual to 6dm high; leaf blades ovate-lanceolate to elliptic-oblanceolate, 2-10cm long, 0.5-4cm wide, the lower ones petioled; flowers usually bisexual, rarely unisexual; calyx 15-30mm long; petals white to pink, 19-35mm long; styles 3; capsule usually 3 celled. Disturbed areas. Natrona and Albany Cos.

Silene _oregana_ Wats., Proc. Am. Acad. 10:343. 1875. Perennial to 5dm high; leaves mostly basal, the blades oblanceolate to linear, 2-10cm long, 1-30mm wide, at least the lower petioled; calyx 10-18mm long; petals pinkish-white, 15-25mm long; styles 3 or rarely 4 or 5; capsule 1 celled above, 3-5 celled below. Sagebrush, slopes, and open woods. NW, NE.

Silene _parryi_ (Wats.) Hitchc. & Maguire, Univ. Wash. Publ. Biol. 13:36. 1947. Perennial to 4dm high; leaf blades linear or linear-oblanceolate to spatulate, 2-8cm long, 1-10mm wide, the lower mostly petioled; calyx 10-16mm long; petals white or green to purplish, 13-18mm long; styles 3 or rarely 4 or 5; capsule 1 celled; carpophore usually 2.5-4mm long. Meadows and slopes in the mountains. NW, SW, SE. _S._ _tetonensis_ E. Nels. This species may not be distinct from _S._ _scouleri_ Hook.

<u>Silene</u> <u>repens</u> Patrin ex Pers., Syn. 1:500. 1805. Perennial with mostly erect stems from a trailing base, to 3dm high; leaves linear to lanceolate or oblanceolate, 1-4cm long, 2-6mm wide, mostly sessile; calyx 10-15mm long; petals white to pink or purple, 8-15mm long; styles 3 or rarely 4; capsule 3 or rarely 4 celled; carpophore 4-7mm long. Rocky slopes in the mountains. Teton and Sublette Cos.

<u>Spergularia</u> (Pers.) J. & K. Presl Sand Spurrey

Mostly annual herbs with linear, opposite, sometimes fascicled leaves with prominent scarious stipules; flowers in leafy cymes; sepals 5, separate; petals 5 or rarely lacking, white or pink, usually shorter than sepals; stamens 2-10; styles 3; capsule 3 valved.

Stamens usually 10, sometimes as few as 6; secondary leaves usually clustered in axils of primary leaves <u>S</u>. <u>rubra</u>
Stamens 2-5; secondary leaves not in axils of primary <u>S</u>. <u>marina</u>

Spergularia marina (L.) Griseb., Spicil. Fl. Rumel. 1:213. 1843. Stems ascending to erect, to 35cm long; leaves usually sparsely glandular-pubescent, 0.5-3cm long, to 1.5mm wide; sepals 2.5-5mm long; petals white to pink, 1.5-4mm long; stamens 2-5. Alkaline areas. SW, SE, NE. *Tissa sparsiflora* Greene, *S. sparsiflora* (Greene) A. Nels.

Spergularia rubra (L.) J. & K. Presl, Fl. Cechica 94. 1819. Stems mostly prostrate, to 3dm long; leaves usually glabrous, 3-15mm long, to 1.5mm wide; sepals 3-5mm long; petals pinkish, 2-4mm long; stamens 6-10. Disturbed areas. Statewide.

Stellaria L. Chickweed

Annual or perennial herbs with opposite leaves without stipules; flowers bisexual, in cymes or solitary in axils; sepals mostly 5, separate; petals 5 or sometimes lacking, white; stamens usually 10, sometimes fewer; styles usually 3; capsules 1 celled, usually dehiscing by 6 valves.

1. Plants annual weeds; stems with longitudinal lines of hairs *S. media*
1. Plants usually perennial; stems glabrous or the hairs uniformly distributed
    2. Inflorescence densely glandular-pubescent; petals about twice as long as sepals *S. jamesiana*

2. Inflorescence glabrous or pubescent but not glandular; petals less than twice as long as sepals

3. Stems with long spreading hairs above — *S. simcoei*

3. Stems glabrous or minutely scaberulous above

4. Leaves ovate to ovate-lanceolate, some usually short petioled, mostly 2cm or less long; flowers usually solitary in leaf axils; petals usually lacking

5. Leaves mostly less than 1cm long and twice as long as wide or less, ciliate-margined at base; sepals 2-2.5mm long, usually obtuse — *S. obtusa*

5. Leaves sometimes over 1cm long and often over twice as long as wide, ciliate-margined or not at base; sepals 2-4mm long, usually acute

6. Leaf margins minutely crisped; stems usually terminated by a pair of young leaves — *S. crispa*

6. Leaf margins flat; stems usually terminated by a flower or appearing so — *S. calycantha*

4. Leaves usually linear to lanceolate or oblanceolate and sessile, sometimes over 2cm long; flowers often few to many in terminal and axillary inflorescences; petals present or lacking

7. Petals lacking or vestigial and very much shorter than sepals

8. Flowers or inflorescence in axils of green bracts similar to stem leaves; pedicels not reflexed in fruit — *S. sitchana*

8. Flowers or inflorescence in axils of scarious or scarious-margined bracts; pedicels usually reflexed in fruit — *S. umbellata*

7. Petals exceeding to slightly shorter than the sepals
   9. Leaf margins finely tuberculate-scaberulous under magnification (30X); flowers usually many in branched cymes ... *S. longifolia*
   9. Leaf margins mostly smooth; flowers often few and axillary
      10. Petals as long as or longer than the sepals
         11. Leaves stiff and sharply acute or acuminate; sepals about 4mm long at anthesis
            12. Flowers mostly 1 or 2 per stem; leaves usually blue-green and glaucous; plants mostly alpine and subalpine and less than 15cm high ... *S. monantha*
            12. Flowers 1 to several per stem; leaves green, not glaucous; plants mostly below subalpine and often over 15cm high ... *S. longipes*
         11. Leaves not stiff, obtuse to acute; sepals 2-3mm long at anthesis ... *S. crassifolia*
      10. Petals shorter than sepals
         13. Bracts of inflorescence foliaceous; flowers usually not solitary ... *S. sitchana*
         13. Bracts of inflorescence, or at least of the lower flowers, membranous, or if not, the flowers solitary at tip of stem ... see #12 above

Stellaria calycantha (Ledeb.) Bong., Mem. Acad. St. Petersb. VI, 2:127. 1832. Perennial to 3dm high; leaves mostly sessile and ovate, 0.5-2cm long, 2-10mm wide; flowers mostly axillary; sepals 2-3mm long; petals usually lacking. Woods, meadows, and stream banks. NW, SW, SE. _S. borealis_ Bigel.

Stellaria crassifolia Ehrh., Hannov. Bot. Mag. 8:116. 1784. Perennial to 15cm high; leaves sessile, mostly lanceolate, 2-10mm long, 0.5-3mm wide; flowers mostly solitary in axils; sepals 2-3mm long; petals 2-5mm long. Moist areas. NW, SW, SE.

Stellaria crispa Cham. & Schlecht., Linnaea 1:51. 1826. Perennial with prostrate or ascending stems to 4dm long; leaves ovate-lanceolate to ovate, sessile to short petioled, 0.5-2cm long, 2-10mm wide; flowers solitary in axils and at tip of stem; sepals 2.5-4mm long; petals usually lacking. Moist areas. Sheridan Co. and Yellowstone Park.

Stellaria jamesiana Torr., Ann. Lyc. N. Y. 2:169. 1828. Perennial to 4dm high; leaves sessile, linear to lanceolate, 2-10cm long, 2-11mm wide; flowers in axillary and terminal cymes; sepals 3.5-6mm long; petals 6-10mm long. Woods, meadows, and slopes. NW, SW, SE. _Arenaria jamesiana_ (Torr.) Shinners.

Stellaria longifolia Muhl. ex Willd., Enum. Pl. Hort. Berol. 479. 1809. Perennial with decumbent to ascending stems to 6dm long; leaves sessile, linear, lanceolate, or oblanceolate, 1-4cm long, 1-5mm wide; flowers mostly in terminal cymes; sepals 2.5-4mm long; petals about as long as sepals. Stream banks and meadows. NW, NE, SE.

Stellaria longipes Goldie, Edinb. New Phil. Journ. 6:327. 1822. Perennial to 3dm high; leaves sessile, linear to lanceolate, 0.5-3.5cm long, 1-4mm wide; flowers in cymes or rarely solitary; sepals about 4mm long; petals 3-5mm long. Woods, meadows, and slopes. Statewide. *Alsine validus* Goodding.

Stellaria media (L.) Vill., Hist. Pl. Dauph. 3:615. 1789. Annual with decumbent stems to 3dm long; leaves sessile or petioled, the blades mostly ovate, 2-25mm long, 1-10mm wide; flowers solitary in axils or rarely 2-7 in bracteate cymes; sepals 2.5-5mm long; petals 2-5mm long. Disturbed areas. Albany Co.

Stellaria monantha Hultén, Bot. Notiser 1943:265. 1943. Perennial to 15cm high; leaves sessile, linear to lanceolate, 0.5-2.5cm long, 1-4mm wide; flowers 1 or 2 or rarely 3 per stem; sepals about 4mm long; petals 3-5mm long. Mostly alpine and subalpine. Statewide. *S. longipes* var. *altocaulis* (Hultén) Hitchc.

Stellaria obtusa Engelm., Bot. Gaz. 7:5. 1882. Perennial to 10cm high; leaves sessile or short-petioled, ovate, 3-10mm long, 2-7mm wide; flowers solitary in axils; sepals 4 or 5, 2-2.5mm long; petals lacking; stamens 8-10; styles 3 or 4. Streambanks and meadows. NW, SW, SE.

Stellaria simcoei (Howell) Hitchc. in Hitchc. et al., Vasc. Pl. Pac. N. W. 2:310. 1964. Annual or perennial with prostrate to ascending stems to 2dm long; leaves ovate to lanceolate, sessile or short petioled, mostly 8-15mm long and 3-12mm wide; flowers solitary in leaf axils; sepals 4 or 5, about 2.5mm long; petals lacking or to 1mm long; stamens 8 or 10; styles 3 or 4. Stream banks and meadows. Fremont Co. and Yellowstone Park.

Stellaria sitchana Steud., Nom. Bot. II, 2:637. 1841. Perennial to 5dm high; leaves sessile, mostly lanceolate, 1-5cm long, 2-10mm wide; flowers usually axillary and in terminal cymes; sepals 2-4mm long; petals lacking or to as long as sepals. Woods, meadows, and stream banks. Statewide. _S. borealis_ Bigel., _S. calycantha_ var. _bongardiana_ (Fern.) Fern. and var. _sitchana_ (Steud.) Fern.

Stellaria umbellata Turcz. ex Kar. & Kir., Bull. Soc. Nat. Mosc. 15:173. 1842. Perennial to 3dm high; leaves sessile, oblong, lanceolate, or linear, 5-20mm long, 1-10mm wide; flowers in umbel-like cymes; sepals 1.5-3mm long; petals vestigial or lacking. Woods, slopes, and meadows. NW, SE, SW.

__Vaccaria__ Medic. Cow Herb

__Vaccaria__ __pyramidata__ Medic., Philos. Bot. 1:96. 1789. Annual to 8dm high; leaves sessile or short petioled, cordate to lanceolate or oblong-lanceolate, 2-12cm long, 3-40mm wide; flowers in a cyme; sepals united, 11-15mm long; petals pink, 16-26mm long; styles 2 or rarely 3; stamens 10; capsule 1 celled, dehiscing by 4 valves. Disturbed areas. Statewide. __Saponaria__ __vaccaria__ L., __V__. __segetalis__ (Neck.) Garcke ex Asch.

CELASTRACEAE Staff Tree Family

Shrubs or vines with simple, alternate or opposite leaves; flowers bisexual or unisexual with rudiments of the opposite sex, regular, in cymes, racemes, panicles, or axillary clusters; sepals 4 or 5, united; petals 4 or 5, separate; stamens 4 or 5; ovary 1, superior; carpels 2-5; style 1; locules 2-5; ovules usually 1 or 2 per carpel; placentation axile; fruit a capsule.

Plants woody vines or twining shrubs; leaves 3-12cm long __Celastrus__

Plants small, upright shrubs; leaves 0.5-3cm long __Pachistima__

Celastrus L. Bittersweet

Reference: Hou, D. 1955. Ann. Mo. Bot. Gard. 42:215-302.

Celastrus scandens L., Sp. Pl. 196. 1753. Vine or twining shrub to 20m long; leaf blades elliptic, ovate, or obovate, 3-12cm long, crenate-serrate, glabrous; plants dioecious but with sterile parts of opposite sex; calyx about 1-2mm long; petals greenish, about 3-4mm long; fruits about 8-12mm long, orange. Thickets and woods. Crook Co.

Pachistima Raf. Mountain Lover

Pachistima myrsinites (Pursh) Raf., Sylva Tell. 42. 1838. Shrub to 1m high; leaves opposite or subopposite, the blades elliptic to obovate or oval, 0.5-3cm long, serrate, leathery and evergreen; flowers in axillary clusters; sepals 1-2mm long; petals maroon, 1-2mm long; capsule about 3-4mm long. Woods and thickets. NW, SW, SE. (Sometimes spelled Paxistima.)

CERATOPHYLLACEAE Hornwort Family

Aquatic herbs without roots; leaves compound, whorled, with linear divisions which are finely toothed; flowers unisexual, regular, solitary in leaf axils; plants monoecious; sepals 7-15, usually united at base; petals lacking; stamens 10-20; ovary 1, superior; carpel 1; style 1; locule 1; ovule 1; placentation parietal; fruit an achene or nutlet.

Ceratophyllum L. Coontail

Ceratophyllum demersum L., Sp. Pl. 992. 1753. Usually submerged aquatic with stems to 3m long; leaves whorled, 1-3cm long, once or twice palmately or dichotomously dissected into linear, finely toothed segments; flowers solitary in leaf axils, mostly 1-7mm long; body of achenes 4-6mm long, with a terminal spine to 12mm long and 2 basal spines to 6mm long. Ponds, lakes, and slow streams. NW, SE.

Figure 5. Chenopodiaceae. A. Salicornia rubra, portion of stem (X 4). B. Bassia hyssopifolia, flower (X 4). C. Kochia scoparia, flower (X 4). D. Chenopodium berlandieri, flower at left, fruit at right (X 10). E. Corispermum, flower (X 4). F-P. Fruits: F. Suckleya (X 4); G. Cycloloma (X 5); H. Grayia spinosa (X 2); I. Atriplex argentea (X 4); J. Atriplex canescens (X 4); K. Atriplex confertifolia (X 2); L. Atriplex dioica (X 7); M. Atriplex gordonii (X 2); N. Atriplex patula (X 4); O. Atriplex powellii (X 5); P. Atriplex truncata (X 7).

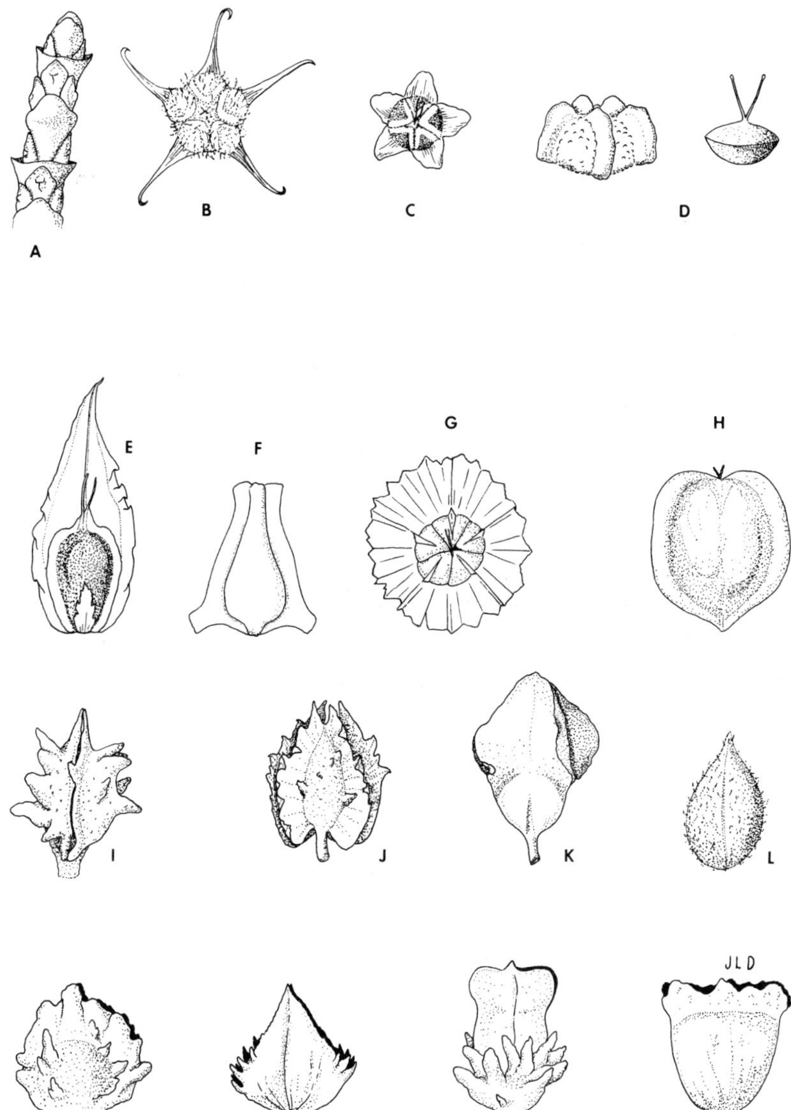
A
B
C
D
E
F
G
H
I
J
K
L
M
N
O
P
JLD

CHENOPODIACEAE Goosefoot Family

Herbs or shrubs with simple, alternate, or rarely opposite, sometimes succulent or gray-scurfy leaves, common on alkaline soils; flowers bisexual, or unisexual and the plants monoecious or dioecious, regular or rarely irregular, solitary or clustered in the leaf axils or in spikes, racemes, or panicles; sepals usually 5, rarely 3 or 4 or fewer, usually united, sometimes lacking; petals lacking; stamens usually 5, rarely 3 or 4 or fewer; ovary 1, superior; carpels 2 or 3; styles or at least stigmas 2 or 3; locule 1; ovule 1; placentation basal; fruit an achene or utricle, the seed enclosed by a free or adherent pericarp.

1. Stems jointed and bearing opposite scale-like leaves; flowers sunken in depressions of the spikes; plants succulent annuals <u>Salicornia</u>
1. Stems not jointed; leaves usually not scale-like, mostly alternate; flowers not as above; plants succulent or not, annual or perennial
   2. Plants either shrubs, or semishrubs that are woody below and herbaceous above
      3. Plants with spine-tipped branches and linear, semiterete leaves <u>Sarcobatus</u>
      3. Plants either lacking spine-tipped branches or with spine-tipped branches and flattened, non-linear leaves

4. Leaves densely stellate-pubescent, linear or slightly wider with revolute margins — *Ceratoides*

4. Leaves glabrous or pubescent with simple hairs, or scurfy, or if sparsely stellate-pubescent, the leaves not linear with revolute margins

  5. Leaves linear, less than 3mm wide, usually fleshy, terete or nearly so (rarely flattened); branches not spine-tipped

    6. Flowers and young stems villous-tomentose — *Kochia*

    6. Flowers and young stems glabrous, puberulent, or farinose — *Suaeda*

  5. Leaves linear or broader, some usually over 3mm wide, or else not terete or fleshy; branches sometimes spine-tipped

    7. Bracts of pistillate flowers completely united to enclose the fruit, laterally compressed, usually winged all around and entire, usually lacking appendages on faces; leaves often succulent; staminate calyx usually 4 lobed — *Grayia*

    7. Bracts of pistillate flowers free at least at tip, sometimes dorsiventrally compressed, usually not winged all around and often toothed or with appendages on faces; leaves usually not succulent; staminate calyx commonly 5 lobed — *Atriplex*

2. Plants herbaceous, mostly annuals

8. Leaves terete to subterete (rarely flattened), succulent, linear or nearly so (some bracts often lanceolate or ovate), entire, glabrous or nearly so or rarely farinose, less than 3mm wide, sometimes tipped with a bristle or spine

9. Leaves with stiff spinulose tips or weak bristle tips, the spines or bristles often 1mm or more long; perianth segments often membranous

10. Leaves with somewhat weak bristle tips, otherwise rounded at tip — Halogeton

10. Leaves with stiff spinulose tips, the leaf often tapering into the spine — Salsola

9. Leaves not as above; perianth segments usually succulent — Suaeda

8. Leaves not with the above combination of characters

11. Perianth lacking or of 1 or rarely 2 or 3 bract-like segments smaller than and not enclosing the fruit, the fruit laterally flattened and subtended by a single bract or no bract (loose pericarp sometimes encloses seed)

12. Inflorescence usually stellate-pubescent; flowers not clustered in leaf axils, each subtended by a usually long-attenuate or acuminate bract; fruit 3-4mm long — Corispermum

12. Inflorescence not stellate-pubescent; flowers usually clustered in leaf axils, the bracts, if present, not long-attenuate or acuminate; fruit less than 2mm long <u>Monolepis</u>

11. Perianth 3-5 lobed or sometimes lacking in pistillate flowers; fruit at least partly enclosed by perianth or 2 large subtending bracts, sometimes dorsiventrally flattened

13. Flowers unisexual, the pistillate usually lacking a perianth and enclosed by 2 partly or wholly connate bracts, the staminate with a 3-5 lobed perianth and no bracts

14. Bracts of pistillate flowers often broad and open at summit, the margins usually not winged at least below; plants usually not succulent but often scurfy and gray; leaves various <u>Atriplex</u>

14. Bracts of pistillate flowers united except at the very narrow summit, the margins winged; plants usually succulent, sparingly if at all scurfy; leaf blades orbicular to broadly ovate, irregularly dentate <u>Suckleya</u>

13. Flowers mostly perfect (sometimes some pistillate or sterile) with a regular 3-5 lobed perianth, lacking enclosing bracts

15. Inflorescence usually densely hairy; leaves linear to lanceolate or oblanceolate, 1-8(12)mm wide, entire, usually hairy, the margins sometimes ciliate

16. Perianth segments, at least in fruit, bearing a usually hooked spine from the back — Bassia

16. Perianth segments without a hooked spine, often with a membranous wing — Kochia

15. Inflorescence glabrous or rarely hairy; leaves often wider or glabrous (or farinose), often lobed or toothed

17. Calyx usually with a broad, horizontal, membranous, circular wing from the middle when mature; leaves with large, very short-spinulose teeth; plants usually pubescent and not farinose — Cycloloma

17. Calyx lacking a wing; leaves usually without short-spinulose teeth; plants usually glabrous or farinose, rarely glandular-pubescent — Chenopodium

<u>Atriplex</u> L. Saltbush; Shadscale

Annual to perennial, monoecious or dioecious herbs or shrubs, glabrous to farinose; leaves mostly alternate, sometimes opposite especially below, entire or occasionally toothed; flowers mostly clustered or in a spike, raceme, or panicle; staminate perianth (3-4)5 parted; stamens (3-4)5; pistillate perianth lacking or rarely of 2-5 segments but each flower usually subtended by 2 at least partly united bracts; stigmas 2 or rarely 3.

References: Brown, G. D. 1956. Am. Midl. Nat. 55:199-210.
Bassett, I. J. & C. W. Crompton. 1973. Canad. Journ. Bot. 51:1715-1723.

1. Plants perennial shrubs or semishrubs
  2. Plants spiny, the spines sharp-pointed twigs or branches, these most common toward base ..... <u>A. confertifolia</u>
  2. Plants not spiny
    3. Plants mostly woody throughout, often over 5dm high; fruiting bracts usually 4 winged lengthwise ..... <u>A. canescens</u>
    3. Plants usually woody only at base, rarely over 5dm high; fruiting bracts not conspicuously winged although sometimes with irregular, flattened appendages

4. Leaves crowded, mostly 3-16mm long; Uinta Co. *A. corrugata*

4. Leaves usually not crowded, often well over 16mm long; widespread *A. gordonii*

1. Plants herbaceous annuals

5. Mature leaves usually green, often glabrate (sometimes gray-farinose when young), often opposite or subopposite below

6. Fruiting bracts ovate to orbicular, mostly 6-15mm wide (lacking in some flowers); plants mostly over 6dm high; some leaf blades often over 3cm wide *A. hortensis*

6. Fruiting bracts often more elongate or else triangular or rhombic, mostly less than 6mm wide; plants often less than 6dm high; leaf blades often less than 3cm wide

7. Staminate flowers mostly mixed with the pistillate; bracts of pistillate flowers united much less than half their length; some leaves often toothed or lobed

8. Main leaf blades, or some of them, triangular or hastate with a truncate base or nearly so *A. triangularis*

8. Main leaf blades sometimes with a pair of lobes toward base but definitely tapering to base, not strictly triangular or hastate *A. patula*

7. Staminate flowers mostly in terminal or axillary clusters or spikes; bracts of pistillate flowers united to tip; leaves entire *A. dioica*

5. Mature leaves usually gray-farinose on both surfaces or at least beneath, usually alternate throughout

9. Fruiting bracts orbicular or nearly so, entire, lacking appendages *A. heterosperma*

9. Fruiting bracts not orbicular, usually with 1 or more teeth or appendages or with a pointed tip with relatively straight sides

10. Fruiting bracts mostly 2-3.5mm long, cuneate to cup shaped at base and truncate to slightly rounded at tip, the faces smooth or with obscure tubercles, the tip often greenish-reticulate with 1 to several teeth *A. truncata*

10. Fruiting bracts not as above (flowers sometimes all staminate)

11. Leaves 2-15(20)mm long, 0.5-3(5)mm wide; fruiting bracts 1.5-3mm long *A. wolfii*

11. Leaves mostly longer or wider; fruiting bracts mostly 3mm or more long

12. Mature bracts of pistillate flowers mostly 3-4mm long, the upper smooth portion broadly oval or horizontally oblong and tipped by a short tooth; leaves mostly entire (flowers sometimes all staminate) *A. powellii*

12. Mature bracts of pistillate flowers often 4mm or more long, the shape various, usually several toothed at tip or long-pointed; leaves often somewhat toothed

13. Leaves mostly sinuate-dentate; fruiting bracts united not quite to middle *A. rosea*

13. Leaves entire to slightly toothed; fruiting bracts usually united to well above middle

14. Leaves, at least the upper, mostly cordate or subcordate; some fruiting bracts usually with a long, hardened, conical base often over 3mm long; Uinta Co. *A. saccaria*

14. Leaves usually truncate to cuneate at base; bracts usually not as above; widespread *A. argentea*

Atriplex argentea Nutt., Gen. Pl. 1:198. 1818. Monoecious annual to 7dm high; leaves mostly farinose, petioled or sessile above, the blades lanceolate to deltoid-ovate, 1-5cm long, 0.5-4cm wide, entire to slightly dentate; flowers in axillary or terminal glomerules; fruiting bracts mostly 4-10mm long with dentate to laciniate margins and often appendaged faces. Plains and hills especially where alkaline. Statewide. *A. volutans* A. Nels.

Atriplex canescens (Pursh) Nutt., Gen. Pl. 1:197. 1818. Four-wing Saltbush. Dioecious shrub to 2m high; leaves sessile or nearly so, elliptic to narrowly spatulate or linear, 1-6cm long, 2-15mm wide, entire, gray-scurfy; flowers in spikes, racemes, or panicles or small clusters; fruiting bracts 4-15mm long, usually 4 winged lengthwise. Plains and hills especially where alkaline. Statewide. *A. aptera* A. Nels., *A. odontoptera* Rydb.

Atriplex confertifolia (Torr. & Frem.) Wats., Proc. Am. Acad. 9:119. 1874. Shadscale. Dioecious shrub to 12dm high with mostly spine-tipped branches; leaves short-petioled to nearly sessile, the blades elliptic or ovate to suborbicular, 0.3-2cm long, 2-17mm wide, entire; flowers in axillary clusters, in spikes, or solitary; fruiting bracts 5-12mm long and about as wide, entire, the faces smooth. Plains and hills especially where alkaline. Statewide.

Atriplex corrugata Wats., Bot. Gaz. 16:345. 1891. Dioecious or monoecious perennial to 2dm high, woody at base and usually mat forming, gray-scurfy; leaves sessile or subsessile, the blades linear to spatulate, 3-16mm long, 1-6mm wide, entire; staminate flowers in terminal, interrupted clusters, the pistillate in axillary clusters; fruiting bracts 4-6mm long, the margins entire, the faces with short flattened appendages. Plains and hills. Uinta Co.

Atriplex dioica (Nutt.) Macbr., Contr. Gray Herb. n. s. 53:11. 1918. Monoecious or dioecious annual to 3dm high, sparsely farinose to glabrate; leaves sessile, lanceolate, elliptic, or ovate, 0.5-3cm long, 2-12mm wide, entire; flowers in axillary or terminal clusters or spikes or occasionally solitary; fruiting bracts 0.5-2mm long and connate to tip, entire. Plains, hills, and disturbed areas, often where alkaline. Statewide. A. suckleyana Rydb., Endolepis ovata Rydb.

Atriplex gordonii (Moq.) Dietr., Syn. Pl. 5:537. 1852; emend. Hook., Journ. Bot. & Kew Misc. 5:261. 1853. Dioecious perennial to 5dm high, the base woody, gray-scurfy; leaves short petioled to sessile, the blades linear-spatulate to obovate, 0.5-5cm long, 2-15mm wide, entire; flowers in axillary and terminal clusters or spikes or panicles; fruiting bracts 3-7mm long, entire or dentate, the faces smooth to appendaged, sometimes beaked at tip. Plains and hills especially where alkaline. Statewide. A. falcata (Jones) Standl., A. nuttallii Wats., A. pabularis A. Nels., A. fruticulosa Osterh., A. eremicola Osterh., A. tridentata Kuntze, A. oblanceolata Rydb., A. buxifolia Rydb. Moquin used the epithet "Gardneri" by mistake.

Atriplex heterosperma Bunge, Reliq. Lehm. 272. 1851. Monoecious annual to 1m high; leaves farinose at least beneath, short petioled, the blades hastate to deltoid-ovate, 7-40mm long, 4-35mm wide, entire or few toothed; flowers in axillary and terminal spikes or panicles; fruiting bracts 1-5mm long, orbicular or nearly so, entire and without appendages, dimorphic, some no larger than the fruit, others much larger. Alkaline areas where moist. Albany Co.

Atriplex hortensis L., Sp. Pl. 1053. 1753. Usually monoecious annual to 25dm high, glabrate in age; leaves petioled, the blades deltoid or deltoid-ovate to broadly lanceolate, 2-20cm long, 1-10cm wide, entire to undulate; flowers in axillary and terminal panicles; fruiting bracts lacking or 6-15mm long with entire or denticulate margins and smooth faces. Disturbed areas. SW, SE.

Atriplex patula L., Sp. Pl. 1053. 1753. Monoecious or semidioecious annual to 1m high, farinose to glabrate; leaves short-petioled, the blades linear-elliptic to ovate, sometimes with a pair of lobes toward base, 1-8cm long, 0.2-5cm wide, entire to dentate; flowers in axillary and terminal spikes or panicles; fruiting bracts 2-8mm long, the margins entire to denticulate, the faces smooth, tuberculate, or appendaged. Plains, hills, banks, and disturbed areas. Statewide. *A. carnosa* A. Nels., *A. subspicata* (Nutt.) Rydb.

Atriplex powellii Wats., Proc. Am. Acad. 9:114. 1874. Monoecious or dioecious annual to 7dm high, mostly farinose; leaves petioled below to sessile above, the blades deltoid, ovate, or rhombic, 5-35mm long, 3-30mm wide, mostly entire; flowers in terminal or axillary glomerules; fruiting bracts 3-4mm long, the margins entire to dentate, the faces smooth to appendaged. Plains, hills, and disturbed areas, often where alkaline. SW, SE, NE. *A. nelsonii* Jones, *A. philonitra* A. Nels.

Atriplex rosea L., Sp. Pl. 2:1493. 1763. Monoecious annual to 12dm high, mostly farinose, rarely glabrate; leaves petioled to sessile, the blades lanceolate to ovate or rhombic, 1-7cm long, 0.5-5cm wide, usually sinuate-dentate; flowers in axillary and terminal clusters or spikes; fruiting bracts 3-8mm long, the margins denticulate, the faces usually tuberculate. Disturbed or alkaline areas. SW, SE. *A. spatiosa* A. Nels.

Atriplex saccaria Wats., Proc. Am. Acad. 9:112. 1874. Monoecious annual to 5dm high, gray-scurfy; leaves petioled or the upper sessile, the blades cordate, at least above, to sometimes deltoid or ovate below, 0.5-2cm long and about as wide, entire; flowers in axillary and terminal clusters; fruiting bracts 3-7mm long, the margins toothed, at least at summit, the faces smooth to appendaged. Plains and hills. Uinta Co.

Atriplex triangularis Willd., Sp. Pl. 4:963. 1806. Monoecious or semi-dioecious annual to 1m high, slightly farinose to glabrous; leaves short-petioled, the blades mostly hastate or deltoid to lanceolate, 1-8cm long, 0.2-7cm wide, entire to dentate; flowers in axillary and terminal spikes or panicles; fruiting bracts 2-7mm long, the margins entire to denticulate, the faces smooth, tuberculate, or appendaged. Plains, hills, and disturbed areas. Statewide. *A. hastata* L. of authors.

Atriplex truncata (Torr.) Gray, Proc. Am. Acad. 8:398. 1872. Monoecious annual to 1m high, gray-scurfy; leaves petioled to sessile, ovate to deltoid or cordate, 0.5-4cm long, 3-25mm wide, entire to undulate; flowers in axillary and terminal clusters or spikes; fruiting bracts 2-3.5mm long, the margins entire or slightly toothed on top, the faces smooth or obscurely tuberculate. Plains and hills especially where alkaline. Statewide.

<u>Atriplex</u> <u>wolfii</u> Wats., Proc. Am. Acad. 9:112. 1874. Monoecious annual to 3dm high, gray-scurfy; leaves sessile, the blades linear to oblong-linear or rarely lance-ovate, 2-15(20)mm long, 0.5-3(5)mm wide, entire; flowers in axillary clusters; fruiting bracts 1.5-3mm long, the margins toothed or entire, the faces usually with tubercles or appendages. Plains and hills especially where alkaline. Sweetwater and Carbon Cos. <u>A</u>. <u>tenuissima</u> A. Nels., <u>A</u>. <u>greenei</u> A. Nels.

<u>Bassia</u> Allion.

<u>Bassia</u> <u>hyssopifolia</u> (Pall.) Kuntze, Rev. Gen. 2:547. 1891. Annual to 2m high, conspicuously pubescent at least in inflorescence; leaves sessile, linear to narrowly lanceolate or oblanceolate, entire, 0.5-4cm long, 1-4mm wide, often with axillary fascicles of smaller leaves; flowers bisexual, pistillate, or sterile, in axillary and terminal clusters or spikes; stamens 5; fruiting perianth about 2mm wide, each lobe with a hooked spine to 2mm long, the fruit dorsiventrally flattened; stigmas 2 or rarely 3. Disturbed areas. Sweetwater and Washakie Cos.

<u>Ceratoides</u> Gagn. Winterfat

Reference: Reveal, J. L. & N. H. Holmgren. 1972. Taxon 21:209.

<u>Ceratoides</u> <u>lanata</u> (Pursh) Howell, Wasmann Journ. Biol. 29:105. 1971. Monoecious subshrub to 7dm high, stellate-pubescent and villous-tomentose; leaves sessile or short petioled, with fascicles of smaller ones in the axils, the blades linear to narrowly lanceolate, oblanceolate, or elliptic, entire, 0.5-4cm long, 1-6mm wide; flowers in axillary clusters or solitary; pistillate flowers subtended by 2 connate, densely hirsute bracteoles about 5-7mm long which enclose the fruit; staminate flowers deciduous, the calyx lobes 4, 1-2mm long; stamens 4; styles 2. Plains and hills. Statewide. <u>Eurotia</u> <u>lanata</u> (Pursh) Moq.

<u>Chenopodium</u> L. Lamb's Quarters; Goosefoot

Annual herbs with alternate, usually petioled, sometimes deciduous, toothed or lobed or entire leaves; flowers bisexual, in terminal and usually axillary clusters, spikes, or panicles; perianth 3-5 cleft; stamens 5 or fewer; stigmas 2 or rarely 3.

References: Wahl, H. A. 1954. Bartonia 27:1-46.
Baranov, A. I. 1964. Rhodora 66:168-171.
Crawford, D. J. 1972. Brittonia 24:118.
1973. Madroño 22:185-195.
Crawford, D. J. & J. F. Reynolds. 1974. Brittonia 26:398-410.

1. Plants glandular and puberulent *C. botrys*

1. Plants not glandular (except in inflorescence of *C. hybridum*), often glabrous or farinose

2. Fruit usually flattened laterally; perianth usually 3 or 4 parted; leaves often somewhat triangular or hastate, green on both sides

3. Glomerules usually not globose, less than 4mm wide, in many, crowded, axillary, simple or compound spikes; calyx not red and fleshy

4. Stems mostly erect and over 15cm high; E Wyo. *C. rubrum*

4. Stems mostly prostrate or nearly so and 10cm or less high; W Wyo. *C. chenopodioides*

3. Glomerules mostly globose, often over 4mm wide at maturity, mostly in an interrupted terminal spike; calyx sometimes red and fleshy *C. capitatum*

2. Fruit usually flattened dorsiventrally, or if laterally, the leaves usually white or gray farinose; perianth usually 5 parted

5. Primary leaf blades entire or nearly so, usually ovate, lanceolate, oblong, linear, or elliptic, rarely over 10(13)mm wide, often with 1-3 somewhat parallel veins

6. Leaf blades often 2-3 times as long as wide, elliptic to ovate; perianth exposing mature fruit laterally and dorsally; fruits maturing unevenly in adjacent parts of inflorescence; often montane *C. atrovirens*

6. Leaf blades often over 3 times as long as wide, lanceolate to oblong or linear; perianth largely covering mature fruit except sometimes dorsally; fruits maturing evenly; rarely montane
 7. Perianth sparingly or not at all farinose
  8. Pericarp readily separable from seed — *C. subglabrum*
  8. Pericarp tightly adherent to seed — *C. pallescens*
 7. Perianth densely farinose
  9. Pericarp readily separable from seed, the seeds mostly 1mm or more wide
   10. Plants mostly branched from near base and bushy; perianth segments usually closely enclosing fruits — *C. desiccatum*
   10. Plants little if at all branched, erect and slender; perianth segments usually somewhat reflexed and exposing fruits — *C. pratericola*
  9. Pericarp somewhat adherent to seed, the seeds mostly 1mm or less wide
   11. Inflorescence leafy to near tip; some leaves often 3 nerved; plants usually erect and slender — *C. hians*
   11. Inflorescence not leafy to tip; leaves 1 nerved; plants sometimes branched and bushy *C. leptophyllum*

5. Primary leaf blades (sometimes deciduous in fruit) mostly lobed or toothed or else deltoid or deltoid-ovate, or both, often over 10mm wide, usually pinnately veined
 12. Leaves white or gray farinose beneath, green above, sinuate-dentate, 12(16)mm or less wide; perianth glabrous — *C. glaucum*

12. Leaves usually not as above; perianth usually farinose
  13. Perianth lobes not dorsally keeled in fruit; leaves glabrous — *C. hybridum*
  13. Perianth lobes usually dorsally keeled in fruit; some leaves often farinose on at least 1 surface
    14. Leaf blades thin and papery when dry, usually barely if at all longer than wide and hastately lobed, often glabrous or nearly so; pericarp loose — *C. fremontii*
    14. Leaf blades somewhat thick, usually about twice or more as long as wide and often toothed throughout, often farinose at least beneath; pericarp various
      15. Seed and pericarp pitted, the latter usually tightly adherent to seed; style not cleft to base, often less than 0.5mm long — *C. berlandieri*
      15. Seed and pericarp smooth or nearly so when mature, the latter usually loose around seed; style essentially cleft to base, often nearly 1mm long
        16. Seeds mostly 1.1-1.5mm wide; fruits largely covered by perianth when mature; widespread — *C. album*
        16. Seeds mostly 0.9-1.2mm wide; fruits not covered by perianth when mature; known only from Albany Co. — *C. strictum*

Chenopodium album L., Sp. Pl. 219. 1753. Lamb's Quarters. Plants to 1m high; leaf blades lanceolate to deltoid-ovate, 0.5-8cm long, irregularly deeply toothed to entire; perianth 1-2mm long; pericarp loose around the dorsiventrally flattened seed. Disturbed areas. NW, NE, SE.

Chenopodium atrovirens Rydb., Mem. N. Y. Bot. Gard. 1:131. 1900. Plants to 8dm high; leaf blades elliptic to ovate, 5-30mm long, entire or rarely with 2 small teeth at base; perianth about 1mm long; pericarp free from or adherent to the dorsiventrally flattened seed. Disturbed areas. Statewide. *C. aridum* A. Nels.

Chenopodium berlandieri Moq., Chenop. Enum. 23. 1840. Plants to 9dm high; leaf blades lanceolate to deltoid-ovate, 0.6-6cm long, entire to few toothed; perianth 1-2.5mm long; pericarp usually tightly adherent to the dorsiventrally flattened seed. Disturbed areas. Statewide. *C. album* L. of authors.

Chenopodium botrys L., Sp. Pl. 219. 1753. Aromatic and glandular-puberulent plants to 5dm high; leaf blades oblong or elliptic to ovate, 1-6cm long, sinuate to pinnatifid; perianth about 1mm long; pericarp thin, the seed dorsiventrally or rarely laterally flattened. Disturbed areas. Converse and Albany Cos.

Chenopodium capitatum (L.) Asch., Fl. Brandenb. 1:572. 1864. Plants glabrous, to 8dm high; leaf blades mostly hastate or deltoid-ovate, 1-8cm long, entire to sinuate-dentate or lobed; perianth about 1mm long; pericarp usually loose, the seed laterally flattened. Disturbed and moist areas. Statewide. *C. overi* Aellen, *Blitum capitatum* L., *B. hastatum* Rydb.

Chenopodium chenopodioides (L.) Aellen, Ostenia 98. 1933. Plants glabrous, to 1dm high, often prostrate; leaf blades mostly deltoid-ovate, 3-25cm long, lobed or sinuately toothed to subentire; perianth about 1mm long; pericarp loose, the seed laterally or rarely dorsiventrally flattened. Streambanks and shores. Teton Co.

Chenopodium desiccatum A. Nels., Bot. Gaz. 34:362. 1902. Plants to 4dm high; leaf blades lanceolate, oblong, elliptic, or linear, 5-30mm long, entire or nearly so; perianth about 1mm long; pericarp free from the dorsiventrally flattened seed. Plains, hills, and disturbed areas. NW, SW, SE.

Chenopodium fremontii Wats., Bot. King Exp. 287. 1871. Plants to 9dm high; leaf blades ovate to deltoid, 5-45mm long, mostly entire or hastately lobed; perianth about 1mm long; pericarp loose on the dorsiventrally flattened seed. Plains, hills, meadows, and open woods. Statewide.

<u>Chenopodium glaucum</u> L., Sp. Pl. 220. 1753. Plants to 5dm high; leaf blades lanceolate to deltoid-ovate, 3-27mm long, sinuate-dentate; perianth 1mm or less long; pericarp free from the dorsiventrally or laterally flattened seed. Moist alkaline areas. Statewide. <u>C. salinum</u> Standl.

<u>Chenopodium hians</u> Standley, N. Am. Fl. 21:16. 1916. Plants to 8dm high; leaf blades elliptic, oblong, lanceolate, or linear, 5-30mm long, entire; perianth about 1mm long; pericarp adherent to the dorsiventrally flattened seed. Plains and hills. SW, SE. <u>C. incognitum</u> Wahl, <u>C. inamoenum</u> Standl.

<u>Chenopodium hybridum</u> L., Sp. Pl. 219. 1753. Plants to 15dm high; leaf blades deltoid to ovate, 2-20cm long, sinuate-dentate or lobed; perianth 1-2mm long; pericarp adhering to the dorsiventrally flattened seed. Moist and disturbed areas. Natrona and Big Horn Cos. <u>C. gigantospermum</u> Aellen.

<u>Chenopodium leptophyllum</u> (Moq.) Wats., Proc. Am. Acad. 9:94. 1874. Plants to 7dm high; leaf blades linear or oblong, 5-30mm long, entire or nearly so; perianth about 1mm long; pericarp somewhat adhering to the dorsiventrally flattened seed. Plains and hills. NW, SW, SE.

Chenopodium pallescens Standley, N. Am. Fl. 21:15. 1916. Plants to 6dm high; leaf blades linear or nearly so, 1-3.5cm long, entire; perianth 1-2.5mm long; pericarp tightly adherent to the dorsiventrally flattened seed. Plains, hills, and disturbed areas. Laramie Co.

Chenopodium pratericola Rydb., Bull. Torrey Club 39:310. 1912. Plants to 8dm high; leaf blades lanceolate, oblong, elliptic, or linear, 5-40mm long, entire or nearly so; perianth about 1mm long; pericarp free from the dorsiventrally flattened seed. Plains, hills, and disturbed areas. Statewide. *C. desiccatum* var. *leptophylloides* (Murr.) Wahl.

Chenopodium rubrum L., Sp. Pl. 218. 1753. Plants to 7dm high; leaf blades mostly deltoid-ovate and often hastate, 1.5-7cm long, entire to sinuate-dentate or lobed; perianth 1mm or less long; pericarp loose, the seeds mostly laterally flattened. Moist alkaline and disturbed areas. NE, SE. *C. succosum* A. Nels.

Chenopodium strictum Roth, Nov. Pl. Sp. 180. 1821. Plants to 8dm high; leaf blades elliptic to deltoid-ovate, 1-6cm long, entire or toothed; perianth about 1-1.5mm long; pericarp tightly adherent to or loose around the dorsiventrally flattened seed. Disturbed areas. Albany Co.

Chenopodium subglabrum (Wats.) A. Nels., Bot. Gaz. 34:362. 1902. Plants to 6dm high; leaf blades linear or nearly so, 1-4.5cm long, entire; perianth about 1mm long; pericarp free from the dorsiventrally flattened seed. Plains and hills. Natrona Co.

Corispermum L. Bugseed

Corispermum hyssopifolium L., Sp. Pl. 4. 1753. Stellate-hirsute to glabrous annual to 6dm high; leaves alternate, sessile or short petioled, the blades mostly linear, 0.5-6cm long, 1-4mm wide, entire; flowers bisexual, in terminal and axillary spikes, each subtended by a bract 3-14mm long, the perianth reduced to 1-3 membranous scales about 1mm long; stamens (1-2)3-5; styles 2; fruit 3-4mm long, usually winged. Disturbed and sandy areas. NW, SW, SE. *C. orientale* Lam. of authors, *C. imbricatum* A. Nels., *C. emarginatum* Rydb., *C. villosum* Rydb., *C. marginale* Rydb.

Cycloloma Moq. Ringwing

Cycloloma atriplicifolium (Spreng.) Coulter, Mem. Torrey Club 5:143. 1894. Pubescent to glabrate annual to 4dm high; leaves alternate, petioled, mostly lanceolate to oblanceolate, 1-8cm long, 2-30mm wide, sinuate-dentate; flowers bisexual or pistillate, in spikes or panicles; perianth with 5 slightly keeled lobes, 1-3mm long, becoming horizontally winged in fruit; stamens 5; stigmas 3. Plains, hills, and disturbed areas. SE.

Grayia H. & A. Hopsage

Shrubs with often exfoliating bark and alternate, sometimes fleshy, entire leaves, gray-scurfy or not; flowers in terminal and axillary clusters or spikes, unisexual and the plants dioecious or monoecious; staminate perianth usually 4 lobed; stamens 4 or rarely 5; pistillate perianth lacking, each ovary enclosed by 2 connate bracts forming a sac which is winged in fruit; stigmas 2 or 3.

Branches usually spine-tipped, at least below; at least the older leaves mostly fleshy and green _G. spinosa_
Branches not spine-tipped; leaves mostly not fleshy, gray-scurfy _G. brandegei_

Grayia brandegei Gray, Proc. Am. Acad. 11:101. 1876. Shrub to 5dm high, not spinose; leaf blades ovate to oblanceolate, 1-6cm long, 3-21mm wide, gray-scurfy; staminate perianth about 2mm long; bracts of pistillate flowers 4-7mm long, usually slightly wider and gray-scurfy. Plains and hills. Carbon Co.

Grayia spinosa (Hook.) Moq. in DC., Prodr. 13(2):119. 1849. Shrub to 15dm high with usually spine-tipped branches at least below; leaves mostly oblanceolate, 0.5-4cm long, 2-10mm wide, usually fleshy and the older ones green; staminate perianth about 1.5-2mm long; bracts of pistillate flowers 7-16mm long in fruit and about as wide or slightly narrower, often red or greenish-white, not gray-scurfy. Plains and hills especially where alkaline. NW, SW, SE.

Halogeton Meyer

Halogeton glomeratus (Bieb.) Meyer in Ledeb., Fl. Alt. 1:378. 1829. Annual herb with spreading to ascending stems to 5dm long; leaves linear or very short oblong, nearly terete, fleshy, 2-15mm long, 0.5-2mm wide, tipped with a weak bristle; flowers solitary or 2 or 3 together, axillary, bisexual or unisexual; pistillate perianth mostly 2-5mm long, segments 5; staminate perianth about 1-2mm long or obscure, segments 2-5; stamens 1-3(5); style 2 parted. Plains, hills, and disturbed areas. NW, SW, SE.

<u>Kochia</u> Roth Summer Cypress

Annual or perennial herbs or subshrubs with entire leaves; flowers bisexual or some unisexual, solitary or clustered in axils; perianth 5 lobed; stamens usually 5; stigmas 2 or 3.

Plants perennial, woody at base; leaves linear, subterete
<u>K</u>. <u>americana</u>
Plants annual; leaves linear to lanceolate or oblanceolate, flat <u>K</u>. <u>scoparia</u>

<u>Kochia</u> <u>americana</u> Wats., Proc. Am. Acad. 9:93. 1874. Perennial with a woody base and erect or ascending stems to 5dm high; leaves linear, nearly terete, usually fleshy, 3-25mm long, 0.5-2mm wide; perianth about 1.5mm long in flower, enlarged in fruit. Plains and hills mostly where alkaline. SW, SE. <u>K</u>. <u>vestita</u> (Wats.) Rydb.

<u>Kochia</u> <u>scoparia</u> (L.) Schrad., Neues Journ. Bot. 3(3):85. 1809. Erect annual to 15dm high; leaves linear to lanceolate or oblanceolate, flat, not fleshy, 0.3-6cm long, 1-12mm wide, ciliate and usually pubescent at least beneath; perianth about 1.5mm long in flower, usually membranous winged in fruit. Disturbed areas. Statewide.

<u>Monolepis</u> Schrad. Poverty Weed

Annual herbs with sessile or petioled, entire or hastate leaves; flowers bisexual or unisexual, in dense axillary clusters or occasionally solitary; perianth segments 1-3 or lacking; stamens 1 or 2; stigmas usually 2.

Leaves usually hastately lobed, the larger mostly 1-5cm long; flowers usually over 5 per axil; seeds mostly over 1mm long, the pericarp pitted ... <u>M. nuttalliana</u>

Leaves entire, mostly 2-12mm long; flowers usually 1-5 per axil; seeds less than 1mm long, the pericarp usually glandular or tuberculate ... <u>M. pusilla</u>

<u>Monolepis nuttalliana</u> (Schultes) Greene, Fl. Francis. 168. 1891. Plants prostrate to ascending with stems to 3dm long; leaf blades hastate or linear, lanceolate, or oblanceolate, 0.5-5cm long, 1-20mm wide; perianth 0.5-2.5mm long; pericarp pitted. Mostly in disturbed areas. Statewide.

<u>Monolepis pusilla</u> Torr. in Wats., Bot. King Exp. 289. 1871. Plants erect to ascending with stems to 2dm long; leaf blades oblong to ovate or oblanceolate, 2-12mm long, 0.5-4mm wide; perianth about 0.5mm long; pericarp glandular or tuberculate. Dry alkaline areas. Sweetwater and Washakie Cos.

Salicornia L. Saltwort

Salicornia rubra A. Nels., Bull. Torrey Club 26:122. 1899. Glabrous, succulent annual to 2dm high with jointed stems, usually reddish; leaves scale-like, opposite, connate, 1-3mm long; flowers bisexual or some unisexual, mostly in groups of 3 sunken in depressions of the spikes; perianth fleshy and sac-like below with a pointed or rounded margin above, about 1.5-2mm long; stamens 1 or 2; stigmas usually 2. Alkaline areas. NW, SW, SE.

Salsola L. Russian Thistle

Salsola kali L., Sp. Pl. 222. 1753. Much branched annual to 8dm high, becoming a tumble-weed; leaves linear, 0.5-6cm long, 0.3-2mm wide, entire, succulent, spinulose at tip, deciduous; flowers bisexual, solitary or clustered in axils of often spiny bracts; perianth of 5 nearly free segments, membranous and about 1.5mm long in flower, somewhat winged and slightly enlarged in fruit; stamens mostly 5; stigmas 2. Disturbed areas. Statewide. S. pestifer A. Nels.

<u>Sarcobatus</u> Nees Greasewood

<u>Sarcobatus</u> <u>vermiculatus</u> (Hook.) Torr. in Emory, Notes Mil. Recon. 149. 1848. Spiny shrub to 2.5m high; leaves alternate, linear, 0.5-4cm long, 0.5-4mm wide, fleshy, semiterete; flowers unisexual, in axillary or terminal spikes or solitary, the staminate above the pistillate; staminate flowers with 2-3 stamens beneath a peltate scale; pistillate perianth cup-like at base and shallowly lobed to entire at tip, about 1-3mm long in flower, becoming winged in fruit. Alkaline areas. Statewide.

<u>Suaeda</u> Forsk. ex Scop. Sea Blite

Annual or perennial herbs or semishrubs with alternate, mostly linear, fleshy, often terete leaves (some bracts often lanceolate or ovate); flowers bisexual or unisexual, solitary or clustered in leaf axils or in spikes; perianth fleshy and 5 lobed; stamens usually 5; stigmas usually 2 or 3.

1. Perianth lobes unequal, 1 or more with a horn-like or wing-like appendage on back at maturity; usually annuals
   2. Perianth lobes at most horned when mature — <u>S</u>. <u>depressa</u>
   2. Perianth lobes broadly thin-winged when mature — <u>S</u>. <u>occidentalis</u>

1. Perianth lobes equal, without appendages or wings; perennials or rarely annual
   3. Leaves mostly contracted to a narrow petiole — *S. nigra*
   3. Leaves not petioled — *S. torreyana*

Suaeda depressa (Pursh) Wats., Bot. King Exp. 294. 1871. Glabrous and glaucous annual to 5dm high; leaves 0.3-3cm long, 0.5-2.5mm wide, semiterete; bracts mostly ovate to lanceolate; perianth lobes unequal, about 0.5-1.5mm long, cucullate. Alkaline areas. Statewide. *S. erecta* (Wats.) A. Nels., *Chenopodium conardii* A. Nels.

Suaeda nigra (Raf.) Macbr., Contr. Gray Herb. n. s. 56:50. 1918. Glabrous annual or occasionally perennial to 7dm high; leaves 3-25mm long, 0.5-2mm wide, somewhat flattened; bracts linear or nearly so; perianth lobes about equal, 1-2mm long, somewhat cucullate. Alkaline areas. NW, SW, SE. *S. diffusa* Wats.

Suaeda occidentalis (Wats.) Wats., Proc. Am. Acad. 9:90. 1874. Glabrous annual to 3dm high; leaves 3-25mm long, 0.5-2mm wide, subterete; bracts similar to leaves or slightly wider; perianth lobes unequal, about 1-2mm long, somewhat cucullate and winged when mature. Alkaline areas. Sweetwater and Park Cos.

<u>Suaeda</u> <u>torreyana</u> Wats., Proc. Am. Acad. 9:88. 1874. Glabrous to puberulent perennial to 7dm high, woody at base; leaves 3-25mm long, 0.5-2.5mm wide, flattened to terete; bracts similar to leaves; perianth lobes all alike, 1-2mm long. Alkaline areas. Statewide. <u>S</u>. <u>fruticosa</u> (L.) Forsk. of authors, <u>S</u>. <u>intermedia</u> Wats., <u>S</u>. <u>moquinii</u> (Torr.) A. Nels.

<u>Suckleya</u> Gray

<u>Suckleya</u> <u>suckleyana</u> (Torr.) Rydb., Mem. N. Y. Bot. Gard. 1:133. 1900. Glabrous or sparsely farinose annual with prostrate stems to 3dm long; leaves petioled, alternate, the blades mostly suborbicular, 5-30mm long and about as wide, sinuate-dentate; flowers unisexual, in axillary clusters, the plants monoecious; staminate perianth 3-4 parted, about 1mm or less long; stamens 3 or 4; pistillate perianth lacking, the pistil enclosed between 2 keeled, connate bracts, in fruit 5-8mm long; stigmas 2. Moist, often alkaline areas. SW, SE.

Herbs with simple, entire, alternate leaves; flowers bisexual, regular, in cymes, the chasmogamous ones maturing at a different time than the cleistogamous ones; sepals 5, separate or united at base; petals 5 and separate, early deciduous, lacking in cleistogamous flowers; stamens numerous or 3-5 in cleistogamous flowers; ovary 1, superior; carpels 3-5; style 1; locule 1; ovules 1 to many; placentation parietal; fruit a capsule.

Helianthemum Miller Frostweed

Reference: Daoud, H. S. & R. L. Wilbur. 1965. Rhodora 67:63-82, 201-216, 255-312.

Helianthemum bicknellii Fern., Rhodora 21:36. 1919. Stellate-pubescent perennial to 5dm high; cauline leaf blades elliptic or elliptic-oblanceolate, 5-35mm long; chasmogamous flowers with sepals 3-7mm long and yellow petals 7-12mm long which usually dry white; cleistogamous flowers 1.5-2.5mm long. Plains, hills, and open woods. Crook Co.

COMMELINACEAE Spiderwort Family

Herbs with simple, linear, alternate, sheathing leaves; flowers bisexual, mostly regular, in a cyme or umbel or solitary; sepals 3, green, separate; petals 3, mostly blue, purple, or rose, usually separate; stamens 6; pistil 1; ovary superior; carpels 3; style 1; locules 1-3; ovules 1 or few per locule; placentation axile; fruit a capsule.

Tradescantia L. Spiderwort

Plants not glaucous; uppermost leaves with ciliate margins; sepals mostly (8)10-13mm long; petals mostly 15-20mm long; pedicels densely pubescent, the hairs often 1mm or more long *T. bracteata*

Plants somewhat glaucous; uppermost leaves without cilia or with obscure cilia on margins; sepals 4-10(12)mm long; petals 7-16(18)mm long; pedicels moderately pubescent to glabrate, the hairs usually much less than 1mm long *T. occidentalis*

Tradescantia bracteata Small ex Britt. & Brown, Ill. Fl. 3:510. 1898. Plants to 4.5dm high; roots fibrous; leaves linear or linear-lanceolate, 4-20cm long, 3-18mm wide, glabrous or the upper ones ciliate on margins; sepals mostly (8)10-13mm long, glandular-pubescent; petals mostly 15-20mm long, usually rose, rarely blue; capsule about 6mm long. Plains, hills, and meadows. Crook Co.

Tradescantia occidentalis (Britt.) Smyth, Trans. Kans. Acad. Sci. 16:163. 1899. Plants to 6dm high; roots fibrous; leaves linear or linear-lanceolate, 0.5-4dm long, 2-20mm wide, glaucous and glabrous; sepals 4-12mm long, glandular-pubescent to glabrate; petals 7-18mm long, blue to purple or rarely rose; capsule about 5-6mm long. Sandy areas and prairies. NE, SE. *T. laramiensis* Goodding.

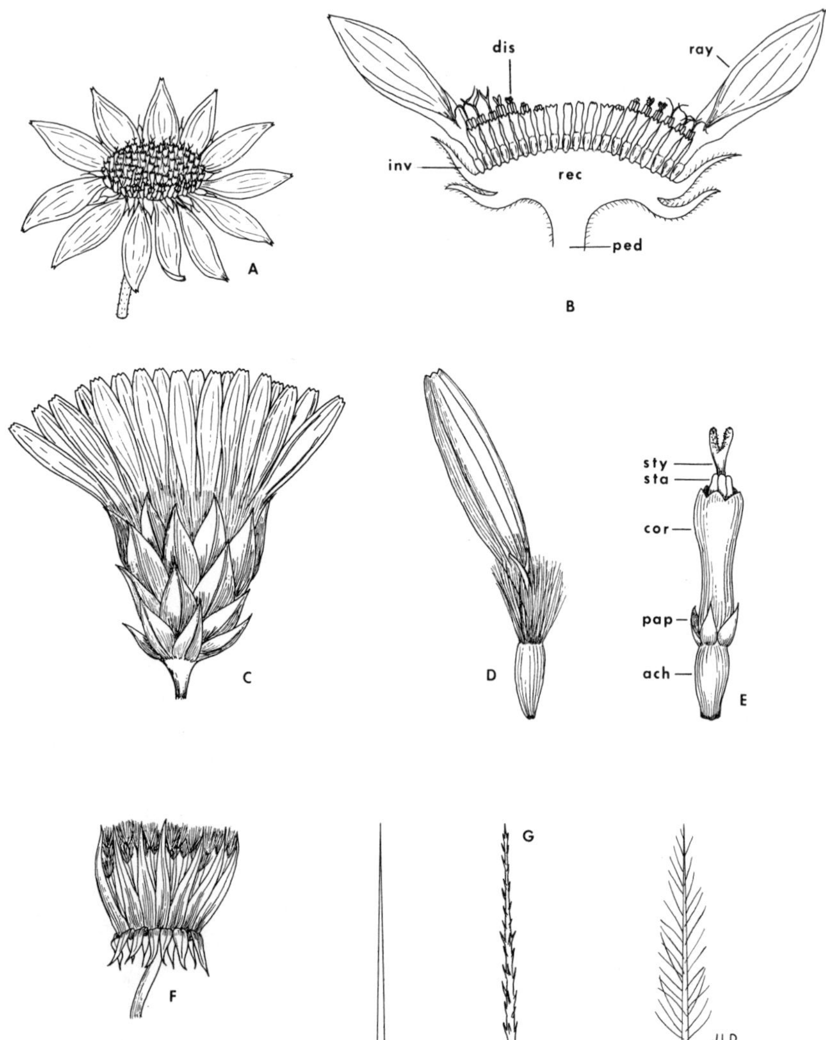
dis
ray
inv
rec
ped
A
B
C
D
sty
sta
cor
pap
ach
E
F
G
JLD

Herbs or shrubs with basal, alternate, or opposite leaves; flowers bisexual or unisexual or sometimes neutral, in a head surrounded by an involucre (rarely 1 flower per head), the head often appearing like a single flower; head often including an outer series of ray flowers bearing a flat, strap-shaped corolla and an inner group of disk flowers; heads of some species may contain all ray flowers and of others may contain all disk flowers; calyx (pappus) usually of scales, awns, a short crown, or bristles or sometimes lacking; petals 5 or rarely 4, united, rarely lacking; stamens mostly 5, the anthers usually united to form a tube, rarely free with the filaments united; ovary 1, inferior; carpels 2; style 1, usually 2 cleft at tip; locule 1; ovule 1; placentation basal; fruit an achene.

References: Rydberg, P. A. 1922. N. Am. Fl. 33:1-46.

1914-27. N. Am. Fl. 34:1-360.

Sherff, E. E. & E. J. Alexander. 1955. N. Am. Fl. II, 2:1-149.

Figure 6. Compositae. A. Head of Helianthus (X 0.7). B. Longitudinal section of head (X 1.3): ray = ray flower, dis = disk flower, rec = receptacle, inv = involucral bract, ped = peduncle. C. Pressed head of Aster with involucral bracts imbricate (X 3). D. Ray flower (X 4) with a pappus of bristles. E. Disk flower (X 5): ach = achene, pap = pappus (of scales), cor = corolla, sta = stamens, sty = style. F. Head of Taraxacum in fruit with biseriate involucral bracts (X 1). G. Capillary bristles (X 8): smooth at left, barbed at center, plumose at right.

1. Plants shrubs GROUP I p. 249
1. Plants herbs or woody only at base (shrubs with flowers will also run here)
  2. Heads of all ray flowers; juice milky
    3. Rays bright yellow or orange (sometimes drying whitish or purplish) GROUP II p. 250
    3. Rays white, blue, purple, pink, or rose GROUP III p. 252
  2. Heads of both ray and disk flowers or of all disk flowers, the ray flowers, when present, marginal in the head; juice watery
    4. Ray flowers lacking, only disk flowers present
      5. Pappus of capillary bristles GROUP IV p. 255
      5. Pappus of scales which are sometimes fringed with hairs, of awns which may be retrorsely barbed, of a short crown, or lacking GROUP V p. 261
    4. Ray flowers and disk flowers both present
      6. Pappus partly or entirely of capillary bristles
        7. Rays yellow or orange when fresh GROUP VI p. 265
        7. Rays white, pink, blue, purple, or rose when fresh GROUP VII p. 268
      6. Pappus of scales, awns (which may be retrorsely barbed), a short crown, or entirely lacking
        8. Receptacle with bristles (hairs) or chaffy scales or bracts among the disk flowers, sometimes reduced to a single row of scales between ray and disk flowers GROUP VIII p. 270
        8. Receptacle naked GROUP IX p. 274

GROUP I

1. Plants with spines or spine-tipped branches
  2. Leaves 3-5 parted, the divisions often with 3 linear subdivisions ... Artemisia
  2. Leaves entire ... Tetradymia
1. Plants without spines or spine-tipped branches
  3. Leaves toothed, lobed, or cleft usually from the tip or else dissected ... Artemisia
  3. Leaves mostly entire or with wavy margins or rarely toothed to base
    4. Leaves and stems glabrous, glandular, or puberulent
      5. Leaf blades mostly ovate to suborbicular, densely glandular-puberulent; SW desert area ... Brickellia
      5. Leaf blades mostly linear, elliptic, lanceolate, or oblanceolate, glandular or not; widespread
        6. Leaves glandular-pubescent; plants of the western mountains ... Haplopappus
        6. Leaves not glandular-pubescent, sometimes glandular-punctate; plants of the plains or mountains
          7. Herbage usually glandular-punctate, the leaves mostly less than 3mm wide ... Gutierrezia
          7. Herbage usually not glandular-punctate, the leaves sometimes over 3mm wide ... Chrysothamnus
    4. Leaves or stems or both somewhat canescent or tomentose

8. Stems mostly densely white-tomentose, the leaves greenish and often glandular-puberulent; rare plants of the NW mountains — Haplopappus

8. Stems not white-tomentose, or if so, the leaves often similarly so; plants of the plains or mountains

 9. Leaves and young stems with a loose tomentum, the leaves usually elliptic or oblanceolate or sometimes linear — Tetradymia

 9. Leaves and young stems not tomentose, sometimes canescent, or if tomentose, the leaves less so or else linear and the stem tomentum close and not loose

  10. Leaves and twigs somewhat canescent — Artemisia

  10. Leaves and twigs either tomentose, glabrous, or rather sparsely puberulent — Chrysothamnus

GROUP II

1. Pappus of plumose bristles, sometimes with a scale-like base

 2. Involucral bracts over 2cm long, rarely shorter; achenes long beaked at tip — Tragopogon

 2. Involucral bracts not over 2cm long; achenes truncate at tip, not beaked — Microseris

1. Pappus not of plumose bristles

3. Heads 1 per stem
  4. Stem leafy, the leaves rarely much reduced
    5. Pappus white; leaves lobed — Crepis
    5. Pappus tawny or brown; leaves entire or minutely toothed — Hieracium
  4. Stem without leaves, the leaves basal or nearly so (rarely with 1 or 2 much reduced stem leaves)
    6. Pappus bristles broader and flattened toward base; achenes not beaked — Nothocalais
    6. Pappus bristles not flattened or broader toward base; achenes often beaked
      7. Outer series of involucral bracts prominently shorter and usually spreading or reflexed, the inner series ascending — Taraxacum
      7. Outer series of involucral bracts about the same size as the inner (rarely shorter), all of them ascending
        8. Achenes often beaked at tip; involucre 10mm or more long — Agoseris
        8. Achenes not beaked at tip; involucre 5-10mm long — Hieracium
3. Heads usually 2 or more per stem
  9. Pappus bristles readily deciduous, united below and all falling together (rarely 1-5 remaining); annuals — Malacothrix

9. Pappus bristles persistent, or later deciduous, mostly separate below; annuals to perennials
   10. Achenes usually flattened; plants sometimes prickly at least on leaf margins or veins
      11. Involucres cylindrical, mostly about twice as long as wide or more; either the achenes beaked or the pappus brownish ... Lactuca
      11. Involucres bell-shaped or hemispherical, little if at all longer than wide; achenes not beaked, the pappus white ... Sonchus
   10. Achenes not flattened; plants usually not prickly
      12. Pappus white; leaves often lobed or coarsely toothed, or if not, the plants usually glabrous or annuals ... Crepis
      12. Pappus tawny or brown; leaves entire or slightly toothed (rarely slightly lobed); plants usually with long hairs (rarely glabrous or with stellate hairs); perennials ... Hieracium

GROUP III

1. Pappus bristles plumose, at least above
   2. Involucral bracts 2cm or less long

3. Pappus plumose to base or nearly so (except in 1 annual or biennial); involucral bracts glabrous or nearly so or sometimes glandular-puberulent or glandular-punctate
Stephanomeria

3. Pappus (at least on mature achenes) with a scaly base about 1-3mm long; plants perennial; involucral bracts often with some hairs which are often black, not glandular
Microseris

2. Involucral bracts over 2cm long Tragopogon

1. Pappus bristles not plumose or the pappus of scales

4. Heads 1 per stem; leaves all basal or nearly so; rays pink or purple

5. Pappus bristles broader and flattened toward base; achenes not beaked Nothocalais

5. Pappus bristles not flattened or broader toward base; achenes often beaked

6. Outer series of involucral bracts prominently shorter and usually spreading or reflexed, the inner series ascending Taraxacum

6. Outer series of involucral bracts about the same size as the inner (rarely shorter), all of them ascending
Agoseris

4. Heads 2 or more per stem, or rarely 1 but then the rays white or the stems definitely leafy, the leaves sometimes much reduced or deciduous

7. Pappus of minute scales — <u>Cichorium</u>

7. Pappus of capillary bristles

  8. Achenes flattened, 1mm or more wide, mostly 4 times or more as wide as deep, often beaked; leaves often pinnatifid and somewhat prickly margined, rarely entire — <u>Lactuca</u>

  8. Achenes not flattened, often less than 1mm wide and rarely over twice as wide as deep, usually not beaked; leaves pinnatifid or not, usually not prickly margined

    9. Corolla white or ochroleucous

      10. Pappus white; leaves often lobed or coarsely toothed, or if not, the plants usually glabrous or annuals — <u>Crepis</u>

      10. Pappus tawny or brown; leaves entire or slightly toothed (rarely slightly lobed); plants usually with long hairs (rarely glabrous or with stellate hairs); perennials — <u>Hieracium</u>

    9. Corolla pink or rose to purplish (very rarely white)

      11. Leaves linear or linear-lanceolate, 5mm or less wide, often deciduous

        12. Plants annual — <u>Shinnersoseris</u>

        12. Plants perennial — <u>Lygodesmia</u>

      11. Leaves broader, mostly well over 1cm wide, rarely deciduous

        13. Stem leaves clasping; SE Wyo. — <u>Prenanthes</u>

        13. Stem leaves lacking or not clasping; NW mountains — <u>Crepis</u>

GROUP IV

1. Leaves with spines often 2mm or more long on their margins; involucre usually spiny at least at tips of outer bracts
  2. Pappus bristles plumose, at least on the inner flowers — Cirsium
  2. Pappus bristles barbellate
    3. Receptacle not fleshy or honeycombed, densely bristly, the bristles much longer than the achenes — Carduus
    3. Receptacle naked or sometimes fleshy or honeycombed, or if slightly bristly, the bristles shorter than the achenes
      4. Corollas 1.5cm or more long, often purple — Onopordum
      4. Corollas 8mm long or less, usually yellow or white — Machaeranthera
1. Leaves without spiny margins or the spines usually less than 1mm long; involucres spiny or not
  5. Receptacle densely bristly; involucral bracts spiny or with hooked tips or strongly fimbriate or fringed toward tip (outer ones sometimes merely with broad scarious tips)
    6. Involucral bracts with hooked tips forming a bur; leaves ovate, deltoid, or cordate, entire or merely toothed — Arctium
    6. Involucral bracts without hooked tips, the spines, if present, straight; leaves not as above — Centaurea

5. Receptacle naked or rarely short hairy; involucral bracts usually not spiny nor with hooked tips nor strongly fimbriate or fringed

7. Flowers unisexual in separate heads (staminate sometimes with a vestigial ovary and entire style); perennials

8. Involucral bracts in 1 series (excluding a few much reduced ones at base of head); basal leaves cordate or sagittate <u>Petasites</u>

8. Involucral bracts in 2 or more series; basal leaves not cordate or sagittate

9. Basal leaves readily deciduous, not much larger than the well developed and numerous stem leaves which are usually less pubescent and greenish on upper surface and white-tomentose beneath; stolons lacking <u>Anaphalis</u>

9. Basal leaves persistent, usually tufted; stem leaves lacking or usually reduced upward, often about equally pubescent on both sides; stolons sometimes present <u>Antennaria</u>

7. Flowers perfect, or if unisexual, the pistillate usually marginal and the staminate or perfect central in the same head; annuals to perennials

10. Involucral bracts with conspicuous, resinous, yellow or orange dots mostly 0.3-1mm long; leaves mostly opposite and dissected <u>Dyssodia</u>

10. Involucral bracts lacking yellow or orange dots or these minute; leaves various
   11. Involucral bracts in 1 series (excluding some occasional short bracts at base)
      12. Basal leaves cordate or sagittate, white-tomentose beneath; stem leaves much reduced or lacking; corollas white to purplish or drying yellowish ... <u>Petasites</u>
      12. Basal leaves not cordate or sagittate, or if so, usually not tomentose beneath and corollas usually not white to purplish; stem leaves well developed or not
         13. Leaves opposite ... <u>Arnica</u>
         13. Leaves alternate or all basal (rarely fascicled)
            14. Plants shrubby; involucral bracts 4-6 per head ... <u>Tetradymia</u>
            14. Plants herbaceous; involucral bracts 7 or more per head
               15. Leaves entire or ciliate margined
                  16. Leaves all less than 14mm wide ... <u>Erigeron</u>
                  16. Leaves, or some of them, 17mm or more wide ... <u>Senecio</u>
               15. Leaves toothed or lobed
                  17. Leaves palmately or ternately divided into mostly linear segments ... <u>Erigeron</u>

17. Leaves not as above *Senecio*

11. Involucral bracts in 2 or more series

18. Leaves opposite or whorled

19. Achenes 5 angled; leaves usually whorled *Eupatorium*

19. Achenes not 5 angled, usually 10 or more ribbed; leaves usually opposite

20. Involucral bracts conspicuously longitudinally striate; stem leaves more than 4 pair *Brickellia*

20. Involucral bracts not striate; stem leaves mostly 2-4 pair *Arnica*

18. Leaves alternate or all basal

21. Involucral bracts scarious and pure white; stems not viscid *Anaphalis*

21. Involucral bracts sometimes scarious but not white (occasionally yellowish or dingy white), or if rarely white, the stems viscid

22. Plants white-woolly at least above, the involucral bracts scarious throughout or sometimes with a green base *Gnaphalium*

22. Plants not as above

23. Plants annual

24. Involucre 5mm or more long, the bracts mostly herbaceous *Aster*

24. Involucre 2-4mm long, the bracts with scarious margins their entire length and nearly 1/3 their width — Conyza

23. Plants perennial or rarely biennial

25. Involucral bracts conspicuously longitudinally striate, greenish and less than 2mm wide — Brickellia

25. Involucral bracts not longitudinally striate (anastomosing veins sometimes prominent), sometimes not green or wider

26. Heads in a spike-like inflorescence, or if not, the involucral bracts mostly obovate and not in vertical rows, usually purplish or pink at least at tip; corollas pink-purple or white — Liatris

26. Heads in an open inflorescence (rarely raceme-like or cymose), the involucral bracts not obovate (rarely so), often in vertical rows, usually not purplish; corollas usually yellow

27. Plants biennial or perennial herbs; involucral bracts linear and usually less than 0.6mm wide, or if wider, the leaves with 3 or more linear divisions toward tip — Erigeron

27. Plants perennial herbs or shrubs; involucral bracts not linear, usually over 0.6mm wide, the leaves entire or minutely toothed (rarely coarsely spinulose-toothed)

28. Leaves with coarse teeth which are spinulose-tipped; involucre 5-11mm long — Machaeranthera

28. Leaves not as above, or if so, the involucre 12-30mm long

29. Heads mostly 10-30mm wide, or if as little as 7mm, the plants in rocky places in the mountains; involucres 9-30mm long — Haplopappus

29. Heads mostly 2-6mm wide, or if as much as 8mm, the plants on the plains or foothills; involucres 4-13mm long

30. Plants herbaceous from a woody caudex, 2dm or less high, usually with a basal tuft of leaves — Petradoria

30. Plants shrubs or subshrubs often over 2dm high, lacking a basal tuft of leaves — Chrysothamnus

GROUP V

1. Anthers not united or only slightly so; flowers all unisexual
   2. Staminate and pistillate flowers in the same head, the pistillate few and marginal
      3. Plants densely white-woolly, less than 10cm high — Psilocarphus
      3. Plants not white-woolly, over 10cm high — Iva
   2. Staminate and pistillate flowers in different heads, the pistillate completely enclosed in an often bur-like involucre
      4. Involucral bracts of staminate head separate; spines of pistillate head hooked — Xanthium
      4. Involucral bracts of staminate head united; spines of pistillate head, if present, not hooked — Ambrosia
1. Anthers united; flowers usually bisexual except sometimes the marginal ones (central ones rarely with ovary aborted)
   5. Involucral bracts with conspicuous yellow or orange, resinous dots mostly 0.3-1mm long; leaves mostly opposite and dissected — Dyssodia
   5. Involucral bracts without conspicuous yellow or orange dots; leaves various
      6. Involucral bracts mostly 4 in 1 series, each enclosing a marginal flower; annuals — Madia

6. Involucral bracts usually in 2 or more series, usually more than 4 and none enclosing marginal flowers; annuals to perennials
  7. Involucral bracts, or some of them, tipped with spines 1cm or more long — Centaurea
  7. Involucral bracts not as above
    8. Receptacle with chaffy scales, these sometimes only between marginal and central flowers or rarely densely white-woolly and enclosing entire flower
      9. Plants densely white-woolly annuals less than 10cm high — Psilocarphus
      9. Plants not as above
        10. Plants densely caespitose and acaulescent, to 5cm high, the heads sessile or nearly so in the leaves — Parthenium
        10. Plants not as above
          11. Receptacle hemispherical to cylindrical; leaves alternate — Rudbeckia
          11. Receptacle flat or nearly so; leaves mostly opposite, rarely all basal
            12. Inner involucral bracts longer than outer, united at least 1/3 their length — Thelesperma
            12. Inner involucral bracts usually shorter than outer, separate or united only at very base — Bidens

8. Receptacle naked or hairy or rarely glandular
  13. Pappus lacking or an obscure crown; involucral bracts usually dry and scarious, at least on the margins
    14. Heads in a spike, raceme, or panicle (rarely subcapitate or solitary in an alpine or subalpine perennial); corollas of central flowers usually 5 toothed; herbs or shrubs — Artemisia
    14. Heads in a corymb, or capitate or subcapitate, or solitary at tips of stem or branches; corollas of central flowers 4 or 5 toothed; herbs
      15. Heads mostly solitary at ends of branches; plants annual — Matricaria
      15. Heads in a definite inflorescence or solitary on an unbranched stem; plants perennial — Tanacetum
  13. Pappus of scales, awns, or lacking; involucral bracts usually herbaceous or occasionally scarious only on the margins or at very tip
    16. Involucral bracts with pectinately arranged filiform processes about 1mm long toward tip — Centaurea
    16. Involucral bracts not as above
      17. Plants annuals; leaves entire; corollas mostly rose or pinkish — Palafoxia
      17. Plants biennials or perennials, or if annuals, the leaves mostly pinnately dissected; corollas various

18. Achenes 5-10 angled or ribbed; leaf blades entire, oblanceolate or linear — Hymenoxys

18. Achenes mostly 4 angled or ribbed; leaf blades lobed or toothed, or if entire, then either lanceolate or ovate to cordate

19. Leaf blades entire or toothed, elliptic or ovate to cordate

20. Leaves along stem — Chrysanthemum

20. Leaves all basal — Chamaechaenactis

19. Leaf blades mostly pinnately dissected

21. Involucral bracts mostly over 3 times as long as wide, not scarious-margined — Chaenactis

21. Involucral bracts mostly broader, usually scarious-margined — Hymenopappus

GROUP VI

1. Plants shrubby — Haplopappus
1. Plants herbaceous but sometimes woody at base
  2. Receptacle bristly; involucral bracts spiny — Centaurea
  2. Receptacle scaly or naked, rarely short bristly or hairy; involucral bracts not spiny (rarely weakly spinulose-tipped)
    3. Involucral bracts with conspicuous, yellow or orange, resinous dots mostly 0.3-1mm long; leaves mostly opposite and dissected — Dyssodia
    3. Involucral bracts without conspicuous resinous dots or these smaller; leaves various
      4. Leaves mostly opposite — Arnica
      4. Leaves alternate or basal
        5. Involucral bracts in 1 series (excluding the few reduced ones at base), not imbricate — Senecio
        5. Involucral bracts in 2 or more series, sometimes imbricate
          6. Involucre sticky, appearing varnished when dry, the outer bracts hooked at tip — Grindelia
          6. Involucre not as above
            7. Pappus in 2 series, the outer of small, often inconspicuous scales or bristles, the inner of longer barbellate bristles; stems leafy, the blades mostly not linear — Heterotheca

7. Pappus of subequal bristles in 1 series (do not mistake hairs of achene for part of pappus), or if not, the leaves mostly basal and linear
  8. Heads solitary on each stem
    9. Involucral bracts 1mm or less wide; leaves linear, less than 2mm wide, entire — Erigeron
    9. Involucral bracts mostly wider; leaves often wider, sometimes toothed — Haplopappus
  8. Heads more than 1 per stem
    10. Leaves pinnatifid or bipinnatifid — Haplopappus
    10. Leaves not as above
      11. Heads about 7mm or more wide
        12. Involucral bracts 1mm or less wide; leaves linear and entire, less than 2mm wide — Erigeron
        12. Involucral bracts mostly wider; leaves often wider or toothed
          13. Heads mostly 2-4 per stem, often over 1cm wide, or if 5 or more per stem, the involucral bracts cuspidate or acuminate — Haplopappus
          13. Heads mostly 5 or more per stem, usually less than 1cm wide; involucral bracts not cuspidate or acuminate

14. Leaves and stems pubescent throughout, or if not, the lower leaves with long multicellular hairs on margins of petioles and often rounded at tip — *Solidago*

14. Leaves and stems glabrous except sometimes the upper stem and the leaf margins, the lower leaves at most with very short cilia on the petiole margins, sometimes sharply acute at tip — *Haplopappus*

11. Heads mostly about 5mm or less wide

15. Involucre 7-10mm long, the bracts cuspidate or acuminate at tip — *Haplopappus*

15. Involucre shorter or the bracts not cuspidate or acuminate, or both

16. Inflorescence of heads a raceme or panicle — *Solidago*

16. Inflorescence of heads corymbose or cymose

17. Ray flowers less than 5 per head; involucral bracts in vertical rows or nearly so — *Petradoria*

17. Ray flowers usually more than 5 per head; involucral bracts not in vertical rows — *Solidago*

GROUP VII

1. Receptacle bristly; involucral bracts spiny or with pectinately arranged processes about 1mm long toward tip — Centaurea
1. Receptacle scaly or naked; involucral bracts not as above
  2. Basal leaf blades cordate or sagittate, white-tomentose beneath — Petasites
  2. Basal leaves lacking or not as above
    3. Rays inconspicuous, little if at all longer than pappus
      4. Plants annual; involucre 2-4mm long, the bracts with scarious margins their entire length and nearly 1/3 their width — Conyza
      4. Plants annual to perennial, if annual, the involucre over 4mm long and the bracts herbaceous
        5. Plants annual; involucral bracts in 2 or more series, the outer series loose and foliaceous — Aster
        5. Plants biennial or perennial; involucral bracts in 1 or 2 series, the outer series not foliaceous — Erigeron
    3. Rays conspicuous and longer than pappus
      6. Involucral bracts often somewhat subequal, long and narrow, entirely green and often with scarious margins or green at base, usually pubescent
        7. Plants densely matted, to 5cm high, the heads mostly sessile or nearly so amongst the entire leaves; involucre mostly 8mm or more long — Townsendia

7. Plants not as above <u>Erigeron</u>

6. Involucral bracts mostly imbricate, the outer definitely shorter than the inner, or if not, then usually foliaceous, usually broadened at tip or base, entirely green, green at tip, or not green, often glabrous or glabrate

8. Leaves, at least the lower ones, once or twice pinnatifid, or if not, the leaves with spine-tipped or mucronate lobes or coarse teeth; plants not rhizomatous, usually taprooted <u>Machaeranthera</u>

8. Leaves not as above, or if so, then with slender rhizomes

9. Heads several per stem or else the plants with well developed cauline leaves, or both

10. Plants densely caespitose perennials from a taproot, to 7(15)cm high, the leaves mostly all basal or else closely subtending the few heads <u>Townsendia</u>

10. Plants not as above

11. Plants taprooted annuals or biennials (rarely short lived perennials)

12. Involucral bracts green only toward tip; pappus of ray and disk flowers similar <u>Machaeranthera</u>

12. Involucral bracts green throughout or toward base; pappus of ray flowers often much shorter than that of disk flowers <u>Townsendia</u>

11. Plants perennials often with rhizomes or fibrous roots
    13. Rays white when fresh (becoming pinkish or brownish in age, rarely purplish); involucre (7)8-15mm long; plants from a stout woody taproot and branched caudex; plains and hills, rarely in the mountains *Xylorhiza*
    13. Rays not white, or if so, either the involucre 3-8(13)mm long, the plants from rhizomes or fibrous roots, or not of the plains and hills *Aster*

9. Heads solitary (sometimes sessile amongst basal leaves); plants with all basal leaves or the cauline leaves few and much reduced
    14. Heads sessile or nearly so *Townsendia*
    14. Heads peduncled
        15. Basal leaves with nearly parallel sides *Aster*
        15. Basal leaves much broadened toward tip and narrowed to base *Townsendia*

GROUP VIII

1. Involucral bracts with conspicuous, yellow or orange, resinous dots mostly 0.3-1mm long; leaves mostly opposite and dissected *Dyssodia*

1. Involucral bracts without resinous dots or these minute; leaves various
  2. Involucral bracts with pectinately arranged processes about 1mm long toward tip — Centaurea
  2. Involucral bracts not as above
    3. Receptacle bristly
      4. Rays 3 toothed at tip; pappus a short crown or lacking — Anthemis
      4. Rays deeply 3 lobed at tip; pappus of awned scales — Gaillardia
    3. Receptacle with chaffy scales or bracts
      5. Involucral bracts dry and scarious throughout except sometimes for a green midrib
        6. Rays commonly 3-5 per head, less than 5(6)mm long; plants perennial — Achillea
        6. Rays commonly 10-30 per head, mostly 5-15mm long; plants annual or perennial — Anthemis
      5. Involucral bracts, at least the outer ones, herbaceous or scarious only on the margins
        7. Rays white, purple, pink, or rose, rarely drying yellowish
          8. Rays white to ochroleucous, not reflexed — Wyethia
          8. Rays usually purple, pink, or rose, usually reflexed

9. Leaves mostly pinnately divided; receptacular bracts about equaling achenes <u>Ratibida</u>

9. Leaves not pinnately divided; receptacular bracts much longer than achenes <u>Echinacea</u>

7. Rays yellow or orange (sometimes purple or brown at base)

10. Involucral bracts mostly 4 in 1 series, enclosing the achenes of the marginal flowers <u>Madia</u>

10. Involucral bracts in more than 1 series, or if in 1, not enclosing the achenes

11. Scales of receptacle few and scattered between ray and disk flowers <u>Helenium</u>

11. Scales of receptacle many or in a definite series between ray and disk flowers

12. Involucral bracts in 2 very dissimilar series, the inner usually longitudinally striate; leaves not hastate or cordate

13. Inner series of involucral bracts longer than the outer and united at least 1/3 their length <u>Thelesperma</u>

13. Inner series of involucral bracts shorter than or equal to the outer, distinct or nearly so <u>Bidens</u>

12. Involucral bracts in 1 or more somewhat similar series, the inner not striate; leaves sometimes hastate or cordate

14. Plants scapose or nearly so, the cauline leaves, if any, much reduced and usually inconspicuous <u>Balsamorhiza</u>
14. Plants leafy stemmed
  15. Cauline leaves all alternate
    16. Receptacle flat or slightly convex <u>Wyethia</u>
    16. Receptacle cylindrical, columnar, or hemispherical
      17. Rays subtended by receptacular bracts; heads, excluding rays, about 1cm wide; leaves mostly pinnately divided <u>Ratibida</u>
      17. Rays not subtended by bracts; heads, excluding rays, mostly 1.5cm or more wide; leaves entire to pinnatifid or palmatifid <u>Rudbeckia</u>
  15. Cauline leaves, at least the lower (sometimes these deciduous), opposite or subopposite
    18. Pappus lacking; rays mostly 7-19mm long; perennial <u>Viguiera</u>
    18. Pappus present, sometimes of only 2 awns and sometimes deciduous; rays mostly 15-50mm long; annual to perennial

19. Ray flowers with achenes; disk achenes corky-winged — Verbesina
19. Ray flowers usually sterile, sometimes with aborted achenes; disk achenes not corky-winged
    20. Pappus persistent; disk achenes strongly compressed and thin-edged; bracts of receptacle often rounded or flat across tip; perennials — Helianthella
    20. Pappus deciduous (at least the 2 main awn-scales); disk achenes usually slightly to moderately compressed, often not thin-edged; bracts of receptacle often pointed at tip; annuals or perennials — Helianthus

GROUP IX

1. Involucre with conspicuous, orange or yellow, resinous dots mostly 0.3-1mm long; leaves mostly opposite and dissected — Dyssodia
1. Involucre without conspicuous resinous dots or these smaller; leaves various

2. Rays white, purple, pink, or rose
 3. Pappus of scales or awns _Townsendia_
 3. Pappus lacking or a minute crown
  4. Leaves toothed to once or twice pinnately compound with broad segments; rays often over 12mm long _Chrysanthemum_
  4. Leaves finely dissected into linear segments; rays 5-12mm long _Matricaria_
2. Rays yellow or orange
 5. Pappus of 2-8 rigid, deciduous, slender awns, not scale-like; involucre sticky, appearing varnished when dry, the outer bracts with hooked tips _Grindelia_
 5. Pappus of scales which may be awned from tip, rarely lacking; involucre rarely sticky, the bracts usually not hooked
  6. Involucral bracts cuspidate or awned at tip; leaves dissected
   7. Involucral bracts all similar and free at base _Bahia_
   7. Involucral bracts in 2 dissimilar series, the outer series connate at base _Hymenoxys_
  6. Involucral bracts usually not cuspidate or awned; leaves rarely dissected
   8. Involucral bracts mostly 4-15, tending to subtend the ray flowers individually; achenes about 3.5 times as long as wide or longer; plants somewhat tomentose (rarely not); leaves on the stems

9. Plants annual
    10. Plants tomentose *Antheropeas*
    10. Plants not tomentose *Madia*
9. Plants perennial *Eriophyllum*

8. Involucral bracts often 20 or more, usually not subtending the ray flowers individually, or if so and less than 20, then the bracts somewhat varnished or resinous or the leaves all basal; achenes various; plants often not tomentose
    11. Heads mostly 1.5-3mm wide, the involucral bracts glabrous or glabrate and resinous or varnished; leaves linear, not divided, 2mm or less wide, well distributed on the stems *Gutierrezia*
    11. Heads often wider, the bracts sometimes pubescent, not resinous or varnished, sometimes glandular-dotted; leaves, if linear and undivided, all basal or nearly so
        12. Leaves opposite at least below *Picradeniopsis*
        12. Leaves alternate or basal
            13. Achenes at least 4 times as long as wide; receptacle flat or nearly so; leaves shallowly pinnately lobed or coarsely toothed *Hulsea*
            13. Achenes mostly 2-3 times as long as wide, rarely longer; receptacle usually convex to subglobose; leaves entire, slightly toothed, or divided into mostly linear segments

14. Plants either scapose and the leaves all 1cm or less wide or with pinnately or ternately dissected leaves; involucral bracts usually appressed and erect

15. Rays well developed, 4mm or more long *Hymenoxys*

15. Rays poorly developed, 2mm or less long *Tanacetum*

14. Plants either leafy stemmed and the leaves entire or toothed or else scapose with entire leaves some of which are 1.5cm or more wide; involucral bracts sometimes loose and eventually reflexed

16. Plants either with rays 3 lobed at tip or else base of involucre tomentose or woolly hairy *Helenium*

16. Plants with rays entire or nearly so at tip and base of involucre scabrous or glandular-puberulent *Platyschkuhria*

<u>Achillea</u> L. Yarrow

References: Mulligan, G. A. & I. J. Bassett. 1959. Canad. Journ. Bot. 37:73-79.

Hiesey, W. M. & M. A. Nobs. 1970. Bot. Gaz. 131: 245-259.

<u>Achillea</u> <u>millefolium</u> L., Sp. Pl. 899. 1753. Perennial, aromatic herb to 8dm high, often rhizomatous; leaves alternate, the blades pinnately dissected, 1-12cm long; heads many in a paniculate or corymb-like inflorescence; involucre 3-6mm long, the bracts imbricate in several series, scarious almost throughout; rays mostly 3-5, white or sometimes pink, 3-6mm long; disk corollas about 3mm long; pappus lacking; receptacle chaffy. Plains, hills, slopes, and disturbed areas. Statewide. <u>A</u>. <u>lanulosa</u> Nutt.

<u>Agoseris</u> Raf. False Dandelion

Taprooted, scapose, perennial herbs, the heads solitary on each scape, the flowers all ligulate with orange, yellow, pink, or purple rays; involucral bracts in 2 to several series, imbricate or not; pappus of capillary bristles; receptacle naked or rarely chaffy.

Corolla yellow, sometimes drying pinkish or purplish; achenes with a beak less than half as long as body of achene *A. glauca*

Corolla orange, often drying purplish or pinkish; achenes with a beak over half as long as body of achene *A. aurantiaca*

Agoseris aurantiaca (Hook.) Greene, Pittonia 2:177. 1891. Plants to 6dm high; leaf blades mostly oblanceolate or oblong, 3-35cm long, entire or with a few teeth or lobes; involucre 1-3cm long; rays usually orange, often drying purplish or pinkish, mostly 1-3cm long. Woods and meadows in the mountains. Statewide. *Troximon aurantiacum* Hook., *T. purpureum* (Gray) A. Nels., *T. montanum* (Osterh.) A. Nels.

Agoseris glauca (Pursh) Raf., Herb. Raf. 39. 1833. Plants to 7dm high; leaf blades linear to oblanceolate or oblong, 2-35cm long, entire to laciniate-pinnatifid; involucre 1-3cm long; rays yellow, sometimes drying pinkish or purplish, mostly 1-3cm long. Plains, hills, meadows, and slopes. Statewide. *A. agrestis* Osterh., *Troximon glaucum* Pursh, *T. parviflorum* Nutt., *T. pubescens* (Rydb.) A. Nels., *T. villosum* (Rydb.) A. Nels., *T. arachnoideum* (Rydb.) A. Nels.

Ambrosia L. Ragweed

Annual or perennial herbs with opposite or alternate, toothed to dissected leaves; heads small, unisexual, lacking ray flowers, the staminate in a spike-like or raceme-like inflorescence, the pistillate below in axils of leaves or bracts; pistillate involucre closed, often bearing tubercles or spines, the pistils usually solitary without corolla or pappus; staminate involucre subherbaceous, the bracts connate about half or more their length; filaments often united, the anthers free or barely united.

Reference: Payne, W. W. 1964. Journ. Arnold Arb. 45:401-438.

1. Plants annual with a taproot
  2. Leaves 3(5) palmately lobed or not at all lobed A. trifida
  2. Leaves mostly 1-2 times pinnatifid
    3. Fruiting involucre bearing several series of coarse spines usually over 3mm long; staminate involucral bracts usually connate about half their length, the lobes somewhat regular A. acanthocarpa
    3. Fruiting involucre bearing 1 series of short spines 0.5mm or less long or spines lacking; staminate involucral bracts usually connate most of their length, the lobes, if present, somewhat irregular A. artemisiifolia

1. Plants perennial with creeping rootstocks, these sometimes deep seated
   4. Fruiting involucre lacking spines, sometimes tuberculate; lower surface of leaves loosely pubescent and glandular, not tomentose; leaves opposite at least below *A. psilostachya*
   4. Fruiting involucre somewhat spiny; lower surface of leaves usually tomentose, glands not apparent; leaves alternate *A. tomentosa*

Ambrosia acanthicarpa Hook., Fl. Bor. Am. 1:309. 1833. Annual to 8dm high; leaves mostly alternate, occasionally opposite below, 1-2 times pinnatifid, the blades deltoid to lanceolate, 1-8cm long; pistillate involucre 5-10mm long including spines; staminate involucre 0.5-2.5mm long. Plains, hills, and riverbanks. SW, SE, NE. *Franseria acanthicarpa* (Hook.) Cov., *F. montana* Nutt.

Ambrosia artemisiifolia L., Sp. Pl. 988. 1753. Annual to 1m high; leaves opposite below, alternate above, the blades 1-2 times pinnatifid, mostly ovate or elliptic, 1-10cm long; fruiting involucre 3-5mm long, bearing several short spines or not spiny; staminate involucre 1-2.5mm long. Disturbed areas. NW, NE, SE. *A. diversifolia* (Piper) Rydb., *A. media* Rydb., *A. longistylis* Nutt., *A. elatior* L.

Ambrosia psilostachya DC., Prodr. 5:526. 1836. Similar to A. artemisiifolia but perennial with creeping rootstocks; fruiting involucre often tuberculate, lacking spines. Plains, hills, and disturbed areas. NE, SE. A. coronopifolia T. & G.

Ambrosia tomentosa Nutt., Gen. Pl. 2:186. 1818. Perennial to 5dm high; leaves alternate, the blades 1-3 times pinnatifid, ovate to lanceolate or elliptic, 2-15cm long; fruiting involucre 3-6mm long including spines; staminate involucre 1-3mm long. Plains, hills, and disturbed areas. NE, SE. Franseria tomentosa (Nutt.) A. Nels., F. discolor Nutt.

Ambrosia trifida L., Sp. Pl. 987. 1753. Annual to 15dm high; leaves opposite or occasionally alternate above, the blades elliptic to cuneate-ovate or suborbicular, usually palmately 3(5) lobed or not lobed, mostly serrate, 2-20cm long; fruiting involucre 5-10mm long, bearing several short spines at tip; staminate involucre 1.5-3mm long. Disturbed areas. NE, SE.

Anaphalis DC. Pearly Everlasting

Anaphalis margaritacea (L.) Benth. & Hook., Gen. Pl. 2:303. 1873. Rhizomatous perennial to 9dm high, white-woolly, dioecious or polygamodioecious; leaves alternate, entire, lanceolate or linear to oblanceolate, 2-12cm long; heads crowded in a short, rounded inflorescence, ray flowers lacking; involucre 5-7mm long, the bracts white, mostly scarious, imbricate in several series; pappus of capillary bristles; receptacle naked. Woods, slopes, and rocky flats. Statewide. _A. subalpina_ (Gray) Rydb., _Nacrea lanata_ A. Nels.

Antennaria Gaertn. Pussy-toes

Dioecious, white-woolly perennial herbs with mostly entire, alternate and basal leaves; heads solitary to numerous in an often compact inflorescence; ray flowers lacking; involucral bracts imbricate in several series, scarious at least at tip; receptacle naked; pappus of capillary bristles, those of the staminate flowers often expanded at tip.

1. Plants usually less than 5cm high, the heads barely if at all exceeding the basal leaves and often solitary
  2. Plants with conspicuous, filiform, naked stolons
     A. flagellaris

2. Plants lacking filiform stolons

3. Leaves mostly spatulate or obovate; involucral bracts mostly all white above the greenish base — *A. rosulata*

3. Leaves mostly oblanceolate or linear; involucral bracts mostly darkened above the greenish base — *A. dimorpha*

1. Plants usually over 5cm high with the heads exceeding the basal leaves, usually several to many heads per stem

4. Plants with conspicuously arching stolons; basal leaves not forming a persistent tuft — *A. arcuata*

4. Plants with trailing or ascending stolons or stolons lacking; basal leaves usually forming a persistent tuft

5. Basal leaves conspicuously less pubescent above than beneath, becoming glabrate and usually green above, white-tomentose beneath

6. Heads in an open, elongate, raceme-like or panicle-like inflorescence, the heads on long peduncles; involucral bracts usually brownish, greenish, or transparent at tip — *A. racemosa*

6. Heads in a crowded or subcapitate cyme; involucral bracts usually whitish at tip — *A. neglecta*

5. Basal leaves about equally pubescent on both sides, usually silvery or gray

7. Plants forming mats; stolons present which are often very leafy

8. Terminal scarious portion of involucral bracts, at least the outer ones, brownish to blackish-green; montane to alpine

 9. Terminal portion of involucral bracts blackish-green throughout, usually sharp-pointed; alpine or subalpine *A. alpina*

 9. Terminal portion of the inner involucral bracts becoming white at tip or the whole scarious portion brownish, usually blunt at tip; montane to alpine *A. umbrinella*

8. Terminal scarious portion of involucral bracts white or pink, sometimes with a basal dark spot; usually not alpine, often below montane

 10. Involucral bracts with a conspicuous dark spot near base of scarious portion; basal leaves mostly oblanceolate *A. corymbosa*

 10. Involucral bracts without a dark spot at base of scarious portion or only slightly darkened; leaves often broader

  11. Involucres mostly (6)7-11mm long; dry pistillate corollas mostly 5-8mm long *A. parvifolia*

  11. Involucres mostly 3-7mm long; dry pistillate corollas mostly 2.5-4.5mm long *A. microphylla*

7. Plants usually not forming mats; stolons lacking
  12. Involucre scarious throughout, glabrous or nearly so except sometimes the outermost bracts *A. luzuloides*
  12. Involucre pubescent and not so scarious on lower portion even on inner bracts
    13. Plants mostly less than 2dm high; involucre blackish except sometimes the inner bracts at tip; alpine and subalpine *A. lanata*
    13. Plants mostly 2-5dm high; involucre blackish, brownish, or whitish; alpine or not
      14. Involucre largely blackish or brownish-green, some of the bracts white-tipped (rarely pinkish) *A. pulcherrima*
      14. Involucre largely whitish, some bracts with a small dark spot at base *A. anaphaloides*

Antennaria alpina (L.) Gaertn., Fruct. 2:410. 1791. Plants mat forming and stoloniferous, to 10cm high; basal leaf blades oblanceolate, 3-25mm long; pistillate involucres 4-7mm long, woolly below, the tips blackish-green or brownish; corollas about 1.5-3mm long. Alpine and subalpine. Statewide. *A. media* Greene.

Antennaria anaphaloides Rydb., Mem. N. Y. Bot. Gard. 1:409. 1900. Plants from a branched caudex or stout rhizome, to 5dm high; basal leaf blades oblanceolate, elliptic, or linear-oblong, 2-15cm long, with usually 3-5 prominent parallel nerves; pistillate involucres 5-8mm long, tomentose toward base, the tip usually white; corollas 2.5-5mm long. Plains, hills, slopes, meadows, and woods. Statewide.

Antennaria arcuata Cronq., Leafl. West. Bot. 6:41. 1950. Plants to 4dm high, with arching stolons; leaf blades oblanceolate, 1-6cm long; heads many; pistillate involucre 4-6mm long, tomentose below, the bracts whitish above. Meadows. Reported from W Wyo.

Antennaria corymbosa E. Nels., Bot. Gaz. 27:212. 1899. Plants mat forming and stoloniferous, to 4dm high; basal leaf blades mostly oblanceolate, 5-30mm long; pistillate involucre 4-7mm long, tomentose at base, the tip usually white, each bract with a black or brown spot below the light tip; pistillate corollas 2.5-4.5mm long. Woods, meadows, and slopes mostly in the mountains. Statewide.

Antennaria dimorpha (Nutt.) T. & G., Fl. N. Am. 2:431. 1843. Plants mat forming from a branched caudex, to 4cm high; leaf blades linear or oblanceolate, 3-25mm long; heads solitary on each stem; pistillate involucre 9-15mm long, glabrate or tomentose at base, the tip mostly brownish; staminate involucre shorter; pistillate corollas about 8mm long. Plains and hills. Statewide.

Antennaria flagellaris (Gray) Gray, Proc. Am. Acad. 17:212. 1882. Plants to 4cm high, with naked, filiform stolons; leaf blades linear, 0.5-3cm long; heads solitary on each stem; pistillate involucres 7-13mm long, brown or reddish tinged, somewhat tomentose below; corollas about 5mm long. Hills and slopes. Reported from N Wyo.

<u>Antennaria</u> <u>lanata</u> (Hook.) Greene, Pittonia 3:288. 1898. Plants from a branched caudex, to 2dm high; basal leaf blades oblanceolate, 1-7cm long; pistillate involucres 5-8mm long, tomentose toward base, the tip sometimes whitish but mostly blackish; corollas 2.5-4mm long. Alpine and subalpine. Park Co.

<u>Antennaria</u> <u>luzuloides</u> T. & G., Fl. N. Am. 2:430. 1843. Plants from a branched caudex, to 7dm high; leaf blades linear, linear-lanceolate, or oblanceolate, 1-13cm long; involucres 4-6mm long, usually glabrous, the tips whitish, the base greenish-brown but still scarious; corollas about 3mm long. Plains, hills, woods, and slopes. NW, SW, SE. <u>A</u>. <u>oblanceolata</u> Rydb.

<u>Antennaria</u> <u>microphylla</u> Rydb., Bull. Torrey Club 24:303. 1897. Plants mat forming and stoloniferous, to 4dm high; basal leaf blades oblanceolate to obovate, 3-30mm long; involucres 3-7mm long, tomentose at base, the tips pink or white; corollas 2.5-4.5mm long. Plains, hills, slopes, meadows, and open woods. Statewide. <u>A</u>. <u>arida</u> E. Nels., <u>A</u>. <u>scariosa</u> E. Nels., <u>A</u>. <u>oxyphylla</u> Greene, <u>A</u>. <u>rosea</u> Greene, <u>A</u>. <u>imbricata</u> E. Nels., <u>A</u>. <u>concinna</u> E. Nels.

Antennaria neglecta Greene, Pittonia 3:173. 1897. Plants stoloniferous, to 4dm high; basal leaf blades oblanceolate, obovate, or suborbicular, 7-35mm long; pistillate involucres 6-9mm long, tomentose below, the tips whitish or pinkish; pistillate corollas about 6-8mm long. Open woods. NW, NE. *A. obovata* E. Nels., *A. oblancifolia* E. Nels.

Antennaria parvifolia Nutt., Trans. Am. Phil. Soc. II, 7:406. 1841. Plants mat forming and stoloniferous, to 25cm high; basal leaf blades oblanceolate to obovate, 5-35mm long; pistillate involucres 7-11mm long, tomentose at base, the tips white to occasionally pink; corollas 5-8mm long. Plains and hills. Statewide. *A. aprica* Greene.

Antennaria pulcherrima (Hook.) Greene, Pittonia 3:176. 1897. Plants from a branched caudex, to 5dm high; basal leaf blades mostly elliptic or oblanceolate, 2-15cm long; involucres 5-8mm long, tomentose toward base, mostly blackish or brownish-green, the tip often whitish; corollas 2.5-4mm long. Hills, woods, and slopes. Fremont and Hot Springs Cos.

Antennaria racemosa Hook., Fl. Bor. Am. 1:330. 1834. Plants stoloniferous, to 6dm high; basal leaf blades elliptic-ovate to elliptic-obovate or suborbicular, 1-7cm long, tomentose beneath, green and glabrous or nearly so above; pistillate involucres 6-8mm long, usually slightly if at all tomentose, the tips brownish or greenish or transparent; corollas about 3-4mm long. Woods. NW, NE.

Antennaria rosulata Rydb., Bull. Torrey Club 24:300. 1897. Plants mat forming, to 4cm high; leaf blades oblanceolate to obovate, 3-20mm long; heads 1-5; involucres 4-8mm long, tomentose at base, the tips mostly white; corollas 2-4mm long. Hills and slopes. Uinta Co.

Antennaria umbrinella Rydb., Bull. Torrey Club 24:302. 1897. Plants mat forming and stoloniferous, to 2dm high; basal leaf blades oblanceolate to obovate, 3-25mm long; pistillate involucres 4-7mm long, woolly below, the tips usually brownish or whitish on the inner bracts; corollas 2-4mm long. Slopes and ridges in the mountains. Statewide. *A. mucronata* E. Nels., *A. fusca* E. Nels., *A. viscidula* (E. Nels.) Rydb., *A. reflexa* E. Nels.

Anthemis L. Dogfennel

Annual or perennial herbs with alternate, pinnatifid or pinnately dissected leaves; heads terminating the branches, with ray and disk flowers, the rays white or yellow; involucral bracts subequal or imbricate in several series, mostly scarious; receptacle chaffy at least near middle or else bristly; pappus a very short crown or lacking.

Rays yellow *A. tinctoria*

Rays white (often drying cream colored)

Rays pistillate, fertile; receptacle chaffy throughout *A. arvensis*

Rays usually neutral, sterile; receptacle chaffy only near middle or else bristly throughout *A. cotula*

Anthemis arvensis L., Sp. Pl. 894. 1753. Annual to 6dm high; leaves pinnately dissected, 1-6cm long; involucre 3-7mm long; rays mostly 10-20, white, 5-11mm long, pistillate and fertile; receptacle chaffy throughout, the receptacular bracts awn-tipped; pappus lacking; achenes not tuberculate. Disturbed areas. Albany Co.

Anthemis cotula L., Sp. Pl. 894. 1753. Similar to above species except the rays sterile and usually neutral, the receptacle chaffy only toward middle or else bristly, and the achenes glandular-tuberculate. Disturbed areas. NE, SE.

Anthemis tinctoria L., Sp. Pl. 896. 1753. Short-lived perennial to 7dm high; leaves 1-2 times pinnatifid, 1-5cm long; involucre 5-7mm long; rays mostly 20-30, yellow, 7-20mm long; receptacle chaffy throughout, the receptacular bracts awn-tipped; pappus a short crown. Disturbed areas. Albany and Teton Cos.

__Antheropeas__ Rydb.

Reference: Carlquist, S. 1956. Madroño 13:226-239.

__Antheropeas__ __wallacei__ (Gray) Rydb., N. Am. Fl. 34:98. 1915. Tomentose annual to 15cm high; leaves alternate, the blades linear to obovate, 2-20mm long, entire or some 3 lobed at tip; heads 1 to few; involucre 3-5mm long, in 1 series; receptacle naked; rays yellow, rarely white, 3-6mm long; pappus of scales, a minute crown, or obsolete; disk corollas 1.5-3mm long. Plains. Albany Co. __Eriophyllum__ __wallacei__ (Gray) Gray.

__Arctium__ L. Burdock

Biennial herbs with alternate, mostly entire or toothed, ovate, deltoid, or cordate leaves; heads several to many, without ray flowers, the corollas pink or purplish with slender lobes; involucral bracts in several series, with a hooked bristle at tip; receptacle bristly; pappus of short, deciduous bristles.

Heads usually 1.5-2.5cm wide including bristles, subsessile or short-peduncled; terminal inflorescence raceme-like __A.__ __minus__

Heads usually 2.5-4cm wide, mostly long-peduncled; terminal inflorescence corymb-like __A.__ __lappa__

Arctium lappa L., Sp. Pl. 816. 1753. Plants to 2m high; leaf blades 4-50cm long; inflorescence corymb-like with long peduncles; involucre about 15-30mm long, 2.5-4cm wide including bristles; corolla 8-15mm long. Disturbed areas. Albany Co.

Arctium minus (Hill) Bernh., Syst. Verz. Erfurt 154. 1800. Similar to above species but the inflorescence raceme-like or thyrsoid with short or no peduncles and the involucre about 15-20mm long and 1.5-2.5(3.5)cm wide. Disturbed areas. SW, SE.

Arnica L.

Perennial herbs with opposite leaves (occasionally alternate above); rays yellow or rarely lacking; involucral bracts herbaceous, somewhat biseriate or uniseriate; receptacle hairy or naked; pappus of capillary bristles.

Reference: Maguire, B. 1947. Am. Midl. Nat. 37:136-145.

1. Stem leaves mostly 5-10 pair
   2. Involucral bracts obtuse to acute with a tuft of long hairs at or just below tip ... A. chamissonis
   2. Involucral bracts sharply acute, the tips about as hairy as the body or not hairy ... A. longifolia

1. Stem leaves, excluding the basal cluster if present, mostly 2-4 pair

  3. Rays usually lacking, rarely vestigial — *A. parryi*

  3. Rays usually present

    4. Pappus subplumose, somewhat tawny; flowering stems usually without tufts of basal leaves — *A. mollis*

    4. Pappus barbellate, usually white; flowering stems often with tufts of basal leaves

      5. Widest leaf blades mostly 1-3 times as long as wide, at least some usually toothed

        6. Leaves with 3-7 somewhat parallel, primary veins, the middle and upper blades rarely broader than lance-ovate

          7. Leaves all entire or rarely denticulate; mostly alpine or subalpine — *A. rydbergii*

          7. Leaves, or some of them, usually coarsely toothed; lower and middle elevations — *A. lonchophylla*

        6. Leaves with mostly pinnate venation, the middle and upper blades often ovate to cordate

          8. Achenes usually glabrous at least below; basal leaves seldom cordate

            9. Involucre long-hairy only toward base; widespread — *A. latifolia*

            9. Involucre long-hairy almost throughout; Carbon Co. — *A. paniculata*

8. Achenes usually short-hairy nearly to base; basal leaves often cordate
   10. Heads mostly solitary, sometimes 2 or 3 — A. cordifolia
   10. Heads several in a corymb or panicle — A. paniculata

5. Widest leaf blades mostly (3)4-10 times as long as wide, entire or nearly so
   11. Heads with mostly 7-10 rays; mostly alpine or subalpine — A. rydbergii
   11. Heads with mostly 10-23 rays; lower and middle elevations
      12. Old leaf bases with tufts of long brown wool in axils; disk corollas usually hairy at least below, often also glandular — A. fulgens
      12. Old leaf bases without tufts of hair, or the hairs few and white; disk corollas glandular, usually not hairy — A. sororia

Arnica chamissonis Less., Linnaea 6:238. 1831. Plants rhizomatous, to 1m high; cauline leaves mostly 5-10 pair, the blades lanceolate, elliptic, or oblanceolate, 1-15cm long, entire or slightly toothed; heads usually several; involucre 7-12mm long; rays 9-20mm long; disk corollas 6-10mm long. Moist places in or near the mountains. Statewide. *A. foliosa* Nutt., *A. ocreata* A. Nels., *A. celsa* A. Nels., *A. rhizomata* A. Nels., *A. exigua* A. Nels., *A. stricta* A. Nels.

Arnica cordifolia Hook., Fl. Bor. Am. 1:331. 1834. Plants rhizomatous, to 6dm high; cauline leaves mostly 2-4 pair, the blades mostly cordate below to ovate above, 1-12cm long, entire or toothed; heads mostly 1-3; involucre 8-24mm long; rays 13-45mm long; disk corollas 6-13mm long. Woods, meadows, and thickets in or near the mountains. Statewide. *A. pumila* Rydb.

Arnica fulgens Pursh, Fl. Am. Sept. 527. 1814. Plants with short rhizomes, to 6dm high; cauline leaves mostly 2-4 pair, the blades lanceolate, oblanceolate, elliptic, or oblong, 1-11cm long, entire or nearly so; heads mostly 1, sometimes 2 or 3; involucre 9-16mm long; rays 12-30mm long; disk corollas 7-10mm long. Plains, hills, slopes, and meadows. Statewide.

Arnica latifolia Bong., Mem. Acad. St. Petersb. VI, 2:147. 1832. Plants rhizomatous, to 6dm high; cauline leaves mostly 2-4 pair, the blades lance-elliptic to ovate or deltoid, 1-10cm long, toothed or entire; heads 1 to several; involucre 6-18mm long; rays 12-30mm long; disk corollas 5-10mm long. Moist places in the mountains. Statewide. *A. gracilis* Rydb., *A. ventorum* Greene, *A. arcana* A. Nels.

Arnica lonchophylla Greene, Pittonia 4:164. 1900. Plants rhizomatous, to 4dm high; cauline leaves mostly 2-4 pair, the blades lance-elliptic to lance-ovate, 1-10cm long, dentate to subentire; heads usually several; involucre 8-13mm long; rays 10-20mm long; disk corollas 6-10mm long. Woods and slopes. Johnson Co.

Arnica longifolia Eaton in Wats., Bot. King Exp. 186. 1871. Plants rhizomatous or from a branching caudex, to 8dm high; leaves mostly 5-7 pair, the blades lanceolate, lance-elliptic, or oblong, 1-16cm long, entire or nearly so, basal leaves lacking; heads 1 to many; involucre 7-15mm long; rays 10-25mm long; disk corollas 6-11mm long. Moist places in the mountains. Statewide. *A. polycephala* A. Nels.

Arnica mollis Hook., Fl. Bor. Am. 1:331. 1834. Plants rhizomatous or from a branched caudex, to 6dm high; cauline leaves mostly 3-4 pair, the blades ovate to obovate, 2-15cm long, entire to dentate; heads 1 or few; involucre 9-20mm long; rays 12-35mm long; disk corollas 6-10mm long. Moist places in the mountains. Statewide. A. subplumosa Greene, A. silvatica Greene.

Arnica paniculata A. Nels. in Coult. & Nels., New Man. Rocky Mts. 572, 610. 1909. Plants rhizomatous, to 6dm high; cauline leaves 3-4 pair, the blades lanceolate to subcordate, 2-12cm long, serrate-dentate to subentire; heads several to many; involucre 9-14mm long; rays 10-15mm long; disk corollas 7-11mm long. Moist woods. Carbon Co.

Arnica parryi Gray in Parry, Am. Nat. 8:213. 1874. Plants rhizomatous, to 6dm high; cauline leaves mostly 2-4 pair, the blades lanceolate or elliptic to oblong, 2-15cm long, entire or denticulate; basal leaves broader; heads 1 to several; involucre 9-17mm long; ray flowers lacking or vestigial; disk corollas 7-9mm long. Open woods, meadows, and slopes. NW, SW, SE.

Arnica rydbergii Greene, Pittonia 4:36. 1899. Plants rhizomatous, to 3dm high; cauline leaves mostly 3-4 pair, the blades oblanceolate to lanceolate, rarely ovate, 1-8cm long, entire or nearly so; heads 1 to few; involucre 8-13mm long; rays 1-2cm long; disk corollas 6-9mm long. Mostly alpine and subalpine. Statewide. *A. caespitosa* A. Nels.

Arnica sororia Greene, Ottawa Nat. 23:213. 1910. Plants rhizomatous, to 6dm high; cauline leaves mostly 2-4 pair, the blades oblanceolate to lanceolate or linear-oblong, 1-12cm long, entire or nearly so; heads mostly 1, sometimes 3; involucre 10-15mm long; rays 15-25mm long; disk corollas 6-10mm long. Plains, hills, and slopes. SW, NW, NE.

Artemisia L. Sagebrush; Wormwood

Annual, biennial, or perennial herbs or shrubs, usually aromatic, with alternate, entire to dissected leaves; heads without ray flowers, in a spike-like, raceme-like, or panicle-like inflorescence, rarely solitary; involucral bracts usually imbricate, dry, often with scarious margins; receptacle naked or long hairy, rarely glandular; pappus usually lacking.

References: Ward, G. H. 1953. Contr. Dudley Herb. 4:155-205.

Beetle, A. A. 1960. Univ. Wyo. Agr. Exp. Sta. Bull. 368:1-83.

Taylor, R. L. et al. 1964. Canad. Journ. Genet. Cytol. 6:42-45.

Holbo, H. R. & H. N. Mozingo. 1965. Am. Journ. Bot. 52:970-978.

Beetle, A. A. & A. Young. 1965. Rhodora 67:405-406.

Marchand, L. S. et al. 1966. Canad. Journ. Bot. 44:1623-1632.

Hanks, D. L. et al. 1973. USDA For. Serv. Res. Paper INT-141:1-24.

1. Plants subshrubs to 2dm high, usually with some spine-tipped branches; achenes and corollas with long cobwebby hairs ... <u>A</u>. <u>spinescens</u>

1. Plants shrubs, subshrubs, or herbs, sometimes over 2dm high, not spiny; achenes and corollas only rarely with long cobwebby hairs

  2. Flowers all perfect; shrubs, rarely appearing like subshrubs

    3. Leaves linear, linear-oblanceolate, or linear-elliptic and entire or a few sometimes irregularly once or twice lobed or toothed

      4. Leaves all filiform and less than 1mm wide ... <u>A</u>. <u>filifolia</u>

      4. Leaves linear or broader, 1mm or more wide ... <u>A</u>. <u>cana</u>

    3. Leaves, or many of them, 3 toothed or 3-6 parted at tip, often cuneate

5. Leaves mostly deeply cleft into 3-6 linear divisions, the basal part of leaf usually about as wide as the divisions and not broadened

6. Involucres 1.5-3mm long; eastern plains *A. filifolia*

6. Involucres 3-4mm long; Laramie Range and westward *A. tripartita*

5. Leaves mostly 3 toothed at tip, or if lobed, the basal part of at least some leaves usually obviously broadened below the lobes

7. Plants mostly 1-4dm high, rarely higher, often glandular-punctate; inflorescence spike-like or a narrow panicle mostly less than 1.5cm wide, rarely wider; rarely above 8000 ft.

8. Involucre at anthesis mostly tomentose or canescent throughout, often not shiny; heads rather few and often somewhat remote *A. arbuscula*

8. Involucre at anthesis (except sometimes the lower bracts) glabrous to rarely loosely pubescent, shiny; heads many and crowded on numerous branches *A. nova*

7. Plants mostly over 4dm high, rarely shorter, glands present or not; inflorescence usually a panicle often over 1.5cm wide; sometimes above 8000 ft.

9. Heads moderate in number, somewhat loosely arranged, averaging 4-5mm long and 3-4mm wide; some leaves usually over 3cm long and irregularly lobed at tip; usually well up in the mountains in Lincoln and Teton Cos. *A. rothrockii*

9. Heads usually many and crowded, mostly smaller; leaves often shorter and usually regularly toothed at tip; widespread *A. tridentata*

2. Flowers pistillate at margin of head, these often few and reduced, the middle ones perfect or sometimes the ovary aborted in flowers at very middle; herbs or subshrubs, very rarely a shrub

10. Leaves all filiform or rarely some trifid, the leaves or segments all less than 1mm wide; shrub or subshrub usually over 2dm high *A. filifolia*

10. Leaves not as above, or if so, then herbaceous, or less than 2dm high and somewhat matted, or with mostly more than 3 leaf segments

11. Flowers at middle of head fertile, the ovary normal

12. Receptacle with long hairs between the flowers

13. Plants mostly 4-12dm high; larger cauline leaf blades 3cm long or more, the ultimate segments (1)1.5-4mm wide; escaped cultivar *A. absinthium*

13. Plants mostly 1-4dm high; cauline leaf blades mostly less than 3cm long, the ultimate segments often less than 1.5mm wide; native

14. Heads numerous, mostly 5mm or less wide; involucral bracts without brown or black margins; mostly low and middle elevations *A. frigida*

14. Heads few, mostly 5-12mm wide; involucral bracts with conspicuous brown or black margins; high elevations

15. Heads often 5-25 per stem; basal leaves mostly twice pinnately divided; widespread *A. scopulorum*

15. Heads 1-4 per stem; basal leaves mostly once pinnately divided or trifid; SE Wyo. *A. pattersonii*

12. Receptacle not hairy

16. Plants annual or biennial with a taproot; leaves often glabrous or nearly so *A. biennis*

16. Plants perennial from a rhizome or caudex or rarely a taproot; leaves usually hairy at least beneath

17. Leaves primarily basal, the stem leaves few and progressively reduced upward (rarely not in a glabrous or glabrate species); lower leaves usually 2-3 times cleft or divided; alpine or subalpine

18. Corollas usually long-hairy; leaves long-hairy on both sides *A. norvegica*

18. Corollas glabrous; leaves glabrous at least above *A. parryi*

17. Leaves mostly cauline, often less divided; mostly not alpine

19. Leaves finely 2-3 times pinnately divided, the rachis and divisions usually 1mm or less wide; escaped cultivar *A. abrotanum*

19. Leaves entire or divided with the rachis and divisions commonly over 1mm wide; native

20. Leaves mostly 1-3 times pinnately dissected, less pubescent on upper surface, very strong-odored; involucres glabrous or nearly so *A. michauxiana*

20. Leaves entire to subbipinnatifid, pubescence various, odor moderate; involucres tomentose to glabrate

21. Plants with deep creeping rhizomes, the stems loosely clustered or solitary; leaves entire to subbipinnatifid, only occasionally narrow with a gradually tapering tip *A. ludoviciana*

21. Plants lacking creeping rhizomes, the stems clustered from a woody caudex; leaves mostly entire, narrow with a gradually tapering tip, rarely pinnately lobed *A. longifolia*

11. Flowers at middle of head sterile, the ovary aborted

22. Plants somewhat matted subshrubs mostly less than 2dm high; inflorescence spike-like or raceme-like with few heads

23. Corollas of central flowers 2.5-3.5mm long; involucre 2-4mm long *A. pedatifida*

23. Corollas of central flowers mostly 4-4.5mm long; involucre 4.5-7mm long *A. porteri*

22. Plants herbaceous, to 15dm high; inflorescence usually panicle-like with many heads

24. Leaves mostly entire, the lower rarely with 3-5 narrow segments *A. dracunculus*

24. Leaves mostly pinnatifid or dissected except the upper ones *A. campestris*

Artemisia abrotanum L., Sp. Pl. 845. 1753. Subshrub to 2m high; leaf blades 2-8cm long, 2-3 times pinnately dissected into linear segments; involucre about 2mm long, pubescent; marginal flowers pistillate; receptacle naked; corollas 0.5-2mm long. Disturbed areas. NW, SE.

Artemisia absinthium L., Sp. Pl. 848. 1753. Perennial herb to 12dm high; leaves 2-3 times pinnatifid at least below, less so above, 2-8cm long, often silvery-sericeous; involucre 2-3mm long, sericeous; marginal flowers pistillate; receptacle hairy; corollas 1-2mm long. Disturbed areas. Albany and Carbon Cos.

Artemisia arbuscula Nutt., Trans. Am. Phil. Soc. II, 7:398. 1841. Low Sagebrush. Shrub to 6dm high; leaves cuneate-oblanceolate, mostly with 3 teeth or lobes at tip, 3-16(20)mm long; involucre 2.5-6mm long, canescent or tomentose; flowers all perfect; receptacle naked; corollas 1.5-3mm long. Plains, hills, and slopes. NW, SW, SE. A. longiloba (Osterh.) Beetle is an ecological race.

Artemisia biennis Willd., Phytogr. 11. 1794. Taprooted annual or biennial to 2m high; leaves mostly 1-2 times pinnatifid, 1-15cm long; involucre 1.5-3mm long, glabrous; marginal flowers pistillate; receptacle glabrous; corollas 0.5-1mm long. Disturbed areas. Statewide.

Artemisia campestris L., Sp. Pl. 846. 1753. Taprooted biennial or perennial to 1m high; leaves mostly 1-3 times pinnate or ternate, 1-10cm long; involucre 2-4mm long, glabrous or not; marginal flowers pistillate; receptacle glabrous; corollas 1-3mm long. Open places. Statewide. A. borealis Pall., A. caudata Michx., A. canadensis Michx.

Artemisia cana Pursh, Fl. Am. Sept. 521. 1814. Silver Sagebrush. Shrub to 2m high; leaves linear, linear-oblanceolate, or linear-elliptic, 1-7cm long, entire or sometimes with 1 or 2 teeth or lobes, rarely tridentate; involucre 3.5-6mm long, canescent or tomentose to glabrate; flowers all perfect; receptacle naked; corollas 2-5mm long. Plains and hills, often where moist. Statewide.

Artemisia dracunculus L., Sp. Pl. 849. 1753. Rhizomatous perennial herb to 15dm high; leaves linear or nearly so, 1-7cm long, entire or rarely a few cleft; involucre 2-4mm long, glabrous or sparsely pubescent; marginal flowers pistillate; receptacle naked or glandular; corollas 0.5-2mm long. Plains, hills, slopes, and streambanks. Statewide. A. dracunculoides Pursh, A. aromatica A. Nels.

Artemisia filifolia Torr., Ann. Lyc. N. Y. 2:211. 1828. Sand Sagebrush. Shrub or subshrub to 12dm high; leaves filiform, 1-7cm long, some often trilobed with filiform divisions; involucre 1.5-3mm long, canescent; marginal flowers pistillate; receptacle naked; corollas 0.5-2mm long. Plains and hills, often where sandy. NE, SE.

Artemisia frigida Willd., Sp. Pl. 3:1838. 1803. Fringed Sagebrush. Subshrub to 4(6)dm high; leaves mostly 2-3 times ternately divided into linear divisions about 1.5mm or less wide, the blades 3-25mm long; involucre 2-3mm long, canescent; marginal flowers pistillate; receptacle hairy; corollas 1-2mm long. Plains, hills, and slopes. Statewide.

Artemisia longifolia Nutt., Gen. Pl. 2:142. 1818. Perennial herb from a woody caudex, to 8dm high; leaf blades linear to lanceolate, 1-10cm long, usually entire or rarely with a few teeth or lobes; involucre 3-5mm long, tomentose or tomentulose; marginal flowers pistillate; receptacle naked; corollas 1-3mm long. Plains and hills, often where alkaline. Statewide. *A. natronensis* A. Nels.

Artemisia ludoviciana Nutt., Gen. Pl. 2:143. 1818. Rhizomatous perennial herb to 1m high; leaf blades mostly lanceolate or lance-elliptic to oblanceolate and entire or toothed, or sometimes deeply lobed to 1-2 times pinnatifid and ovate or elliptic in outline, 1-10cm long; involucre 2.5-5mm long, tomentose to glabrate; marginal flowers pistillate; receptacle naked; corollas 1-3mm long. Plains, hills, slopes, and disturbed areas. Statewide. *A. incompta* Nutt., *A. gnaphalodes* Nutt., *A. brittonii* Rydb., *A. argophylla* Rydb., *A. floccosa* Rydb., *A. flodmanii* Rydb., *A. pabularis* (A. Nels.) Rydb., *A. paucicephala* A. Nels., *A. gracilenta* A. Nels., *A. rhizomata* A. Nels.

Artemisia michauxiana Bess. in Hook., Fl. Bor. Am. 1:324. 1834. Perennial herb with either rhizomes, a caudex, or a taproot, to 7dm high; leaves mostly 1-3 times pinnately dissected, sometimes entire above, 1-5cm long; involucre 2-4mm long, glabrous or nearly so; marginal flowers pistillate; receptacle glabrous; corollas 1-3mm long. Rocky places in the mountains. SW, NW, NE. *A. discolor* Dougl. ex Bess., *A. graveolens* Rydb., *A. tenuis* Rydb., *A. subglabra* A. Nels.

<u>Artemisia</u> <u>norvegica</u> Fries, Nov. Fl. Suec. 56. 1817. Perennial herb from a branched caudex, to 6dm high; basal leaves mostly 2-3 times pinnately dissected, the blades 1-10cm long; involucre 3-7mm long, villous or glabrous in age; marginal flowers pistillate; receptacle glabrous; corollas 2-3.5mm long. Alpine and subalpine. Fremont and Sublette Cos. <u>A</u>. <u>saxicola</u> Rydb.

<u>Artemisia</u> <u>nova</u> A. Nels., Bull. Torrey Club 27:274. 1900. Black Sagebrush. Shrub to 4dm high; leaves cuneate-oblanceolate, mostly with 3 teeth or lobes at tip, 3-16(20)mm long; involucre 2.5-5mm long, glabrous or rarely loosely hairy; flowers all perfect; receptacle naked; corollas 1.5-3mm long. Plains and hills. NW, SW, SE.

<u>Artemisia</u> <u>parryi</u> Gray, Proc. Am. Acad. 7:361. 1868. Perennial herb with a branched caudex, to 4dm high; lower leaves mostly 2-3 times pinnately divided, 1-5cm long, glabrous, or glabrate beneath; involucre 2-4mm long, usually glabrous; marginal flowers pistillate; receptacle naked; corollas 1.5-3mm long. Alpine and subalpine. Fremont Co.

<u>Artemisia</u> <u>pattersonii</u> Gray, Syn. Fl. 2nd ed. 1(2):453. 1886. Perennial herb from a branched caudex or rhizomes, to 2dm high; leaves, at least the lower, pinnately or palmately divided into mostly linear segments, 1-5cm long, canescent to glabrate; involucre 4-7mm long, usually villous; marginal flowers pistillate; receptacle hairy; corollas about 2-3mm long. Alpine and subalpine. Reported from the Medicine Bow Mts. (Wiens & Richter, Am. Journ. Bot. 53:981. 1966).

<u>Artemisia</u> <u>pedatifida</u> Nutt., Trans. Am. Phil. Soc. II, 7:399. 1841. Somewhat mat forming subshrub to 15cm high; lower leaves mostly 1-2 times ternate with linear segments, becoming entire above, 0.3-3cm long; involucre 2-4mm long, tomentose or tomentulose; marginal flowers pistillate; receptacle naked; corollas 2-3.5mm long. Plains and hills. Statewide.

<u>Artemisia</u> <u>porteri</u> Cronq., Madroño 11:145. 1951. Mat forming perennial with a woody taproot, to 15cm high; leaves entire and linear or ternate with linear divisions, 1-5cm long; involucre 4.5-7mm long, tomentulose or canescent; marginal flowers pistillate; receptacle naked; corollas 2-4.5mm long. Plains and hills. Endemic in Fremont Co.

<u>Artemisia</u> <u>rothrockii</u> Gray, Bot. Calif. 1:618. 1876. Shrub to 15dm high; leaves cuneate or flabelliform, 0.5-6cm long, mostly 3 toothed or lobed or sometimes entire and rounded at tip; involucres mostly 4-5.5mm long, canescent or tomentose or glabrate in age; flowers all perfect; receptacle naked; corollas 2-3.5mm long. High mountain slopes. Teton and Lincoln Cos.

Artemisia scopulorum Gray, Proc. Acad. Phila. 1863:66. 1863. Perennial herb from a branching caudex, to 3dm high; leaves pinnatifid or ternate-pinnatifid below, with mostly linear segments, often becoming entire above, the blades 5-35mm long; involucre about 4mm long, usually woolly-villous; marginal flowers pistillate; receptacle hairy; corollas 1.5-3mm long. Slopes, meadows, and ridges in the high mountains. Statewide.

Artemisia spinescens Eaton in Wats., Bot. King Exp. 180. 1871. Subshrub to 2dm high, the old persistent branches usually spine-tipped; leaf blades 3-15mm long, 3-5 divided, the divisions often again lobed or toothed; involucre 2-4mm long, densely villous; marginal flowers pistillate; receptacle glabrous; corollas 1-3mm long, bearing cobwebby hairs. Plains and hills. Statewide. *Picrothamnus desertorum* Nutt.

Artemisia tridentata Nutt., Trans. Am. Phil. Soc. II, 7:398. 1841. Big Sagebrush. Shrub to 3m high; leaves cuneate-oblanceolate, mostly 3 toothed at tip, the upper becoming linear and entire, 0.5-4(5)cm long; involucre 2-5mm long, pubescent to glabrate; flowers all perfect; receptacle glabrous; corollas 2-3.5mm long. Plains, hills, and slopes. Statewide. *A. vaseyana* Rydb.

Artemisia tripartita Rydb., Mem. N. Y. Bot. Gard. 1:432. 1900. Three-tip Sagebrush. Shrub to 1m high; leaves 0.5-6cm long, deeply cleft at tip into mostly 3-6 linear segments; involucres mostly 3-4mm long, usually canescent; flowers all perfect; receptacle glabrous; corollas 2-3mm long. Plains and hills. NW, SW, SE. A. trifida Nutt.

Aster L.

Annual to perennial herbs with alternate and/or basal leaves; heads solitary to numerous; rays blue, purple, pink, or white, rarely lacking; involucral bracts mostly herbaceous, occasionally dry, usually imbricate; receptacle usually naked; pappus of capillary bristles.

References: Wiegand, K. M. 1933. Rhodora 35:16-38.
Cronquist, A. 1943. Am. Midl. Nat. 29:429-468.
1948. Leafl. West. Bot. 5:73-82.

1. Plants annual with inconspicuous rays barely exceeding the pappus, or the rays lacking
   2. Rays about 2-4mm long; outer involucral bracts usually widest above the middle — A. frondosus
   2. Rays lacking or rudimentary; outer involucral bracts usually widest at or below the middle — A. brachyactis

1. Plants perennial, the rays well developed
   3. Plants with a taproot; heads solitary on each stem; leaves entire, mostly basal, usually over 10 times longer than wide; alpine or subalpine, rarely lower — A. alpigenus
   3. Plants often fibrous rooted or with rhizomes; heads only occasionally solitary; leaves various; mostly not alpine
      4. Pappus of an outer series of inconspicuous short bristles and an inner series of long bristles; heads solitary on each stem; leaves 4mm wide or less, 15mm or less long; plants from a branched caudex — A. scopulorum
      4. Pappus not as above; heads usually not solitary; leaves often larger; plants only occasionally from a branched caudex
         5. Involucral bracts mostly dry, not herbaceous, sometimes keeled, often purplish or colorless, sometimes greenish tinged, the margins usually strongly fringed or erose
            6. Involucral bracts glabrous or nearly so (except margins); rays sometimes white or ochroleucous
               7. Involucres averaging 8mm or more long; outer involucral bracts acute; rays usually white — A. engelmannii
               7. Involucres 4-8mm long; outer involucral bracts sometimes obtuse; rays white or not
                  8. Basal leaves well developed or at least with clusters of old leaf bases; leaves mostly all 8mm or less wide; rays white or ochroleucous — A. ptarmicoides

8. Basal leaves usually lacking; leaves often over 8mm wide; rays mostly purplish or pinkish but occasionally white

9. Peduncles densely pubescent; leaves often toothed, mostly less than 4 times as long as wide *A. meritus*

9. Peduncles often glabrous or glabrate; leaves not toothed, mostly over 4 times as long as wide *A. glaucodes*

6. Involucral bracts conspicuously pubescent; rays purple or lavender

10. Leaves mostly 10mm or less wide, not toothed *A. perelegans*

10. Leaves, or some of them, often over 10mm wide, often toothed *A. meritus*

5. Involucral bracts mostly herbaceous, not keeled, green at least in part, the margins various

11. Peduncles and involucres glandular

12. Leaves mostly less than 10mm wide

13. Leaves, except sometimes the basal, mostly linear and much reduced upward, some lower ones usually 5cm or more long; Carbon and Albany Cos. *A. pauciflorus*

13. Leaves mostly broader or else not reduced upward or much shorter; widespread

14. Leaves mostly less than 12mm long; SW Wyo. *A. arenosus*

14. Leaves, or some of them, over 12mm long; widespread

15. Leaves strongly clasping the stem, little if at all reduced upward; plants often over 5dm high *A. novae-angliae*

15. Leaves barely or not at all clasping, usually much reduced upward; plants mostly less than 4dm high but occasionally higher

16. Leaves often strongly dimorphic, the lower (sometimes deciduous) much larger than the upper; Black Hills *A. oblongifolius*

16. Leaves not dimorphic, the upper gradually reduced in size; widespread *A. campestris*

12. Leaves, or some of them, 15mm or more wide

17. Leaves mostly toothed *A. conspicuus*

17. Leaves entire or merely ciliate

18. Plants mostly less than 5(7)dm high, the leaves reduced upward with some lower ones usually 8cm or more long *A. integrifolius*

18. Plants over 5dm high, the leaves little if at all reduced upward (rarely less than 5dm high but then with all the leaves shorter than 8cm) *A. novae-angliae*

11. Peduncles and involucres not glandular

19. Plants pubescent throughout with stiff, straight hairs, the outer involucral bracts and some leaves spinulose-tipped, the main leaves 7cm or less long and often with fascicles of smaller leaves in their axils; rays white

20. Plants with well developed creeping rhizomes; involucral bracts often subequal — *A. falcatus*

20. Plants with a very short rhizome or caudex; involucral bracts unequal, the inner much longer and narrower — *A. pansus*

19. Plants not as above

21. Lower leaves usually ovate to cordate, rarely deciduous — *A. ciliolatus*

21. Lower leaves mostly oblanceolate, lanceolate, or linear

22. Plants with slender rhizomes and stems mostly less than 2mm thick, without a tuft of basal leaves; leaves all linear or lance-linear, usually 8mm or less wide and mostly 6 or more times as long as wide — *A. junciformis*

22. Plants not as above

23. Leaves mostly all 8mm or less wide, the basal tuft well developed or with old persistent bases; rays white or ochroleucous; involucre often shiny as if varnished — *A. ptarmicoides*

23. Leaves either wider or without a basal tuft, or if not, then without the other characters

24. Peduncles glabrous to sparsely pubescent, usually glaucous; involucral bracts acute; some bracts often cordate-clasping; E Wyo. *A. laevis*

24. Peduncles pubescent or glabrous, not glaucous; involucral bracts often obtuse; bracts various; widespread

25. Leaves often toothed, mostly 4 times longer than wide or less, the venation prominently pinnate; basal tuft of leaves lacking; involucre often pubescent, the bracts often with purplish margins at least toward tip; N Wyo. *A. meritus*

25. Leaves usually entire to ciliate, often over 4 times longer than wide, the venation often only reticulate or somewhat parallel; basal tuft of leaves present or not; involucre usually glabrous (except sometimes the margins of bracts), the bracts only occasionally with purplish margins; widespread

26. Involucral bracts only green on midrib or at very tip, the outer mostly ovate or ovate-oblong, usually not narrowed in lower half except at very base *A. glaucodes*

26. Involucral bracts not as above

27. Pubescence of stems mostly in lines from leaf bases; inflorescence usually long, broad, and leafy with many heads *A. hesperius*

27. Pubescence only rarely in lines, usually somewhat uniform or lacking; inflorescence often compact and sparingly leafy with few heads

28. Rays white, pink, or lavender; inflorescence usually scattered along 1/3 to 1/2 the plant; tuft of basal leaves usually lacking *A. eatonii*

28. Rays mostly blue or purple; inflorescence often compact at top of plant; tuft of basal leaves often present (the following are distinguished with difficulty)

29. Reticulations of leaf veins mostly longer than wide; middle leaves mostly 7 or more times as long as wide; tuft of basal leaves usually lacking; often in rather dry places *A. adscendens*

29. Reticulations of leaf veins averaging about as long as wide; middle leaves often less than 7 times as long as wide; tuft of basal leaves often present; often in wet places

30. Involucre with long foliaceous bracts at base *A. foliaceus*

30. Involucre lacking long foliaceous bracts at base

31. Middle and lower (not basal) leaves rarely over 15mm wide *A. occidentalis*

31. Middle and lower leaves often well over 15mm wide *A. foliaceus*

Aster adscendens Lindl. in Hook., Fl. Bor. Am. 2:8. 1834. Perennial to 1m high, from a rhizome or branching caudex; leaf blades linear or oblong to oblanceolate, 1-15cm long, mostly with minute spinulose teeth; heads several to many; involucre 5-9mm long; rays mostly blue, purple, or pinkish, 5-17mm long. Plains, hills, slopes, and woods. Statewide. *A. nelsonii* Greene, *A. chilensis* Nees ssp. *adscendens* (Lindl.) Cronq.

Aster alpigenus (T. & G.) Gray, Proc. Am. Acad. 8:389. 1872. Subscapose perennial to 2dm high, from a taproot and caudex; leaves linear, linear-elliptic, or linear-oblanceolate, 1-15cm long, entire; heads solitary on each stem; involucre 6-13mm long; rays lavender or purple, 7-17mm long. Mostly alpine and subalpine, rarely lower. SW, NW, NE. *Oreastrum alpigenum* (T. & G.) Greene, *A. haydenii* Porter.

Aster arenosus Blake, Journ. Wash. Acad. Sci. 30:471. 1940. Perennial to 15cm high with a rhizome or taproot and branching caudex; leaf blades linear to oblanceolate, 3-12mm long, the margins with coarse cilia; heads solitary on branches; involucre 5-8mm long; rays white to lavender, 5-10mm long. Plains and hills. Sweetwater Co. *Leucelene ericoides* (Torr.) Greene.

Aster brachyactis Blake in Tidestrom, Contr. U. S. Nat. Herb. 25:564. 1925. Taprooted annual to 7dm high; leaf blades linear or nearly so, 1-9cm long, entire or scabrous-margined; heads several to many; involucre 5-11mm long; rays lacking. Moist places, often where alkaline. Johnson and Albany Cos. *A. angustus* (Lindl.) T. & G.

Aster campestris Nutt., Trans. Am. Phil. Soc. II, 7:293. 1840. Rhizomatous perennial to 5dm high; leaf blades mostly linear or oblong, 1-5(7)cm long, the margins ciliate; heads several to many; involucre 5-9mm long; rays purplish, 6-15mm long. Plains, hills, meadows, and slopes. Statewide.

Aster ciliolatus Lindl. in Hook., Fl. Bor. Am. 2:9. 1834. Rhizomatous perennial to 12dm high; leaf blades mostly cordate or ovate below, lanceolate or elliptic above, 1.5-17cm long, mostly serrate; heads several to many; involucre 5-8mm long; rays blue, 8-15mm long. Woods and openings. NE, SE. *A. lindleyanus* T. & G.

Aster conspicuus Lindl. in Hook., Fl. Bor. Am. 2:7. 1834. Rhizomatous perennial to 1m high; leaf blades ovate to obovate, 3-18cm long, serrate or rarely subentire; heads few to many; involucre 8-12mm long; rays blue or purple, 10-25mm long. Woods and slopes in the mountains. NW, NE.

Aster eatonii (Gray) Howell, Fl. N. W. Am. 310. 1900. Rhizomatous perennial to 1m high; leaf blades linear, oblong, or lanceolate, 1-15cm long, usually scaberulous-margined, very rarely toothed; heads several to many; involucre 4.5-10mm long; rays white, lavender, or pink, 5-13mm long. Moist places. NW, NE, SE. *A. oregonus* (Nutt.) T. & G. of authors, *A. proximus* Greene, *A. mearnsii* Rydb.

Aster engelmannii (Eaton) Gray, Syn. Fl. 1(2):199. 1884. Perennial to 15dm high, from a caudex or stout rhizome; leaf blades lanceolate, elliptic, or lance-ovate, 2-12cm long, ciliate-margined, very rarely toothed; heads few to several; involucres mostly 8-13mm long; rays white, sometimes drying pinkish, 15-30mm long. Woods and slopes. Statewide.

Aster falcatus Lindl. in Hook., Fl. Bor. Am. 2:12. 1834. Rhizomatous perennial to 7dm high; leaf blades linear or oblong, 0.3-7cm long, the margins entire or stiff-ciliate; heads few to many; involucre 4-8mm long; rays white, 3-12mm long. Plains, hills, meadows, and streambanks. Statewide. *A. commutatus* (T. & G.) Gray, *A. cordineri* A. Nels.

Aster foliaceus Lindl. ex DC., Prodr. 5:228. 1836. Perennial to 1m high, from a rhizome or caudex; leaf blades oblanceolate or obovate below to lanceolate or ovate above, 1-20cm long, mostly entire to ciliate-margined; heads 1 to many; involucre 6-12mm long; rays blue, purple, or rose-purple, 1-2cm long. Moist places in or near the mountains. Statewide. A. canbyi (Gray) Vasey ex Rydb., A. glastifolius Greene, A. apricus (Gray) Rydb., A. frondeus (Gray) Greene, A. incertus A. Nels.

Aster frondosus (Nutt.) T. & G., Fl. N. Am. 2:165. 1841. Taprooted annual to 14dm high; leaf blades linear to oblanceolate, 1-6cm long, entire or ciliate-margined; heads several to many; involucre 5-9mm long; rays mostly pinkish, about 2-4mm long. Moist places, often where alkaline. Lincoln and Albany Cos.

Aster glaucodes Blake, Proc. Biol. Soc. Wash. 35:174. 1922. Rhizomatous perennial to 6dm high; leaf blades lanceolate or oblong to oblanceolate, 1-9cm long, entire or scabrous-margined; heads few to many; involucre 5-8mm long; rays white to violet or pinkish, 7-15mm long. Woods, slopes, and canyons. Statewide. A. glaucus (Nutt.) T. & G.

Aster hesperius Gray, Syn. Fl. 1(2):192. 1884. Rhizomatous perennial to 15dm high; leaf blades linear or oblanceolate to lanceolate, 0.5-15cm long, entire or toothed; heads few to many; involucre 5-8mm long; rays white, pink, lavender, or blue, 6-15mm long. Streambanks, shores, and meadows. Statewide. A. laetevirens Greene, A. coerulescens DC. of authors.

Aster integrifolius Nutt., Trans. Am. Phil. Soc. II, 7:291. 1840. Perennial to 7dm high, from a caudex or short rhizome; leaf blades oblanceolate or elliptic below to lance-ovate above, 2-16cm long, the margins entire or ciliate; heads few to several; involucre 7-14mm long; rays purplish or pinkish, 1-2cm long. Woods, slopes, and meadows in the mountains. SW, NW, NE.

Aster junciformis Rydb., Bull. Torrey Club 37:142. 1910. Perennial to 8dm high, with slender rhizomes; leaf blades linear or lance-linear, 1-9cm long, the margins entire or scabrous; heads 1 to several; involucre 4-8mm long; rays white, blue, lavender, or pinkish, 7-15mm long. Swamps, bogs, and wet meadows. Fremont and Albany Cos.

Aster laevis L., Sp. Pl. 876. 1753. Perennial to 12dm high, from a short rhizome or caudex; leaf blades obovate to ovate but often narrower, 0.5-14cm long, entire or toothed; heads several to many; involucre 5-9mm long; rays blue or purple, 6-17mm long. Plains, hills, woods, and slopes. NE, SE. A. geyeri (Gray) Howell.

Aster meritus A. Nels., Bot. Gaz. 37:268. 1904. Rhizomatous perennial to 5dm high; leaf blades lanceolate to oblanceolate or obovate, 1-10cm long, entire to toothed; heads 1 to several; involucre 6-9mm long; rays purple, 8-17mm long. Woods and slopes in the mountains. NW, NE. A. sibiricus L. var. meritus (A. Nels.) Raup.

Aster novae-angliae L., Sp. Pl. 875. 1753. Rhizomatous perennial to 15dm high; leaf blades mostly lanceolate or oblong, 2-12cm long, the margins entire to ciliate; heads few to several; involucre 8-13mm long; rays rose to purplish, 9-20mm long. Streambanks and disturbed areas. Weston and Albany Cos.

Aster oblongifolius Nutt., Gen. Pl. 2:156. 1818. Rhizomatous perennial to 8dm high; leaf blades elliptic or oblong to oblanceolate, 0.3-5cm long, the margins scabrous or ciliate; heads few to several; involucre 5-8mm long; rays purplish, 5-15mm long. Plains and hills. Crook Co.

Aster occidentalis (Nutt.) T. & G., Fl. N. Am. 2:164. 1841. Perennial to 1m high, from a rhizome or sometimes a caudex; leaf blades oblanceolate below to elliptic, linear, or lance-linear above, 1-15cm long, the margins entire to ciliate; heads 1 to many; involucre 5-10mm long; rays blue or purplish, 6-18mm long. Meadows, thickets, and slopes in the mountains. Statewide. A. andinus Nutt., A. fremontii (T. & G.) Gray, A. williamsii Rydb.

Aster pansus (Blake) Cronq., Leafl. West. Bot. 6:45. 1950. Perennial to 15dm high, from a caudex or very short rhizome; leaf blades mostly linear or oblong, 0.2-7cm long, the margins entire or stiff-ciliate; heads many; involucre 3-7mm long; rays white, 3-8mm long. Plains, hills, and streambanks. NW, NE, SE. A. hebecladus DC., A. multiflorus Ait.

Aster pauciflorus Nutt., Gen. Pl. 2:154. 1818. Usually rhizomatous perennial to 5dm high; leaf blades linear to narrowly oblanceolate, 1-10cm long, entire or nearly so; heads few to several; involucre 4-7mm long; rays blue, purplish, or whitish, 4-12mm long. Plains and hills, often where alkaline. Carbon and Albany Cos.

Aster perelegans Nels. & Macbr., Bot. Gaz. 56:477. 1913. Perennial to 6dm high, from a caudex; leaf blades linear-oblong to lanceolate or elliptic, 1-6cm long, the margins entire to ciliate; heads few to many; involucre 5-10mm long; rays purple or lavender, 7-20mm long. Woods, hills, and slopes. SW, NW, NE. A. elegans (Nutt.) T. & G.

Aster ptarmicoides (Nees) T. & G., Fl. N. Am. 2:160. 1841. Perennial to 5dm high, from a rhizome or caudex; leaf blades linear, linear-oblanceolate, or oblong, 2-15cm long, the margins entire or scabrous; heads 1 to several; involucre 4-7mm long; rays white or ochroleucous, 5-13mm long. Woods, hills, and slopes. NW, NE.

Aster scopulorum Gray, Proc. Am. Acad. 16:98. 1880. Perennial to 2dm high, from a branched woody caudex; leaf blades linear, linear-oblanceolate, or oblong, 3-15mm long, spinulose-tipped, the margins scabrous; heads solitary on each stem; involucre 7-11mm long; rays blue or purplish, 6-15mm long. Plains and hills. Yellowstone Park. *Ionactis alpina* (Nutt.) Greene.

Bahia Lag.

Reference: Ellison, W. L. 1964. Rhodora 66:67-86, 177-215, 281-311.

Bahia dissecta (Gray) Britt., Trans. N. Y. Acad. Sci. 8:68. 1889. Annual or biennial to 8dm high; leaves alternate and basal, the blades 1-5cm long, ternately divided, the primary segments usually again divided; heads with ray and disk flowers; involucre 5-8mm long, the bracts subequal; receptacle naked; pappus lacking; rays yellow, 5-9mm long. Woods, slopes, and hills. Albany Co. *Amauriopsis dissecta* (Gray) Rydb.

Balsamorhiza Nutt. Balsamroot

Scapose or subscapose perennials from a thick taproot or caudex; heads usually solitary on each stem, with ray and disk flowers, the rays yellow; involucral bracts herbaceous, imbricate or subequal in several series; receptacle chaffy, the bracts clasping the achenes; pappus usually lacking. The species sometimes hybridize.

References: Weber, W. A. 1953. Madroño 12:47-49.
Helton, N. et al. 1972. Madroño 21:526-535.

1. Leaves entire or nearly so — *B. sagittata*
1. Leaves deeply cleft or pinnatifid
  2. Plants with long, soft, tangled hairs especially on the leaves and involucres — *B. incana*
  2. Plants with short stiff hairs or sometimes with long glassy hairs
    3. Leaves, or some of them, over 8cm wide; involucre usually 3cm or more wide — *B. macrophylla*
    3. Leaves less than 8cm wide; involucre usually about 2.5cm or less wide — *B. hispidula*

Balsamorhiza hispidula Sharp, Ann. Mo. Bot. Gard. 22:137. 1935. Plants to 4dm high; leaf blades 1-2 times pinnatifid, 0.5-2dm long; involucre 12-20mm long; rays 15-30mm long; disk corollas about 8mm long. Plains and hills. Sweetwater Co. B. hookeri Nutt. var. hispidula (Sharp) Cronq.

Balsamorhiza incana Nutt., Trans. Am. Phil. Soc. II, 7:350. 1840. Plants to 6dm high; leaf blades pinnatifid, 0.5-3dm long; involucre 10-22mm long; rays 2-6cm long; disk corollas about 7mm long. Plains, hills, and slopes. Statewide except not in W, S, and E border Cos.

Balsamorhiza macrophylla Nutt., Trans. Am. Phil. Soc. II, 7:350. 1840. Plants to 1m high; leaf blades mostly pinnatifid, 1-5dm long; involucre 25-50mm long; rays 35-60mm long; disk corollas about 11mm long. Meadows, hills, woods, and slopes. NW, SW.

Balsamorhiza sagittata (Pursh) Nutt., Trans. Am. Phil. Soc. II, 7:350. 1840. Plants to 8dm high; leaf blades mostly cordate or hastate, 4-30cm long, entire or nearly so; involucre 15-30mm long; rays 2-5cm long; disk corollas about 8mm long. Plains, hills, slopes, and open woods. Statewide.

Bidens L. Beggar Ticks

Annuals with opposite, simple to ternately or pinnately compound leaves; heads usually several; rays yellow or lacking; involucral bracts biseriate, the outer somewhat foliaceous, the inner shorter and usually longitudinally striate; receptacle chaffy; pappus of 2-5 awns which are usually retrorsely barbed.

1. Primary leaves simple
   2. Rays usually well developed, rarely lacking; achenes lacking a cartilaginous margin along top ... B. cernua
   2. Rays usually lacking, rarely to 4mm long; achenes with a cartilaginous margin along top set off from the body by color or constriction ... B. comosa
1. Primary leaves compound with mostly 3-5 leaflets
   3. Outer involucral bracts mostly 5-8; disk corollas orange, 2.5-3mm long ... B. frondosa
   3. Outer involucral bracts of larger heads mostly 10 or more; disk corollas pale yellow, 2.5-4mm long ... B. vulgata

Bidens cernua L., Sp. Pl. 832. 1753. Plants to 1m high; leaves simple, lance-linear to elliptic, 1-15cm long, coarsely serrate to subentire, at least the upper sessile; inner involucral bracts 6-13mm long; rays to 15mm long or lacking; disk corollas about 3.5mm long. Wet places. NW, NE, SE. B. glaucescens Greene.

Bidens comosa (Gray) Wieg., Bull. Torrey Club 24:436. 1897. Plants to 15dm high; leaves simple, lanceolate, elliptic, or ovate, 2-15cm long, usually coarsely serrate, sessile or petioled; inner involucral bracts 6-13mm long; rays lacking or to 4mm long; disk corollas about 3.5mm long. Moist areas. Hot Springs and Goshen Cos.

Bidens frondosa L., Sp. Pl. 832. 1753. Plants to 12dm high; leaves pinnately compound with mostly 3-5 lanceolate, serrate leaflets 0.5-10cm long; inner involucral bracts 5-8mm long; rays lacking or nearly so; disk corollas 2.5-3mm long. Wet areas. NE, SE.

Bidens vulgata Greene, Pittonia 4:72. 1899. Similar to *B. frondosa* but with mostly 10-16 outer involucral bracts rather than 5-8 and averaging more robust and more leafy with yellow disk corollas rather than orange. Disturbed areas. Sheridan and Platte Cos.

Brickellia Ell.

Perennial herbs or shrubs with alternate or occasionally some opposite leaves; heads several, lacking ray flowers; involucral bracts imbricate in several series, longitudinally striate; receptacle naked; disk corollas white to pink-purple; pappus of capillary bristles.

References: Shinners, L. H. 1946. Wrightia 1:122-144.
Shinners, L. H. & R. McVaugh. 1968. Taxon 17:732-734.

Leaf blades mostly 15mm or less long; SW Wyo. *B. microphylla*
Leaf blades, or some of them, 20mm or more long; widespread
Pappus bristles plumose; leaf blades linear, lanceolate, or elliptic *B. eupatorioides*
Pappus bristles barbellate; leaf blades deltoid-ovate to subcordate *B. grandiflora*

Brickellia eupatorioides (L.) Shinners, Sida 4:274. 1971. Perennial herb to 1m high, from a woody caudex; leaf blades linear, lanceolate, or elliptic, 1-6cm long, toothed to entire; involucre 8-14mm long; disk corollas 4.5-7mm long. Plains and hills. NE, SE. *Kuhnia eupatorioides* L., *K. glutinosa* Ell., *K. reticulata* A. Nels., *K. hitchcockii* A. Nels.

Brickellia grandiflora (Hook.) Nutt., Trans. Am. Phil. Soc. II, 7:287. 1840. Perennial herb to 7dm high, from a woody caudex; leaf blades mostly deltoid-ovate to subcordate, 1-8cm long, toothed to entire; involucre 7-14mm long; disk corollas 5-8mm long. Rocky places in or near the mountains. NW, NE, SE. *Coleosanthus congestus* A. Nels.

<u>Brickellia</u> <u>microphylla</u> (Nutt.) Gray, Pl. Wright. 1:85. 1852. Shrub or subshrub to 6dm high; leaf blades mostly ovate to suborbicular, 1-15mm long, somewhat toothed to entire; involucre 7-12mm long; disk corollas 5-7mm long. Rocky or sandy hills and washes. Sweetwater Co. B. scabra (Gray) A. Nels. ex Robins.

<u>Carduus</u> L. Musk Thistle

<u>Carduus</u> <u>nutans</u> L., Sp. Pl. 821. 1753. Annual or biennial to 25dm high; leaves lanceolate, elliptic, or oblanceolate, pinnately lobed, 5-40cm long, spiny, the base usually decurrent along stem; heads mostly solitary at ends of branches, without ray flowers; involucre 2-4cm long, the middle and outer bracts with spiny tips; receptacle bristly; pappus of barbellate capillary bristles; corollas purple or reddish, 2-3cm long. Disturbed areas. Statewide.

<u>Centaurea</u> L. Star-thistle

Annual to perennial herbs with alternate, simple to pinnately compound leaves; heads 1 to several, without ray flowers or these poorly developed; involucral bracts imbricate in several series, either spine-tipped or with a fimbriate or broad hyaline tip; receptacle bristly; corollas purple, lavender, pinkish, yellow, or rarely white, with long narrow lobes; pappus of bristles or scales or lacking.

Reference: Moore, R. J. 1972. Rhodora 74:331-346.

1. Involucral bracts, at least some, spine-tipped, some spines usually over 1cm long
   2. Stem winged by decurrent leaf bases — *C. solstitialis*
   2. Stem not winged although often angled — *C. calcitrapa*
1. Involucral bracts not spine-tipped although often lacerate or fringed
   3. Pappus mostly 6mm long or more; deep creeping rhizomes present — *C. repens*
   3. Pappus mostly less than 3mm long or lacking; rhizomes lacking — *C. maculosa*

Centaurea calcitrapa L., Sp. Pl. 917. 1753. Biennial to 8dm high; leaves pinnatifid or some merely toothed, 1-9cm long; involucre 1-2cm long excluding the spines which terminate the bracts; corollas purple or lavender, 9-20mm long. Disturbed areas. Converse Co.

Centaurea maculosa Lam., Encyc. Meth. 1:669. 1785. Taprooted biennial or perennial to 15dm high; leaves pinnately compound or pinnatifid at least below, 1-11mm long; involucre 10-14mm long, the bracts with fimbriate tips; corollas pink-purple or rarely white, 1-2cm long. Disturbed areas. Natrona Co.

Centaurea repens L., Sp. Pl. 2:1293. 1763. Perennial from a deep creeping rhizome, to 8dm high; leaves slightly lobed to entire, oblong, elliptic, or oblanceolate or sometimes linear or lanceolate, 1-12cm long; involucre 9-15mm long, the outer bracts with broad hyaline tips, the inner much narrower with somewhat plumose-hairy tips; corollas purple or pinkish, 1-2cm long. Disturbed areas. NW, SE.

Centaurea solstitialis L., Sp. Pl. 917. 1753. Taprooted annual or biennial to 8dm high; lower leaves often pinnatifid or lyrate, the upper becoming entire, elliptic or oblanceolate to linear, 1-10cm long, the base decurrent along stem; involucre 8-15mm long excluding spines which terminate the bracts; corollas yellow, 1-2cm long. Disturbed areas. Platte and Hot Springs Cos.

Chaenactis DC. False Yarrow

Annual to perennial herbs with taproots and alternate or basal, mostly pinnately dissected leaves; heads 1 to several, lacking ray flowers; involucral bracts herbaceous or nearly so, equal or somewhat imbricate; receptacle naked; pappus of scales; corollas white, yellow, or pink.

Reference: Stockwell, P. 1940. Contr. Dudley Herb. 3:89-168.

Plants alpine or subalpine, perennial, with 1 or more rosettes; plants 15cm or less high, the stems little if at all leafy — *C. alpina*

Plants usually not alpine or subalpine, annual to perennial, the rosette solitary or lacking; plants often over 15cm high, the stems usually leafy

- Pappus scales mostly 8-16; biennial or perennial — *C. douglasii*
- Pappus scales 4; annual — *C. stevioides*

Chaenactis alpina (Gray) Jones, Proc. Calif. Acad. Sci. II, 5:699. 1895. Perennial to 15cm high; leaves mostly near base, the blades 0.5-5cm long; heads 1 to few; involucre 8-13mm long; corollas 5-7mm long. Alpine and subalpine slopes and talus. NW, SW, SE.

Chaenactis douglasii (Hook.) Hook. & Arn., Bot. Beechey Voy. 354. 1839. Biennial or perennial to 6dm high; leaves along the stem as well as basal, the blades 1-12cm long; heads usually several; involucre 7-16mm long; corollas 5-10mm long. Plains, hills, and slopes. Statewide. *C. angustifolia* Greene, *C. achilleaefolia* Hook. & Arn., *C. cineria* Stockw., *C. humilis* Rydb.

<u>Chaenactis</u> <u>stevioides</u> Hook. & Arn., Bot. Beechey Voy. 353. 1839. Annual to 3dm high; leaves basal and along stem, the blades 0.5-5cm long; heads 1 to several; involucre 5-11mm long; corollas 4-6mm long. Plains and hills. Carbon Co.

<u>Chamaechaenactis</u> Rydb.

Reference: Preece, S. J., Jr. & B. L. Turner. 1953. Madroño 12:97-103.

<u>Chamaechaenactis</u> <u>scaposa</u> (Eastw.) Rydb., Bull. Torrey Club 33:156. 1906. Hirsute-canescent perennial to 12cm high, from a branching caudex; leaves all basal, the blades ovate to orbicular or cordate, 3-16mm long, the margins somewhat revolute and mostly entire; heads solitary on each stem, lacking ray flowers; involucre 10-17mm long, the bracts biseriate; receptacle naked; pappus of scales; disk corollas white to yellowish, 5.5-8mm long. Plains and hills. Sweetwater Co. <u>Actinella</u> <u>carnosa</u> A. Nels.

Chrysanthemum L.

Perennial herbs with alternate, toothed to irregularly pinnately lobed leaves; heads 1 to many; rays white or lacking; involucral bracts imbricate, somewhat dry and usually scarious at least at tip; receptacle naked; pappus a short crown or lacking.

Reference: Mulligan, G. A. 1968. Nat. Canad. 95:793-795.

Rays usually lacking; leaves merely crenate-serrate or lobed at very base _C. balsamita_

Rays well developed; leaves usually irregularly lobed _C. leucanthemum_

Chrysanthemum balsamita L., Sp. Pl. 2:1252. 1763. Perennial to 12dm high; upper leaf blades elliptic to ovate, 1-15cm long, crenate-serrate, sometimes lobed near base; heads many; involucre 3-5mm long; rays lacking or very rarely present and under 1cm long; pappus minute or lacking; disk corollas about 2mm long. Disturbed areas. Albany Co.

Chrysanthemum leucanthemum L., Sp. Pl. 888. 1753. Rhizomatous perennial to 8dm high; leaf blades mostly oblanceolate or obovate, usually irregularly lobed or cleft, 1-10cm long; heads solitary at ends of branches; involucre 7-11mm long; rays 12-23mm long; pappus lacking; disk corollas about 3mm long. Disturbed areas. NW, SE.

Chrysothamnus Nutt. Rabbitbrush

Shrubs or subshrubs with alternate, narrow, entire to ciliate-margined leaves and many heads; ray flowers lacking; involucral bracts imbricate, usually in vertical ranks, often chartaceous at least below; receptacle naked; pappus of capillary bristles.

References: Anderson, L. C. 1964. Madroño 17:222-227.
1970. Madroño 20:337-342.

1. Branches covered with a close felt-like tomentum (scrape with sharp instrument under magnification)
   2. Heads in a cyme or corymb at ends of branches; involucre 10mm or less long, or if more, the achenes glabrous ..... *C. nauseosus*
   2. Heads in a terminal raceme or sometimes a panicle; involucre often 10mm or more high; achenes hairy ..... *C. parryi*

1. Branches glabrous or puberulent, lacking tomentum
   3. Mature achenes glabrous or nearly so, longitudinally 10 striate, often over 4mm long; SE Wyo. — *C. vaseyi*
   3. Mature achenes usually densely hairy, not striate, mostly 3-4mm long; widespread
      4. Involucral bracts, or some of them, narrowed to a subulate tip; corollas 5mm or less long; leaves 1.2mm or less wide, 1 nerved; Sweetwater Co. — *C. greenei*
      4. Involucral bracts obtuse or acute; corollas often 5mm or more long; leaves often over 1.2mm wide, often more than 1 nerved; widespread
         5. Plants shrubs mostly over 4dm high, usually in alkaline areas, with adventitious shoots from lateral roots; leaves mostly elliptic, some often over 3mm wide, mostly flat; corollas lacking a bulge or dark band — *C. linifolius*
         5. Plants shrubs or subshrubs, often less than 4dm high, usually in non-alkaline areas, without adventitious shoots; leaves elliptic or not, the width variable, often twisted; corollas often with a bulge or dark band on lower half — *C. viscidiflorus*

Chrysothamnus greenei (Gray) Greene, Erythea 3:94. 1895. Plants to 4dm high, the twigs glabrous or puberulent; leaves linear or linear-filiform, 0.5-3.5cm long; involucre 5-8mm long; corollas 4-5mm long. Plains and hills. Sweetwater Co.

Chrysothamnus linifolius Greene, Pittonia 3:24. 1896. Plants to 24dm high, the twigs glabrous or puberulent; leaves mostly elliptic, 1-5cm long; involucre 4-7mm long; corollas 4-7mm long. Mostly in alkaline areas. NW, SW, SE.

Chrysothamnus nauseosus (Pall. ex Pursh) Britt. in Britt. & Brown, Ill. Fl. 3:326. 1898. Plants to 2m high, the twigs with a very close, smooth tomentum; leaves linear, 1-10cm long; involucre 6-11mm long; corollas 5-11mm long. Plains, hills, and slopes. Statewide. C. oreophilus A. Nels., C. graveolens (Nutt.) Greene, C. frigidus Greene, C. pulcherrimus A. Nels., C. plattensis (Greene) Greene, C. pallidus A. Nels.

Chrysothamnus parryi (Gray) Greene, Erythea 3:113. 1895. Plants to 1m high, the twigs with a very close, smooth tomentum; leaves linear, 1-8cm long; involucre (6)10-13mm long; corollas (6)8-11mm long. Plains and hills. NW, SW, SE. C. wyomingensis A. Nels., C. howardii (Parry) Greene, C. affinis A. Nels.

Chrysothamnus vaseyi (Gray) Greene, Erythea 3:96. 1895. Plants to 3dm high, the twigs glabrous or minutely puberulent; leaves mostly linear, 0.5-5cm long; involucre 4-8mm long; corollas 4-6.5mm long. Plains and hills. Albany and Carbon Cos.

Chrysothamnus viscidiflorus (Hook.) Nutt., Trans. Am. Phil. Soc. II, 7:324. 1840. Plants to 2m high, the twigs glabrous or puberulent; leaves linear to narrowly elliptic, 1-6cm long, often twisted; involucre 4-8mm long; corollas 4-7mm long. Plains, hills, and slopes. Statewide. *C. pumilus* Nutt., *C. lanceolatus* Nutt., *C. glaucus* A. Nels.

Cichorium L. Chicory

Cichorium intybus L., Sp. Pl. 813. 1753. Taprooted biennial or perennial to 17dm high; leaves alternate, the blades oblanceolate to lanceolate, toothed to pinnatifid below, becoming entire above, 1-25cm long; heads 1-3 in each axil of the reduced upper leaves or terminating short branches; flowers all ligulate, the rays blue or occasionally white, 15-25mm long; involucre 8-15mm long, the bracts biseriate; receptacle naked; pappus of minute scales. Disturbed areas. NE, SE.

<u>Cirsium</u> Mill. Thistle

Biennial or perennial herbs with alternate, spiny, subentire to pinnatifid leaves; heads 1 to many, ray flowers lacking; involucral bracts in several series, mostly imbricate, at least the outer spine-tipped; receptacle bristly; pappus of plumose bristles (except sometimes outer flowers); corollas purplish, pink, or whitish.

Reference: Gardner, R. C. 1974. Madroño 22:239-265.

1. Involucre of largest heads 1-1.8(2.2)cm long; heads in loose corymbiform clusters, some heads usually with pappus longer than corolla, some with pappus shorter than corolla; plants with deep creeping rhizomes ..... <u>C</u>. <u>arvense</u>
1. Involucre of largest heads usually over 1.8cm long; heads solitary or in compact terminal clusters or axillary; pappus mostly similar; plants often merely taprooted
   2. Plants with involucres of mature heads 3.5-5cm long; inner involucral bracts dilated and lacerate near tip, often with a glutinous dorsal ridge; achenes with a yellow apical collar 0.4-0.8mm long; Black Hills ..... <u>C</u>. <u>drummondii</u>
   2. Plants without the above combination of characters

3. Heads often in terminal clusters, occasionally axillary, usually exceeded by 1 to several leaves; leaves often not much reduced upward; involucral bracts lacking a glutinous dorsal ridge

4. Bases of most leaves decurrent along stem for 6-30mm; leaves glabrous or rarely thinly tomentose beneath or with multicellular hairs along midrib ..... *C. eatonii*

4. Bases of leaves usually clasping the stem and not decurrent, sometimes all basal; leaves tomentose beneath, sometimes thinly so

5. Pappus usually equaling or exceeding corolla; involucral bracts not dilated and lacerate at tip ..... *C. foliosum*

5. Pappus shorter than corolla; inner involucral bracts sometimes dilated and lacerate at tip

6. Pappus 20-24mm long; style 26-32mm long; corolla 22-30mm long; plants sometimes acaulescent; mostly in S Wyo. ..... *C. acaulescens*

6. Pappus 14-21mm long; style 19-26mm long; corolla 16-24mm long; plants rarely acaulescent; mostly in N Wyo. ..... *C. scariosum*

3. Heads solitary or in loose clusters, 1 always terminal on stem and branches, occasionally axillary, never much exceeded by leaves; leaves much reduced upward; involucral bracts with or without a glutinous dorsal ridge

7. Involucral bracts with coarsely lacerate margins, usually dilated in upper half *C. centaureae*

7. Involucral bracts with entire or subentire margins, tapering to tip, rarely slightly expanded at base of spine

8. Upper surface of leaves with many short spines, otherwise glabrous or glabrate; outer involucral bracts reflexed near middle *C. vulgare*

8. Upper surface of leaves lacking spines, glabrous or tomentose; outer involucral bracts not reflexed except sometimes the tips

9. Leaves somewhat clasping the stem, or if decurrent, the wings less than 12mm long

10. Yellow apical collar of achenes 0.3-0.7mm long; upper leaf surface usually greenish, white-tomentose beneath *C. flodmanii*

10. Yellow apical collar of achenes 0.2mm or less long; upper leaf surface usually gray, white-tomentose or not beneath

11. Leaves rather densely whitish-tomentose beneath; involucral bracts often lacking a glutinous dorsal ridge *C. undulatum*

11. Leaves sparsely tomentose beneath, gray-green; involucral bracts usually with a well developed glutinous dorsal ridge *C. canovirens*

9. Leaves decurrent along stem, the wings of middle leaves mostly 15mm or more long

12. Decurrent wings of lower leaves usually longer than those of uppermost leaves; involucral bracts pubescent with cobwebby hairs which tend to connect apparently alternate bracts and cross over the bract between

13. Leaves tomentose beneath — *C. subniveum*

13. Leaves glabrous or with multicellular hairs along midrib beneath — *C. eatonii*

12. Decurrent wings of lower leaves shorter or subequal to those of uppermost leaves; involucral bracts glabrous, of if not, usually not as above

14. Lower leaves often with some lobes 3.5-5.5 times as long as wide; upper leaf surface slightly tomentose; involucre usually as wide as long or wider — *C. canescens*

14. Lower leaves usually with all lobes 3 times as long as wide or less, or if longer, the upper leaf surface often glabrous; involucre usually longer than wide

15. Involucre of mature heads 30-45mm long; upper leaf surface slightly to densely tomentose — *C. ochrocentrum*

15. Involucre of mature heads 18-28mm long; upper leaf surface glabrous or rarely slightly tomentose

16. Lower leaf surface white-tomentose — *C. pulcherrimum*

16. Lower leaf surface glabrous or with a few scattered hairs along midrib — *C. eatonii*

Cirsium acaulescens (Gray) Schum., Just's Bot. Jahresb. 29(1):566. 1903. Taprooted biennial or perennial to 8dm high; leaves sometimes all basal, the blades elliptic, oblong, or oblanceolate, 4-30cm long, mostly pinnatifid, whitish-tomentose beneath, green above; heads 1 to few; involucre 2-3.3cm long; corollas white to pink or lavender, 22-30mm long. Meadows, slopes, and disturbed areas. SW, SE, NE. C. coloradense (Rydb.) Cock. ex Daniels. Very similar to C. tioganum (Congd.) Petr. if not synonymous.

Cirsium arvense (L.) Scop., Fl. Carn. 2nd ed. 2:126. 1772. Canada Thistle. Perennial with deep rhizomes, to 15dm high; leaf blades lanceolate to oblanceolate, 1-15cm long, toothed to pinnatifid, white-tomentose to glabrate beneath, glabrous or glabrate above; heads few to many, usually unisexual; involucre 1-2.2cm long; corollas pink-purple or rarely white, 8-18mm long. Disturbed areas. Statewide. Carduus arvensis (L.) Robs.

Cirsium canescens Nutt., Trans. Am. Phil. Soc. II, 7:420. 1841. Taprooted biennial to 9dm high; leaf blades lanceolate, oblong, or oblanceolate, 2-32cm long, undulate to pinnatifid, white-tomentose beneath, green above; heads 1 to few; involucre 18-30mm long; corollas white or rarely pink or purple, 22-30mm long. Plains, hills, and disturbed areas. SW, SE, NE. Cnicus nelsonii Pammel, Carduus nelsonii (Pammel) A. Nels., Carduus plattensis Rydb.

<u>Cirsium</u> <u>canovirens</u> (Rydb.) Petr., Beih. Bot. Centralbl. 35, Abt. 2:540. 1917. Taprooted biennial or short lived perennial to 1m high; leaf blades mostly elliptic or oblong, 2-30cm long, pinnatifid, sparsely tomentose beneath, less so above; heads few to many; involucre 18-30mm long; corollas ochroleucous, 20-25mm long. Hills and meadows. Reported from NW Wyo.

<u>Cirsium</u> <u>centaureae</u> (Rydb.) Schum., Just's Bot. Jahresb. 29(1):566. 1903. Taprooted perennial to 12dm high; leaf blades elliptic, oblong, or oblanceolate, 3-27cm long, pinnatifid, white-tomentose beneath, green above; heads few; involucre 18-27mm long; corollas pink-purple to white, 15-25mm long. Woods and openings. Carbon and Albany Cos. <u>Carduus</u> <u>americanus</u> (Gray) Greene.

<u>Cirsium</u> <u>drummondii</u> T. & G., Fl. N. Am. 2:459. 1843. Taprooted perennial to 7dm high; leaf blades elliptic, oblong, or oblanceolate, 5-23cm long, mostly pinnatifid, sparsely long-hairy on both surfaces; heads 1 to few; involucre 3.5-5cm long; corollas purple, 35-43mm long. Plains, hills, and woods. Weston Co. <u>C</u>. <u>coccinatum</u> Osterh., <u>Carduus</u> <u>drummondii</u> (T. & G.) Cov.

Cirsium eatonii (Gray) Robins., Rhodora 13:240. 1911. Taprooted perennial to 8dm high; leaf blades mostly oblong or elliptic, 4-23cm long, pinnatifid, glabrous or glabrate on both surfaces; heads few; involucre 2-3cm long; corollas purple, pink, or white, 15-25mm long. Mountain slopes and talus. NW, NE. *C. tweedyi* (Rydb.) Petr., *C. polyphyllum* (Rydb.) Petr.

Cirsium flodmanii (Rydb.) Arthur, Torreya 12:34. 1912. Somewhat rhizomatous and taprooted perennial to 8dm high; leaf blades lanceolate to oblanceolate or oblong, 3-33cm long, pinnatifid or some merely toothed, white-tomentose beneath, greenish above; heads few; involucre 18-30mm long; corollas purple, 22-35mm long. Moist places. NW, NE, SE.

Cirsium foliosum (Hook.) DC., Prodr. 6:654. 1838. Taprooted perennial to 6dm high; leaf blades oblong, elliptic, or oblanceolate, 5-25cm long, toothed to pinnatifid, somewhat tomentose beneath to glabrate above; heads few to many; involucre 19-25mm long; corollas mostly white, 18-24mm long. Meadows and slopes. Yellowstone Park and Sheridan Co. *Carduus foliosus* Hook.

Cirsium ochrocentrum Gray, Pl. Fendl. 110. 1849. Taprooted or somewhat rhizomatous perennial to 8dm high; leaf blades mostly oblong or elliptic, 3-30cm long, pinnatifid, white-tomentose beneath, less so above; heads few; involucre 3-4.5cm long; corollas white to pink or lavender, 28-38mm long. Plains, hills, and disturbed areas. NW, SW, SE. *Carduus ochrocentrus* (Gray) Greene.

Cirsium pulcherrimum (Rydb.) Schum., Just's Bot. Jahresb. 29(1):566. 1903. Taprooted perennial to 8dm high; leaf blades oblong, elliptic, or oblanceolate, 3-22cm long, toothed to pinnatifid, white-tomentose beneath, green above; heads few; involucre 18-27mm long; corollas white, pink, or purple, 18-25mm long. Plains, hills, slopes, and disturbed areas. Statewide. Carduus pulcherrimus Rydb.

Cirsium scariosum Nutt., Trans. Am. Phil. Soc. II, 7:420. 1841. Elk Thistle. Taprooted perennial to 1m high; leaf blades oblong or oblanceolate, 5-20cm long, toothed to pinnatifid, somewhat tomentose beneath, glabrate or with long multicellular hairs above; heads few to several; involucre 2-3cm long; corollas whitish, pink, or purplish, 16-24mm long. Meadows in or near the mountains. SW, NW, NE. C. foliosum (Hook.) DC. of authors.

Cirsium subniveum Rydb., Fl. Rocky Mts. 1006, 1068. 1917. Taprooted perennial to 13dm high; leaf blades mostly elliptic or oblong, 3-25cm long, pinnatifid, tomentose beneath, less so above; heads few; involucre 17-25mm long; corollas white, pink, or purplish, 20-26mm long. Slopes and flats. Teton and Sublette Cos.

Cirsium undulatum (Nutt.) Spreng., Syst. Veg. 3:374. 1826. Taprooted perennial to 12dm high; leaf blades oblanceolate to lanceolate, 3-30cm long, pinnatifid or toothed, white-tomentose beneath, less so above; heads few; involucre 2-3.5cm long; corollas pink-purple or whitish, 27-40mm long. Plains, hills, slopes, and disturbed areas. Statewide. *Carduus undulatus* Nutt., *Carduus megacephalus* (Gray) Smythe.

Cirsium vulgare (Savi) Tenore, Fl. Nap. 5:209 (No. 2860). 1836. Bull Thistle. Biennial to 15dm high; leaf blades lanceolate to oblanceolate, 3-20cm long, mostly pinnatifid, spiny on upper surface, somewhat tomentose beneath; heads several; involucre 2-4cm long; corollas purple or rarely white, 25-35mm long. Disturbed areas. Statewide. *Carduus lanceolatus* L.

Conyza Less.

Reference: Cronquist, A. 1943. Bull. Torrey Club 70:629-632.

Conyza canadensis (L.) Cronq., Bull. Torrey Club 70:632. 1943. Annual to 1m high; leaves alternate, the blades oblanceolate to linear, 0.5-10cm long, entire or few toothed; heads many; involucre 2-4mm long, the bracts somewhat imbricate; receptacle naked; rays white, barely longer than pappus of capillary bristles; disk corollas about 2.5mm long. Plains, hills, streambanks, and disturbed areas. Statewide. *Leptilon canadense* (L.) Britt.

<u>Crepis</u> L. Hawksbeard

Perennial or rarely annual herbs with mostly basal, entire to bipinnatifid leaves, the reduced cauline leaves alternate; heads few to many, the flowers all ligulate, yellow or drying pink or whitish; involucral bracts mostly in 1 or 2 series; receptacle naked; pappus of capillary bristles.

1. Plants annual — *C. tectorum*
1. Plants perennial
  2. Stems and leaves glabrous or hispid (rarely tomentulose on upper stem)
    3. Flowers mostly 20 or more per head — *C. runcinata*
    3. Flowers mostly 6-12 per head
      4. Plants alpine or subalpine; achenes very short beaked or not beaked — *C. nana*
      4. Plants usually not alpine or subalpine; achenes with a slender beak nearly 1/4 as long as the body — *C. elegans*
  2. Stems and leaves somewhat tomentose or puberulent at least when young
    5. Heads mostly 5-10 flowered and with 5-8 inner involucral bracts; plants often with over 40 heads
      6. Involucre glabrous or glabrate; heads mostly 20-100 or more per plant — *C. acuminata*
      6. Involucre obviously pubescent; heads mostly 10-60 per plant — *C. intermedia*

5. Heads mostly 10-60 flowered and with 8 or more inner involucral bracts; plants rarely with over 40 heads

7. Involucre and/or lower part of stem with black, bristly hairs in addition to the whitish hairs, not glandular; leaf segments often toothed; achenes usually not ribbed *C. modocensis*

7. Involucre and lower part of stem not black bristly, or if so, either the bristles gland-tipped or the leaf segments entire or the achenes ribbed

8. Leaf segments mostly linear or lance-linear and entire; achenes usually greenish *C. atrabarba*

8. Leaf segments mostly lanceolate to deltoid, often toothed; achenes mostly yellowish or brownish

9. Involucre often with gland-tipped, black, bristly hairs in addition to the whitish hairs; heads mostly 10-40 flowered; inner involucral bracts mostly 8-13 *C. occidentalis*

9. Involucre lacking gland-tipped, black hairs; heads mostly 7-12 flowered; inner involucral bracts mostly 8 *C. intermedia*

Crepis acuminata Nutt., Trans. Am. Phil. Soc. II, 7:437. 1841. Tomentulose to glabrate perennial to 7dm high; basal leaf blades elliptic, usually irregularly pinnatifid, 5-25cm long; involucre 8-14mm long; corollas 8-18mm long, yellow but fading to white. Plains, hills, and slopes. Statewide.

Crepis atrabarba Heller, Bull. Torrey Club 26:314. 1899. Tomentulose to glabrate perennial to 7dm high; basal leaf blades linear to elliptic, deeply pinnatifid, 5-25cm long; involucre 8-15mm long; corollas 10-18mm long, yellow but fading to whitish. Plains, hills, slopes, and open woods. NW, SW, SE. *C. gracilis* (Eaton) Rydb.

Crepis elegans Hook., Fl. Bor. Am. 1:297. 1833. Glabrous perennial to 3dm high; basal leaf blades ovate or elliptic to suborbicular, deciduous, 1-5cm long, entire to dentate; involucre 7-10mm long; corollas 7-9mm long, yellow or drying whitish or pinkish. Hills, slopes, woods, and streambanks. Yellowstone Park and Fremont Co.

Crepis intermedia Gray, Syn. Fl. 1(2):432. 1884. Tomentulose perennial to 7dm high; basal leaf blades mostly elliptic, irregularly pinnatifid, 5-25cm long; involucre 10-16mm long; corollas 14-25mm long, yellow but fading to whitish. Plains, hills, and slopes. NW, SW, SE.

Crepis modocensis Greene, Erythea 3:48. 1895. Tomentulose to glabrate perennial to 4dm high; basal leaf blades mostly elliptic, deeply irregularly pinnatifid, 3-25cm long; involucre 10-16mm long; corollas 11-22mm long, yellow but fading to whitish. Plains, hills, and slopes. Statewide. C. scopulorum Cov.

Crepis nana Richards. in Frankl., Narr. 1st Journ. 746. 1823. Glabrous perennial to 2dm high; basal leaf blades ovate to orbicular or sometimes obovate, 0.4-4cm long, entire or rarely toothed or somewhat lobed; involucre 7-12mm long; corollas 7-9mm long, yellow but drying white or pink. Alpine and subalpine. Park and Sublette Cos.

Crepis occidentalis Nutt., Journ. Acad. Phila. 7:29. 1834. Tomentulose, glandular-hirsute, or glabrate perennial to 4dm high; basal leaf blades mostly elliptic, 5-25cm long, mostly irregularly pinnatifid; involucre 11-19mm long; corollas 15-24mm long, yellow but fading to whitish. Plains and hills. Statewide.

Crepis runcinata (James) T. & G., Fl. N. Am. 2:487. 1843. Glabrous or hispid perennial to 7dm high; basal leaf blades oblanceolate, elliptic, or obovate, entire to slightly pinnatifid, 1-25cm long; involucre 7-16mm long; corollas 9-20mm long, yellow but fading to whitish. Moist places. Statewide. C. glauca (Nutt.) T. & G., C. riparia A. Nels., C. petiolata Rydb.

Crepis tectorum L., Sp. Pl. 807. 1753. Annual to 1m high; leaf blades lanceolate, linear, or oblanceolate, 1-15cm long, entire or sometimes pinnatifid; involucre 6-9mm long; corollas 8-13mm long, yellow but fading to whitish. Streambanks and disturbed areas. Teton and Yellowstone Parks.

Dyssodia Cav. Dogweed

Reference: Strother, J. L. 1969. Univ. Calif. Publ. Bot. 48:1-88.

Dyssodia papposa (Vent.) Hitchc., Trans. Acad. Sci. St. Louis 5:503. 1891. Foul-smelling annual to 4dm high; leaves mostly opposite, pinnately dissected, 1-5cm long; heads few to many; involucre 6-9mm long, the bracts biseriate and dotted with large glands; receptacle naked or nearly so; rays to about 4mm long or lacking, yellow or orange; pappus of scales divided to near base into bristles; disk corollas about 3mm long. Plains, hills, and disturbed areas. SW, SE, NE.

Echinacea Moench Purple Coneflower

Reference: McGregor, R. L. 1968. Univ. Kans. Sci. Bull. 48:113-142.

Echinacea _angustifolia_ DC., Prodr. 5:554. 1836. Taprooted perennial herb to 6dm high, stiff-hirsute; leaves alternate, the blades lanceolate to oblanceolate or linear, 2-15cm long, entire; heads solitary on each stem or branch; involucre 8-18mm long, the bracts slightly imbricate; receptacle with somewhat spinescent bracts clasping the achenes; rays purple, pink, or rose or rarely white, 2-4cm long; pappus a short crown; disk corollas 6-8.5mm long. Plains and hills. NE, SE. _Brauneria_ _angustifolia_ (DC.) Heller.

Erigeron L. Daisy; Fleabane

Annual to perennial herbs with alternate or basal, entire to dissected leaves; heads 1 to many, usually with ray and disk flowers, the former rarely lacking or inconspicuous; involucral bracts subequal to rarely somewhat imbricate, herbaceous throughout or only near base; receptacle naked; pappus of capillary bristles, occasionally also with scales.

1. Rays yellow; leaves linear *E. linearis*
1. Rays blue, purple, pink, or white; leaves various
   2. Pappus of disk flowers of bristles as well as short outer setae; pappus of ray flowers of only short setae, the bristles lacking *E. strigosus*
   2. Pappus of disk and ray flowers similar, of bristles or sometimes also with outer setae (rays rarely lacking)
      3. Leaves, or some of them, lobed, divided, parted, or coarsely toothed
         4. Plants annuals or fibrous-rooted biennials or perennials
            5. Leaves linear or narrowly oblanceolate, the lower rarely broadly oblanceolate
               6. Involucre 3-5mm long; rays 4-6mm long *E. bellidiastrum*
               6. Involucre 5-9mm long; rays 8-18mm long *E. glabellus*
            5. Leaves much broader than narrowly oblanceolate *E. philadelphicus*
         4. Plants perennials with deep rhizomes or a well developed woody caudex
            7. Stems scapose or with a few much reduced linear leaves
               8. Leaves pinnately lobed or divided *E. pinnatisectus*
               8. Leaves ternately or palmately lobed, divided, or toothed
                  9. Leaves cuneate-flabellate with broad lobes or teeth *E. flabellifolius*

9. Leaves 1-4 times ternately or palmately lobed or divided, the divisions usually linear or nearly so — *E. compositus*

7. Stems leafy

10. Stem leaves well developed, usually lanceolate or broader, merely toothed

11. Leaves pubescent, sometimes sparsely so

12. Plants 10cm or less high; leaves cuneate-flabellate — *E. flabellifolius*

12. Plants over 10cm high; leaves not cuneate-flabellate — *E. coulteri*

11. Leaves glabrous or glandular — *E. superbus*

10. Stem leaves much reduced, mostly linear or oblanceolate, often lobed or divided or entire

13. Plants usually alpine, densely glandular, lacking stiff, spreading, non-glandular hairs or nearly so — *E. flabellifolius*

13. Plants not alpine, with stiff, spreading, non-glandular hairs, often minutely glandular-puberulent also — *E. allocotus*

3. Leaves entire to ciliate or rarely slightly toothed

14. Rays erect, white, pink, or lavender, 8mm or less long and 0.4mm or less wide, usually barely if at all exceeding pappus, sometimes lacking; involucre glandular, hirsute, or with multicellular hairs

15. Rayless pistillate flowers present between outer ray and hermaphroditic disk flowers; inflorescence a corymb or panicle or with a solitary head *E. acris*

15. Rayless pistillate flowers lacking; inflorescence a raceme or with a solitary head *E. lonchophyllus*

14. Rays mostly spreading and well developed, sometimes blue or purple, usually larger than above and conspicuously exceeding pappus, or if erect and reduced, the involucre woolly-villous

16. Involucre woolly-villous with tangled, sometimes glassy or blackish, multicellular hairs; plants alpine or subalpine with 1 head per stem

17. Middle stem leaves mostly ovate to lance-ovate, 1cm or more wide, at least as large as the basal ones *E. elatior*

17. Middle stem leaves linear or narrowly lanceolate or oblanceolate, mostly less than 1cm wide, smaller than the basal ones

18. Hairs of involucre with black or purple cross-walls or all purple

19. Rays 3-7mm long, 0.3-1mm wide *E. humilis*

19. Rays 7-15mm long, some usually 1mm or more wide *E. melanocephalus*

18. Hairs of involucre with clear cross-walls or the lowermost sometimes reddish-purple

20. Lower leaves all linear or narrowly oblanceolate, usually acute at tip, rarely over 3mm wide; stem hairs appressed or ascending *E. ochroleucus*

20. Lower leaves, or some of them, oblanceolate to obovate and rounded or obtuse at tip, sometimes over 3mm wide; stem hairs spreading

21. Stem usually glandular; some leaves often over 3mm wide *E. simplex*

21. Stem not glandular; leaves 3mm or less wide *E. rydbergii*

16. Involucre usually not woolly-villous (rarely so at very base), sometimes densely hirsute; plants only occasionally alpine, often with more than 1 head per stem

22. Plants annual, biennial, or short lived perennials, lacking rhizomes or a well developed woody caudex

23. Pappus simple, with only bristles; rays mostly 4-6mm long *E. bellidiastrum*

23. Pappus double, with long bristles and usually short, inconspicuous, narrow scales; rays often longer

24. Disk corollas mostly 4-5.5mm long *E. glabellus*

24. Disk corollas mostly 3.5mm long or less

25. Hairs of stem, or some of them, appressed or closely ascending; plants often with leafy stolons *E. flagellaris*

25. Hairs of stem mostly all spreading; plants without stolons *E. divergens*

22. Plants perennial with a rhizome or well developed woody caudex

26. Stem leaves usually well developed except sometimes the very uppermost ones, lanceolate or broader; plants usually tall and erect

27. Involucre woolly-villous with tangled, glassy, multicellular hairs *E. elatior*

27. Involucre not woolly-villous, often strongly hirsute

28. Rays mostly 1.5-3mm wide; pappus simple with bristles, rarely double *E. peregrinus*

28. Rays about 0.5-1.5mm wide; pappus double with bristles and minute scales, rarely simple

29. Leaves glandular and pubescent; stems long-hairy throughout *E. uintahensis*

29. Leaves not both glandular and pubescent (rarely slightly glandular on uppermost ones); stems sometimes glabrous or glabrate below

30. Upper leaves glabrous or nearly so except for ciliate margins; stems glabrous or glabrate below

31. Upper leaves with glandular-puberulent margins, not ciliate *E. superbus*

31. Upper leaves with ciliate margins

32. Leaves conspicuously reduced upward, the upper mostly linear or lance-linear *E. formosissimus*

32. Leaves little if at all reduced upward, the upper mostly ovate or lanceolate *E. speciosus*

30. Upper leaves pubescent; stems hairy below

33. Upper stem viscid, or if not, the leaves conspicuously reduced upward *E. formosissimus*

33. Upper stem not viscid, the leaves, except the very uppermost, only slightly reduced upward *E. subtrinervis*

26. Stem leaves usually much reduced upward, mostly linear, oblong, or oblanceolate, sometimes broader in a few low species; plants mostly low, often spreading

34. Stems usually over 3dm high, mostly simple below (a few depauperate alpine specimens may key here)

35. Involucral bracts spreading or reflexed; rays mostly 1.5-3mm wide; pappus usually simple *E. peregrinus*

35. Involucral bracts mostly appressed; rays 0.5-1.2mm wide; pappus often double

36. Basal leaves, or some of them, obtuse or rounded at tip, sometimes over 1cm wide *E. formosissimus*

36. Basal leaves acute at tip, rarely over 1cm wide *E. corymbosus*

34. Stems usually less than 3dm high, either branched near base or somewhat caespitose

37. Lower leaves mostly obovate to suborbicular, usually silvery-hairy, the blades 25mm or less long; Yellowstone Park *E. tweedyi*

37. Lower leaves narrower, often longer; widespread

38. Pubescence of stem mostly spreading

39. Caudex of slender, somewhat rhizomatous branches with many fibrous roots, usually lacking a stout taproot; plants of the mountains *E. ursinus*

39. Caudex simple or with stout branches without many fibrous roots, the taproot usually well developed; plants of the plains or mountains

40. Plants alpine or subalpine

41. Pappus definitely double with bristles and shorter setae *E. ochroleucus*

41. Pappus simple with bristles or rarely with very inconspicuous setae *E. rydbergii*

40. Plants of the plains, foothills, or lower mountain slopes

42. Plants somewhat sparsely hairy, usually glandular; SE Wyo. *E. vetensis*

42. Plants usually copiously hairy, often not glandular; widespread

43. Plants 10cm or less high; leaves densely clustered at base, stem leaves lacking or few and very much reduced; heads 1 per stem, the rays purple or pink; Lincoln Co. *E. nanus*

43. Plants not as above

44. Hairs of leaves rather dense and curly; leaves often well developed upward and usually oblong, the basal often prominently 3 nerved

45. Rays white; basal leaves not 3 nerved, 4mm or less wide; S Wyo. *E. engelmannii*

45. Rays blue, purple, pink, or rarely white; basal leaves often 3 nerved, some often over 4mm wide; widespread

46. Basal leaves usually rounded or obtuse at tip; stem usually green at base; involucral bracts usually with a raised dorsal ridge *E. caespitosus*

46. Basal leaves acute at tip; stem usually purplish at base; involucral bracts usually lacking a raised dorsal ridge *E. corymbosus*

44. Hairs of leaves either sparse, or straight and stiff or appressed, or both; leaves usually much reduced and linear above, the basal only rarely 3 nerved

47. Hairs of involucre somewhat sparce, long and spreading *E. pumilus*

47. Hairs of involucre usually dense and tangled or appressed *E. ochroleucus*

38. Pubescence of stem appressed or occasionally ascending

48. Caudex of slender, somewhat rhizomatous branches with many fibrous roots, usually lacking a stout taproot; plants of the mountains

49. Involucre usually glandular and with spreading hairs; widespread *E. ursinus*

49. Involucre mostly with appressed or somewhat loose hairs, scarcely if at all glandular; NW Wyo. *E. gracilis*

48. Caudex simple or with stout branches without many fibrous roots, the taproot usually well developed; plants of the plains or mountains

50. Plants somewhat sparsely hairy and sparingly glandular; SE Wyo. (not known from Carbon Co.) *E. vetensis*

50. Plants pubescent to glabrous, not glandular except rarely above; widespread

51. Hairs of leaves usually rather dense and curly; basal leaves sometimes prominently 3 nerved, or if not, then often with white rays

52. Rays white; basal leaves not 3 nerved, 4mm or less wide; S Wyo. *E. engelmannii*

52. Rays blue, purple, pink, or white; basal leaves often 3 nerved, some often over 4mm wide; widespread *E. caespitosus*

51. Hairs of leaves lacking or sparse or mostly stiff and straight and often appressed; basal leaves rarely 3 nerved; rays various

53. (moved to left margin)

53. Plants alpine or subalpine, rarely lower on cliffs or talus; leaves glabrous or glabrate; involucre usually glandular; rays mostly blue or purple; (not known from Big Horn Mts.) *E. leiomerus*

53. Plants not with the above combination of characters

54. Involucre 3-5mm long; rocky slopes and cliffs in mountains of Sublette Co. *E. tener*

54. Involucre (4)5-9mm long; habitat and distribution various

55. Basal leaves usually 3 nerved; rays white, occasionally drying pink or lavender; pappus mostly of 1 series, rarely with a few short bristles *E. eatonii*

55. Basal leaves rarely 3 nerved; rays white or not; pappus of 1 or 2 series

56. Leaves mostly all linear, 2mm or less wide, often sparsely pubescent; involucre 4-7mm long; plains and foothills of SE Wyo. or rarely in SW Wyo.

57. Leaves at base usually somewhat dilated and with spreading cilia; achenes short pubescent; SE Wyo. *E. nematophyllus*

57. Leaves usually not as above; achenes glabrous except for long cilia on the margins; Sweetwater Co. *E. consimilis*

56. Leaves often wider, often densely pubescent; involucre 5-9mm long; plains and mountains, widespread

58. Involucre usually densely pubescent with somewhat tangled hairs toward base; leaves often sparsely pubescent; plants mostly of the mountains, only occasional on the plains *E. ochroleucus*

58. Involucre often sparsely pubescent with mostly straight hairs; leaves often densely pubescent; plants of the plains and hills

59. Plants usually from a simple or few branched caudex; involucre 5-7mm long; achenes glabrous or nearly so *E. canus*

59. Plants usually from a much branched caudex; involucre 6-9mm long; achenes densely pubescent *E. pulcherrimus*

Erigeron acris L., Sp. Pl. 863. 1753. Biennial or perennial to 8dm high; leaf blades oblanceolate or obovate below to lance-ovate or linear-oblong above, 0.5-10cm long, the margins entire or ciliate; heads 1 to many; involucre (4)5-11mm long; rays pink, purple, or white, to 8mm long; disk corollas 4-6.5mm long. Open woods and slopes in the mountains especially where rocky. Statewide. *E. yellowstonensis* A. Nels., *E. lapiluteus* A. Nels.

Erigeron allocotus Blake, Journ. Wash. Acad. Sci. 27:379. 1937. Perennial to 15cm high from a branched woody caudex or rhizome; leaf blades obovate below to linear above, 0.5-3cm long, the lower mostly 3 cleft at tip, sometimes again cleft; heads several to many; involucre 4-5mm long; rays white or drying pink or lavender, 3-6mm long; disk corollas 2-3.5mm long. Endemic on rocky slopes in Big Horn Co.

Erigeron bellidiastrum Nutt., Trans. Am. Phil. Soc. II, 7:307. 1840. Annual or biennial to 5dm high; leaf blades linear or oblanceolate, 0.5-4cm long, entire to toothed at tip or rarely pinnatifid; heads many; involucre 3-5mm long; rays white, lavender, or pink, 4-6mm long; disk corollas 2.5-3mm long. Plains and hills. Platte Co.

<u>Erigeron</u> <u>caespitosus</u> Nutt., Trans. Am. Phil. Soc. II, 7:307. 1840. Perennial to 3dm high from a usually branched caudex; leaf blades oblanceolate below to oblong or linear above, 0.5-12cm long, entire; heads 1 to many; involucre 4-7mm long; rays blue, purplish, pink, or white, 5-15mm long; disk corollas 3-4.5mm long. Plains, hills, and slopes. Statewide.

<u>Erigeron</u> <u>canus</u> Gray, Pl. Fendl. 67. 1849. Perennial to 35cm high from a branching or simple caudex; leaf blades oblanceolate below to linear above, 0.5-10cm long, entire; heads 1-4 per stem; involucre 5-7mm long; rays blue, purplish, or white, 7-12mm long; disk corollas 3.5-5.5mm long. Plains, hills, and slopes. NW, NE, SE. <u>Wyomingia</u> <u>cana</u> (Gray) A. Nels.

<u>Erigeron</u> <u>compositus</u> Pursh, Fl. Am. Sept. 535. 1814. Perennial to 25cm high from a branching caudex; leaf blades trifid to ternately or palmately dissected below, lacking or few, linear, and entire above, 0.5-3cm long; heads solitary on each stem; involucre 5-10mm long; rays white, pink, lavender, or blue, to 15mm long, rarely lacking; disk corollas 3-5mm long. Plains, hills, and slopes especially where rocky. Statewide. <u>E</u>. <u>trifidus</u> Hook., <u>E</u>. <u>multifidus</u> Rydb.

Erigeron consimilis Cronq., Brittonia 6:186. 1947. Perennial to 10cm high from a caudex and taproot; leaves mostly basal, the blades linear, 4-20mm long, entire; heads solitary on each stem; involucre 5-7mm long; rays white, pink, or lavender, 7-14mm long; disk corollas 3.5-4.5mm long. Plains, hills, and juniper woodlands. Sweetwater Co.

Erigeron corymbosus Nutt., Trans. Am. Phil. Soc. II, 7:308. 1840. Perennial to 5dm high from a simple or branched caudex; leaf blades linear, elliptic, or oblanceolate, 0.5-15cm long, entire; heads 1 to several; involucre 5-7mm long; rays blue, purple, or pink, 7-15mm long; disk corollas 3.5-5.5mm long. Plains, hills, and slopes. SW, NW, NE. *E. nelsonii* Greene.

Erigeron coulteri Porter in Porter & Coult., Syn. Fl. Colo. 61. 1874. Fibrous rooted perennial from a rhizome or branched caudex, to 6dm high; leaf blades obovate or oblanceolate below to ovate or lanceolate above, 1-10cm long, entire or toothed; heads 1-4; involucre 7-10mm long; rays white or sometimes drying lavender, 9-24mm long; disk corollas 3-4.5mm long. Meadows and streambanks in the mountains. Reported from Wyoming.

Erigeron divergens T. & G., Fl. N. Am. 2:175. 1841. Taprooted annual, biennial, or perennial to 7dm high; leaf blades oblanceolate to oblong or linear, 0.5-5cm long, entire; heads few to many; involucre 3-5mm long; rays blue, purple, pink, or rarely white, 3-10mm long; disk corollas 2-3mm long. Plains, hills, and slopes. NW, NE, SE.

Erigeron _eatonii_ Gray, Proc. Am. Acad. 16:91. 1880. Taprooted perennial to 3dm high; leaf blades oblanceolate to linear, 1-15cm long, entire; heads 1 to few; involucre 5-7mm long; rays white, rarely drying pink or lavender, 5-10mm long; disk corollas 3-5mm long. Plains, hills, and slopes. NW, SW, SE. _E_. _microlonchus_ Greene.

Erigeron _elatior_ (Gray) Greene, Pittonia 3:163. 1897. Perennial to 7dm high from a short caudex; leaf blades oblanceolate or elliptic below to lanceolate or ovate above, 1-12cm long, the margins entire or ciliate; heads 1 to few; involucre 9-15mm long; rays white, pink, or lavender, 12-25mm long; disk corollas 4-5.5mm long. Woods and slopes well up in the mountains. Carbon and Albany Cos.

Erigeron _engelmannii_ A. Nels., Bull. Torrey Club 26:247. 1899. Perennial to 3dm high from a simple or branched caudex; leaf blades linear or linear-oblanceolate, 0.5-8cm long, entire; heads 1 to few per stem; involucre 3-7mm long; rays white, 5-13mm long; disk corollas 2.5-4.5mm long. Plains and hills. SW, SE.

Erigeron _flabellifolius_ Rydb., Bull. Torrey Club 26:545. 1899. Taprooted or rhizomatous perennial to 10cm high; leaf blades mostly cuneate-flabelliform, 3-5 cleft and/or toothed at tip, 0.5-3cm long; heads solitary; involucre 7-9mm long; rays pink or white, 6-12mm long; disk corollas 3-5mm long. Alpine or talus slopes in the mountains. Park Co.

Erigeron _flagellaris_ Gray, Pl. Fendl. 68.1849. Taprooted biennial or perennial, often stoloniferous, to 4dm high; leaf blades oblanceolate to linear, 0.5-4cm long, entire; heads mostly solitary on each stem; involucre 3-5mm long; rays white, pink, blue, or purple, 5-10mm long; disk corollas 2.5-3.5mm long. Woods, slopes, and streambanks especially where rocky. NW, SE.

Erigeron _formosissimus_ Greene, Bull. Torrey Club 25:121. 1898. Somewhat fibrous rooted perennial to 4dm high; leaf blades oblanceolate or obovate below to linear or lanceolate above, 0.5-12cm long, entire; heads 1 to few; involucre 5-9mm long; rays blue, purple, pink, or rarely white, 8-18mm long; disk corollas 3-4.5mm long. Plains, hills, meadows, and slopes. Albany and Carbon Cos. _E_. _eximius_ Greene.

Erigeron glabellus Nutt., Gen. Pl. 2:147. 1818. Fibrous rooted biennial or perennial to 6dm high; leaf blades oblanceolate to linear, 0.5-15cm long, entire or rarely a few toothed; heads 1 to several; involucre 5-9mm long; rays blue, purple, pink, or white, 8-18mm long; disk corollas 4-5.5mm long. Meadows, plains, hills, and slopes. Statewide. *E. asper* Nutt.

Erigeron gracilis Rydb., Mem. N. Y. Bot. Gard. 1:404. 1900. Fibrous rooted or rhizomatous perennial to 25cm high; leaf blades linear to oblanceolate, 0.5-6cm long, entire; heads solitary on each stem; involucre 5-7mm long; rays blue, pink, or purple, 7-15mm long; disk corollas 4-5.5mm long. Hills and slopes. Sublette Co. and Yellowstone Park.

Erigeron humilis Grah., Edinb. New Phil. Journ. 1828:175. 1828. Taprooted or fibrous rooted perennial to 10cm high; leaf blades oblanceolate below to linear or lance-linear above, 0.3-3cm long, the margins entire or ciliate; heads solitary; involucre 5-9mm long; rays white or purplish, 3-7mm long; disk corollas 3-5mm long. Alpine and subalpine. Big Horn Co.

Erigeron leiomerus Gray, Syn. Fl. 1(2):211. 1884. Perennial to 2dm high, the caudex often with rather long branches; leaf blades obovate or oblanceolate below to linear above, 0.5-7cm long, entire; heads solitary on each stem; involucre 4-6mm long; rays blue, purple, or white, 6-15mm long; disk corollas 3-4.5mm long. Meadows, slopes, and ridges in the mountains especially where rocky. NW, SW, SE. E. spathulifolius Rydb.

Erigeron linearis (Hook.) Piper, Contr. U. S. Nat. Herb. 11:567. 1906. Perennial to 3dm high from a simple or branched caudex; leaf blades linear, 1-9cm long, entire; heads 1 to few; involucre 4-7mm long; rays yellow, 4-12mm long; disk corollas 3.5-5.5mm long. Plains, hills, and slopes. Yellowstone Park. E. luteus A. Nels.

Erigeron lonchophyllus Hook., Fl. Bor. Am. 2:18. 1834. Somewhat fibrous rooted biennial or perennial to 6dm high; leaf blades oblanceolate to linear, 0.5-15cm long, entire; heads 1 to few; involucre 4-10mm long; rays white or lavender, 2-8mm long; disk corollas 3-5mm long. Meadows and slopes in or near the mountains. Statewide.

Erigeron melanocephalus (A. Nels.) A. Nels., Bull. Torrey Club 26:246. 1899. Fibrous rooted perennial to 15cm high; leaves mostly basal, the blades oblanceolate or obovate, becoming linear above, 0.5-5cm long, the margins entire or ciliate; heads solitary on each stem; involucre 5-9mm long; rays white or pinkish, 7-14mm long; disk corollas 2.5-3.5mm long. High mountains. Albany Co.

Erigeron nanus Nutt., Trans. Am. Phil. Soc. II, 7:308. 1840. Perennial to 10cm high from a branched caudex; leaves mostly or all basal, the blades linear to oblanceolate, 1-5cm long, the margins entire or ciliate; heads solitary on each stem; involucre 5-8mm long; rays purple or pinkish, 5-13mm long; disk corollas 3.5-5.5mm long. Plains and hills. Lincoln Co. *E. inamoenus* A. Nels.

Erigeron nematophyllus Rydb., Bull. Torrey Club 32:124. 1905. Perennial to 2dm high from a branched caudex; leaves mostly basal, the blades linear or linear-oblanceolate, 0.5-8cm long, entire; heads solitary on each stem; involucre 4-7mm long; rays white or pinkish or rarely purplish, 4-10mm long; disk corollas 3.5-4.5mm long. Plains and hills. SE.

Erigeron ochroleucus Nutt., Trans. Am. Phil. Soc. II, 7:309. 1840. Perennial to 3dm high from a simple or branched caudex; leaf blades linear or narrowly oblanceolate, 0.5-12cm long, entire; heads 1 to few; involucre 5-9mm long; rays blue, purple, pink, or white, 4-15mm long; disk corollas 2.5-4.5mm long. Plains, hills, and slopes. NW, NE, SE. *E.* *laetevirens* Rydb., *Wyomingia tweedyana* (Canby & Rose) A. Nels.

Erigeron peregrinus (Banks ex Pursh) Greene, Pittonia 3:166. 1897. Fibrous rooted perennial from a short rhizome or caudex, to 7dm high; leaf blades obovate or oblanceolate below to ovate or lanceolate or rarely linear above, 1-15cm long, the margins entire or ciliate; heads 1 to few; involucre 6-11mm long; rays blue or purple or occasionally white, 8-25mm long; disk corollas 3-6mm long. Moist places in the mountains. Statewide. *E.* *callianthemus* Greene, *E.* *glacialis* (Nutt.) A. Nels., *E.* *salsuginosus* (Richards.) Gray of authors.

Erigeron philadelphicus L., Sp. Pl. 863. 1753. Mostly fibrous rooted biennial or perennial, or rarely appearing annual, to 7dm high; leaf blades obovate or oblanceolate below to oblong-lanceolate or ovate above, 1-10cm long, some entire and some toothed; heads 1 to several; involucre 3.5-6mm long; rays pink, lavender, or white, 5-10mm long; disk corollas 2.5-4mm long. Moist places. Weston and Crook Cos.

Erigeron _pinnatisectus_ (Gray) A. Nels., Bull. Torrey Club 26:246. 1899. Perennial to 15cm high from a stout caudex; leaves mostly basal and pinnatifid, 1-7cm long; heads solitary on each stem; involucre 5-8mm long; rays blue or purple, 7-13mm long; disk corollas 3.5-4.5mm long. Alpine and subalpine. Albany Co.

Erigeron _pulcherrimus_ Heller, Bull. Torrey Club 25:200. 1898. Perennial to 3dm high from a branched caudex; leaf blades linear or oblanceolate, 0.5-6cm long, entire; heads solitary on each stem; involucre 6-9mm long; rays white or occasionally pink or purplish, 8-15mm long; disk corollas 3-6mm long. Plains and hills. Carbon and Washakie Cos. _Wyomingia_ _pulcherrima_ (Heller) A. Nels., _W_. _cinerea_ A. Nels., _E_. _wyomingia_ Rydb.

Erigeron _pumilus_ Nutt., Gen. Pl. 2:147. 1818. Perennial to 3(4)dm high from a simple or branched caudex; leaf blades oblanceolate to linear, 0.5-8cm long, the margins entire or ciliate; heads 1 to several; involucre 4-7mm long; rays blue, purplish, pink, or white, 6-15mm long; disk corollas 3-5mm long. Plains, hills, and slopes. Statewide. _E_. _concinnus_ (H. & A.) T. & G., _E_. _wyomingensis_ A. Nels.

Erigeron rydbergii Cronq., Brittonia 6:191. 1947. Perennial to 8cm high from a simple or branched caudex; leaf blades oblanceolate to linear, 0.5-4cm long, the margins entire or ciliate; heads solitary on each stem; involucre 4-6mm long; rays white or purplish, 6-11mm long; disk corollas 3-5mm long. High mountain slopes. Yellowstone Park and Big Horn Co.

Erigeron simplex Greene, Fl. Francis. 387. 1897. Perennial to 25cm high from a caudex; leaf blades oblanceolate or obovate below to linear above, 0.5-6cm long, the margins entire or ciliate; heads solitary on each stem; involucre 5-8mm long; rays blue, purplish, pink, or rarely white, 6-15mm long; disk corollas 2.5-3.5mm long. High mountain slopes. Statewide. *E. uniflorus* L. of authors. I am following Cronquist in considering *E. grandiflorus* Hook. as not present in the state.

Erigeron speciosus (Lindl.) DC., Prodr. 5:284. 1836. Perennial from a woody caudex, to 8dm high; leaf blades oblanceolate below to lanceolate or ovate above, 1-15cm long, the margins entire or ciliate; heads 1 to several; involucre 6-9mm long; rays blue or purple or rarely white, 9-20mm long; disk corollas 4-6mm long. Open woods, meadows, and slopes. Statewide. *E. macranthus* Nutt.

<u>Erigeron</u> <u>strigosus</u> Muhl. ex Willd., Sp. Pl. 3:1956. 1803. Annual or rarely biennial to 7dm high; leaf blades oblanceolate to linear, oblong, or lanceolate, 0.5-12cm long, entire; heads several to many; involucre 2.5-5mm long; rays usually white, rarely lavender or pinkish, 3-8mm long; disk corollas 1.5-2.5mm long. Disturbed and open areas. NW, NE. <u>E</u>. <u>ramosus</u> (Walt.) B. S. P.

<u>Erigeron</u> <u>subtrinervis</u> Rydb. ex Porter & Britt., Mem. Torrey Club 5:328. 1894. Perennial from a woody caudex, to 8dm high; leaf blades oblanceolate below to lanceolate or ovate above, 1-13cm long, entire; heads 1 to several; involucre 6-10mm long; rays blue, purple, pink, or white, 9-18mm long; disk corollas 4-5mm long. Open woods, meadows, and slopes. Statewide.

<u>Erigeron</u> <u>superbus</u> Greene ex Rydb., Fl. Colo. 361, 364. 1906. Perennial to 6dm high from a caudex or rhizome; leaf blades oblanceolate to oval below to lanceolate, ovate, or oblong above, 1-12cm long, entire to toothed; heads 1 to several; involucre 7-10mm long; rays blue, purplish, or white, 12-20mm long; disk corollas 4.5-6mm long. Woods and slopes. NE, SE. Weber (Southw. Nat. 18:319. 1973) has taken up the name <u>E</u>. <u>eximius</u> Greene for this species assuming that Cronquist did not see the type, but Cronquist states (Bull. Torrey Club 70:268. 1943) that he did see the type.

Erigeron tener Gray, Proc. Am. Acad. 16:91. 1880. Perennial to 15cm high from a simple or branched caudex; leaf blades oblanceolate or obovate below to linear above, 3-20mm long, entire; heads 1-3 per stem; involucre 3-5mm long; rays blue or purple, 4-8mm long; disk corollas 2.5-4mm long. Rocky places in the mountains. Sublette Co.

Erigeron tweedyi Canby, Bot. Gaz. 13:17. 1888. Perennial to 2dm high from a simple or branched caudex; leaf blades obovate, elliptic, or suborbicular below to linear above, 4-25mm long, entire; heads 1 to few; involucre 3.5-6mm long; rays blue, purple, or white, 5-10mm long; disk corollas 3-4.5mm long. Mountain slopes. Yellowstone Park.

Erigeron uintahensis Cronq., Bull. Torrey Club 70:270. 1943. Perennial to 5dm high from a woody caudex; leaf blades oblanceolate below to lanceolate or ovate above, 0.5-9cm long, the margins entire or ciliate; heads 1-5; involucre 5-8mm long; rays blue or purple, 9-16mm long; disk corollas 3-5mm long. Hills and slopes. Sweetwater Co.

Erigeron ursinus Eaton in Wats., Bot. King Exp. 148. 1871. Fibrous rooted or rhizomatous perennial to 3dm high; leaf blades oblanceolate below to linear above, 0.5-10cm long, the margins entire or ciliate; heads solitary on each stem; involucre 4-7mm long; rays blue or purple, 7-15mm long; disk corollas 3-4.5mm long. Meadows and slopes in the mountains. Statewide.

Erigeron vetensis Rydb., Bull. Torrey Club 32:126. 1905. Perennial to 25cm high from a simple or branched caudex; leaf blades oblanceolate to linear, 0.5-10cm long, the margins entire or ciliate; heads solitary on each stem; involucre 4-8mm long; rays blue, pink, or purple, rarely white, 6-16mm long; disk corollas 3-5.5mm long. Woods, hills, and slopes. SE. E. glandulosus Porter.

Eriophyllum Lag.

Eriophyllum lanatum (Pursh) Forbes, Hort. Woburn. 183. 1833. Tomentose perennial herb to 5dm high; leaves alternate or opposite, the blades mostly oblanceolate to linear, entire to pinnately or ternately lobed; heads 1 to few; involucre 6-9mm long, the bracts in 1 or apparently 2 series; receptacle naked; rays yellow, 6-13mm long; pappus of scales, a crown, or obsolete; disk corollas about 4mm long. Plains, hills, woods, and slopes. NW, SW. E. integrifolium (Hook.) Greene.

<u>Eupatorium</u> L. Joe Pye Weed

<u>Eupatorium</u> <u>maculatum</u> L., Cent. Pl. 1:27. 1755. Fibrous rooted perennial to 15dm high; leaves whorled, the blades lance-elliptic, lanceolate, or lance-ovate, 5-20cm long, serrate; heads several to many, lacking ray flowers; involucre 6-10mm long, the bracts imbricate; receptacle naked; pappus of capillary bristles; disk corollas about 5mm long, pink, purple, or whitish. Swamps, shores, and moist thickets at lower elevations. NE, SE. <u>E</u>. <u>atromontanum</u> A. Nels.

<u>Gaillardia</u> Foug. Blanketflower

<u>Gaillardia</u> <u>aristata</u> Pursh, Fl. Am. Sept. 573. 1814. Taprooted perennial herb to 7dm high; leaves alternate, the blades mostly lanceolate to oblanceolate, 2-15cm long, entire or toothed to somewhat pinnatifid; heads 1 to few; involucre 1-2cm long, the bracts in 2 or 3 series; receptacle bristly; pappus of awned scales; rays yellow or purplish at base, 3 cleft at tip, 1-4cm long; disk corollas about 8mm long. Plains, hills, slopes, and meadows. Statewide.

Gnaphalium L. Cudweed

Annual to perennial, usually white-woolly herbs with alternate, entire leaves; heads few to many, lacking ray flowers; outer flowers pistillate with slender corollas, the few inner flowers with wider corollas and perfect; involucral bracts imbricate, scarious; receptacle naked; pappus of capillary bristles.

1. Involucre mostly 2-4mm long; plants less than 2(3)dm high, usually much branched
   2. Leaves linear or narrowly oblanceolate; involucral bracts often darkened to tip; tomentum sometimes close *G. uliginosum*
   2. Leaves oblanceolate to oblong; involucral bracts usually whitish at tip; tomentum loose *G. palustre*
1. Involucre mostly 4-7mm long; plants often over 2dm high, usually not much branched
   3. Plants glandular-hairy, sparsely if at all tomentose below the inflorescence *G. viscosum*
   3. Plants not glandular, somewhat tomentose throughout
      4. Plants annual or biennial; leaves often adnate-auriculate; heads usually in a compact inflorescence *G. chilense*
      4. Plants biennial or perennial; leaves usually narrowly decurrent; heads in a somewhat loose inflorescence *G. microcephalum*

Gnaphalium chilense Spreng., Syst. Veg. 3:480. 1826. Annual or biennial to 7dm high; leaf blades lanceolate to oblanceolate or linear, 1-8cm long; involucre 4-7mm long; disk corollas about 2-3mm long. Moist, often disturbed places or around hot springs. Yellowstone Park and Teton Co. *G. proximum* Greene, *G. sulphurescens* Rydb.

Gnaphalium microcephalum Nutt., Trans. Am. Phil. Soc. II, 7:404. 1841. Taprooted biennial or perennial to 7dm high; leaf blades linear or oblanceolate, 1-5cm long; involucre 4-7mm long; disk corollas about 3-4mm long. Open places, often around geysers or hot springs. Yellowstone Park. *G. thermale* E. Nels.

Gnaphalium palustre Nutt., Trans. Am. Phil. Soc. II, 7:403. 1841. Annual to 2(3)dm high; leaf blades mostly oblanceolate or oblong, 3-35mm long; involucre 2-4mm long; disk corollas about 1.5-2mm long. Moist places especially on shores. Statewide.

Gnaphalium uliginosum L., Sp. Pl. 856. 1753. Similar to *G. palustre* but the leaf blades linear or linear-oblanceolate and to 45mm long, the tomentum sometimes close rather than loose, and the involucral bracts often greenish or brownish to tip rather than with whitish tips. Streambanks and disturbed areas. Albany and Natrona Cos. *G. strictum* Gray, *G. angustifolium* A. Nels., *G. exilifolium* A. Nels.

Gnaphalium viscosum H. B. K., Nov. Gen. & Sp. 4:82. 1820. Annual or biennial to 9dm high, the stem glandular-hairy; leaf blades oblanceolate to linear or oblong, 1-9cm long; involucre 4-7mm long; disk corollas about 2-3mm long. Open woods and slopes. NW, NE, SE. G. decurrens Ives, G. ivesii Nels. & Macbr.

Grindelia Willd. Gumweed

Biennial or perennial herbs with alternate, entire or toothed leaves; heads several to many, with both ray and disk flowers; involucral bracts resinous, mostly imbricate, the outer with hooked tips; receptacle naked; pappus of deciduous awns which are smooth or barbellate.

Reference: Steyermark, J. A. 1934. Ann. Mo. Bot. Gard. 21:433-608.

Stem leaves often oblanceolate and narrowed to a petiole-like base, sometimes clasping; pappus awns barbellate; Albany Co. G. subalpina

Stem leaves mostly oblong with a somewhat broad clasping base; pappus awns apparently smooth; widespread G. squarrosa

Grindelia squarrosa (Pursh) Dunal, Mem. Mus. Par. 5:50. 1819. Plants to 1m high; leaf blades oblanceolate to oblong or rarely lanceolate, 0.5-20cm long; involucre 5-12mm long; rays yellow, 7-16mm long; disk corollas about 5-7mm long. Plains, hills, and disturbed areas. Statewide. *G. perennis* A. Nels.

Grindelia subalpina Greene, Pittonia 3:297. 1898. Plants to 5dm high; leaf blades mostly oblanceolate or oblong, 1-13cm long; involucre 7-13mm long; rays yellow, 7-20mm long; disk corollas about 5-7mm long. Hills, slopes, and disturbed areas. Albany Co. *G. erecta* A. Nels.

Gutierrezia Lag. Snakeweed; Matchbrush

References: Solbrig, O. T. 1960. Contr. Gray Herb. 188:1-63.
1964. Contr. Gray Herb. 193:67-115.

Gutierrezia sarothrae (Pursh) Britt. & Rusby, Trans. N. Y. Acad. Sci. 7:10. 1887. Shrub or subshrub to 6dm high; leaves alternate, the blades linear, 0.3-4cm long, the margins entire or scabrous; involucre 3-5mm long, resinous, the bracts imbricate; receptacle naked; pappus of scales; rays mostly 3-8, yellow, 2-5mm long; disk corollas 2-3.5mm long. Plains and hills. Statewide. *G. diversifolia* Greene, *G. myriocephala* A. Nels.

<u>Haplopappus</u> Cass. Goldenweed

Perennial herbs or shrubs with mostly alternate or basal leaves; heads 1 to many, with or without ray flowers, the rays yellow; involucral bracts imbricate; receptacle naked or rarely slightly chaffy or hairy; pappus of capillary bristles.

1. Leaves, or most of them, pinnatifid or bipinnatifid, the teeth or lobes usually bristle-tipped ... *H. spinulosus*
1. Leaves entire or toothed
  2. Plants shrubby, never caespitose; heads on leafy peduncles or in leafy-bracted inflorescences
    3. Twigs closely tomentose; ray flowers lacking ... *H. macronema*
    3. Twigs not tomentose; ray flowers usually present ... *H. suffruticosus*
  2. Plants herbaceous, often caespitose, or if shrubby at base, the heads solitary on naked peduncles
    4. Rays lacking or nearly so
      5. Stems and branches closely white-tomentose ... *H. macronema*
      5. Stems and branches not tomentose
        6. Leaves usually strongly reduced upward; heads 1-4 per stem; NW Wyo. ... *H. carthamoides*
        6. Leaves little if at all reduced upward; heads often more than 4 per stem; SE Wyo. ... *H. wardii*
    4. Rays present, usually conspicuous

7. Plants alpine or subalpine; stems usually relatively leafy, the leaves not much reduced; heads solitary on each stem; involucral bracts mostly herbaceous throughout
   8. Plants conspicuously glandular; W. Wyo. _H. lyallii_
   8. Plants not glandular but with multicellular hairs; SE Wyo. _H. pygmaeus_
7. Plants usually not alpine, or if so, the stems not leafy or the heads not solitary or the involucral bracts not herbaceous throughout
   9. Involucre 10-20mm long; disk corollas mostly 7-14mm long
      10. Involucral bracts mostly broadly obtuse or rounded at tip
         11. Leaves mostly over 1cm wide _H. croceus_
         11. Leaves all less than 1cm wide _H. armerioides_
      10. Involucral bracts mostly acute at tip
         12. Plants from rhizomes _H. parryi_
         12. Plants from a stout taproot or caudex
            13. Rays less than 8mm long _H. carthamoides_
            13. Rays 10mm or more long
               14. Involucral bracts green their whole length or nearly so _H. clementis_
               14. Involucral bracts green no more than about their upper half _H. integrifolius_
   9. Involucre 5-10(11)mm long; disk corollas 5-7(7.5)mm long

15. Heads solitary on each stem (rarely 2); leaves mostly all basal or nearly so; involucres, stems, and leaves puberulent or scabrous at most
  16. Involucral bracts mostly acute — *H. acaulis*
  16. Involucral bracts mostly obtuse or rounded — *H. armerioides*
15. Heads several per stem, or if solitary, either the leaves not all basal or the plants pubescent usually with somewhat woolly hairs
  17. Involucral bracts cuspidate or abruptly acuminate; leaves mostly less than 5mm wide; not known from Uinta Co. — *H. multicaulis*
  17. Involucral bracts mostly acute to blunt (rarely acuminate if in Uinta Co.); some leaves often over 5mm wide
    18. Plants from rhizomes — *H. parryi*
    18. Plants from a stout taproot or caudex
      19. Heads mostly over 4 per stem, rarely fewer; involucral bracts green only toward tip — *H. lanceolatus*
      19. Heads 1-4 per stem; involucral bracts often green throughout
        20. Leaves mostly lanceolate, oblong, or elliptic, sometimes toothed; involucral bracts mostly acute; widespread — *H. uniflorus*
        20. Leaves mostly oblanceolate or linear-oblanceolate, entire; involucral bracts acuminate; Uinta Co. — *H. contractus*

Haplopappus acaulis (Nutt.) Gray, Proc. Am. Acad. 7:353. 1868. Plants densely caespitose from a branched caudex and taproot, to 15cm high; leaves all basal or nearly so, the blades oblanceolate or linear, 0.5-6cm long, scabrous or glabrous, entire; heads solitary on each stem; involucre (5)6-10mm long; rays 6-15mm long; disk corollas 5-7.5mm long. Plains, hills, and slopes. Statewide. *Stenotus acaulis* (Nutt.) Nutt., *S. caespitosus* (Nutt.) Nutt., *S. rudis* A. Nels.

Haplopappus armerioides (Nutt.) Gray, Syn. Fl. 1(2):132. 1884. Plants densely caespitose from a branched caudex and taproot, to 2dm high; leaves mostly basal, the blades mostly oblanceolate, 1-9cm long, glabrous or scabrous, the margins entire or scabrous; heads solitary (rarely 2) on each stem; involucre 7-13mm long; rays 10-20mm long; disk corollas 6.5-9mm long. Plains and hills. SW, SE, NE. *Stenotus armerioides* Nutt.

Haplopappus carthamoides (Hook.) Gray, Proc. Acad. Phila. 1863:65. 1863. Plants from a stout taproot, to 6dm high; leaf blades oblanceolate to oblong, 1-20cm long, entire to toothed; heads 1-4; involucre 12-30mm long; rays to 7mm long or lacking; disk corollas 10-14mm long. Plains, hills, meadows, and slopes. Park Co. *Pyrrocoma subsquarrosa* Greene.

Haplopappus clementis (Rydb.) Blake in Tidestrom, Contr. U. S. Nat. Herb. 25:543. 1925. Plants from a stout taproot, to 4dm high; leaf blades oblanceolate or elliptic to oblong-lanceolate, 2-15cm long, entire or slightly toothed; heads 1 to few; involucre 10-20mm long; rays 10-20mm long; disk corollas about 7mm long. Meadows and slopes in the mountains. NW, NE. There appears to be some intergradation with *H. integrifolius*. *Pyrrocoma clementis* Rydb., *P. villosa* Rydb.

Haplopappus contractus Hall, Carn. Inst. Wash. Publ. 389:155. 1928. Plants caespitose from a woody root, to 15cm high; leaf blades oblanceolate or linear-oblanceolate, 1-4cm long, entire; heads solitary on each stem; involucre 9-10mm long; rays 10-12mm long; disk corollas about 6mm long. Plains. Known only from the type collected in 1873 at Ft. Bridger, Uinta Co. *Pyrrocoma acuminata* Rydb.

Haplopappus croceus Gray, Proc. Acad. Phila. 1863:65. 1863. Plants from a thick taproot, to 7dm high; leaf blades oblanceolate to lanceolate or ovate, 2-25cm long, entire or undulate; heads 1-4; involucre 14-20mm long; rays 13-30mm long; disk corollas 7-11mm long. Meadows, slopes, and open woods. Carbon Co. *Pyrrocoma crocea* (Gray) Greene.

Haplopappus integrifolius Porter ex Gray, Syn. Fl. 1(2):128. 1884. Plants from a taproot, to 4dm high; leaf blades oblanceolate to lanceolate, 1-12cm long, entire or nearly so; heads 1 to few; involucre 10-15mm long; rays 1-2cm long; disk corollas 7-10mm long. Hills, slopes, and meadows. Yellowstone Park and Fremont Co. Pyrrocoma integrifolia (Porter ex Gray) Greene.

Haplopappus lanceolatus (Hook.) T. & G., Fl. N. Am. 2:241. 1842. Plants from a taproot, to 5dm high; leaf blades oblanceolate to lanceolate or oblong, 1-15cm long, entire to somewhat spiny-toothed; heads several; involucre 5-10mm long; rays 5-10mm long; disk corollas 5-7mm long. Hills and meadows, often where alkaline. SW, SE. Pyrrocoma lanceolata (Hook.) Greene.

Haplopappus lyallii Gray, Proc. Acad. Phila. 1863:64. 1863. Plants often rhizomatous or with a weakly developed taproot, to 15cm high; leaf blades mostly oblanceolate or spatulate, 1-7cm long, entire; heads solitary on each stem; involucre 7-13mm long; rays 6-15mm long; disk corollas 6-9mm long. Alpine or subalpine. NW, SW. Tonestus lyallii (Gray) A. Nels.

Haplopappus macronema Gray, Proc. Am. Acad. 6:542. 1865. Subshrub to 4dm high; leaf blades mostly oblong-oblanceolate or linear, 1-5cm long, entire or wavy-margined and glandular; heads 1 to few per branch; involucre 9-16mm long; rays lacking; disk corollas 8-11mm long. Rocky places in the mountains. Yellowstone Park. Macronema discoideum Nutt., M. lineare Rydb.

Haplopappus multicaulis (Nutt.) Gray in Parry, Am. Nat. 8:213. 1874. Plants from a branched caudex and taproot, to 10cm high; leaf blades linear or linear-oblanceolate, 0.5-9cm long, entire; heads 1 to several; involucre 7-10mm long; rays 5-13mm long; disk corollas 5-6mm long. Plains and hills. Statewide. Oonopsis multicaulis (Nutt.) Greene, O. argillacea A. Nels.

Haplopappus parryi Gray in Parry, Am. Journ. Sci. II, 33:239. 1862. Plants from spreading rhizomes, to 5dm high; leaf blades oblanceolate or obovate below to lanceolate or ovate above, 1-15cm long, the margins entire or ciliolate; heads 2 to many; involucre 8-12mm long; rays 5-12mm long; disk corollas 7.5-8.5mm long. Woods and slopes. SW, SE. Oreochrysum parryi (Gray) Rydb., Solidago parryi (Gray) Greene.

Haplopappus pygmaeus (T. & G.) Gray in Parry, Am. Journ. Sci. II, 33:239. 1862. Plants from a taproot and caudex, to 6cm high; leaf blades oblanceolate to oblong, 0.5-5cm long, the margins entire or ciliate; heads solitary on each stem; involucre 7-10mm long; rays 6-11mm long; disk corollas 5-7mm long. Alpine and subalpine. Albany Co. Tonestus pygmaeus (T. & G.) A. Nels.

Haplopappus spinulosus (Pursh) DC., Prodr. 5:347. 1836. Herbs or subshrubs from a woody caudex, to 6dm high; leaf blades 1-2 times pinnatifid, 0.5-5cm long; heads few to many; involucre 4-8mm long; rays 6-12mm long; disk corollas 4-6mm long. Plains and hills. SW, SE, NE. Sideranthus spinulosus (Pursh) Sweet.

Haplopappus suffruticosus (Nutt.) Gray, Proc. Am. Acad. 6:542. 1865. Shrub or subshrub to 4dm high; leaf blades oblanceolate or oblong, 0.5-3cm long, glandular, the margins wavy; heads 1 to few per branch; involucre 9-16mm long; rays 1-2cm long; disk corollas 7-11mm long. Rocky places in the mountains. NW, SW. Macronema grindelifolium Rydb.

Haplopappus uniflorus (Hook.) T. & G., Fl. N. Am. 2:241. 1842. Plants from a taproot, to 3dm high; leaf blades mostly lanceolate, oblong, or elliptic, 0.5-13cm long, toothed or entire; heads solitary or occasionally 2-4; involucre 6-11mm long; rays 6-12mm long; disk corollas 5-7mm long. Plains, hills, meadows, and slopes. Statewide. Pyrrocoma uniflora (Hook.) Greene.

Haplopappus wardii (Gray) Dorn, comb. nov. Plants from a woody, branched caudex, to 4dm high; leaf blades narrowly lanceolate to narrowly oblanceolate or oblong, 2-10cm long, entire; heads usually several; involucre 12-20mm long; rays lacking; disk corollas 6-9mm long. Plains and hills. Endemic in SE Wyo. Aplopappus Fremonti Gray var. Wardi Gray, Syn. Fl. 1(2):128. 1884; Aster Wardii (Gray) Kuntze, Rev. Gen. Pl. 317. 1891; Oonopsis Wardi (Gray) Greene, Pittonia 3:46. 1896; O. condensata (A. Nels.) A. Nels.

Helenium L. Sneezeweed

Perennial herbs with alternate, usually glandular-punctate leaves; heads several to many, with ray and disk flowers, the rays yellow or orange; involucral bracts in 1-3 series, often united at base; receptacle naked; pappus of scales.

References: Bierner, M. W. 1972. Brittonia 24:331-355.
1974. Brittonia 26:385-392.

Stem leaves decurrent forming wings on the stem; rays 6-15mm long; pappus 1-2mm long — H. autumnale

Stem leaves not decurrent; rays 12-30mm long; pappus 2-4mm long — H. hoopesii

Helenium autumnale L., Sp. Pl. 886. 1753. Plants to 12dm high; leaf blades mostly lanceolate to oblanceolate, 2-12cm long, toothed or subentire; involucre 7-12mm long, the bracts eventually reflexed; rays 6-15mm long, eventually reflexed, 3 lobed at tip; disk corollas about 3-4mm long. Streambanks and other moist places. SW, SE.

Helenium hoopesii Gray, Proc. Acad. Phila. 1863:65. 1863. Plants to 1m high; leaf blades oblanceolate below to lanceolate or ovate above, 1-20cm long, entire or nearly so; involucre 6-11mm long; rays 12-30mm long; disk corollas about 5mm long. Hills, meadows, slopes, and woods. NW, SW. Dugaldia hoopesii (Gray) Rydb.

Helianthella T. & G. False Sunflower

Taprooted perennial herbs with opposite, or the upper alternate, entire to ciliate leaves; heads 1 to several, with ray and disk flowers, the rays yellow; involucral bracts herbaceous, subequal or imbricate; receptacle chaffy; pappus of fimbriate scales and 2 awns.

Involucral bracts ovate to lanceolate; bracts of receptacle soft ..... H. quinquenervis

Involucral bracts lance-linear to oblong; bracts of receptacle firm ..... H. uniflora

Helianthella quinquenervis (Hook.) Gray, Proc. Am. Acad. 19:10. 1883. Plants to 15dm high; leaf blades obovate or oblanceolate to lanceolate or ovate, 3-30cm long; involucre 12-30mm long, the bracts mostly ovate or lanceolate; rays 2-5cm long; disk corollas about 5-6mm long. Meadows, slopes, and woods in the mountains. NW, NE, SE.

Helianthella uniflora (Nutt.) T. & G., Fl. N. Am. 2:334. 1842. Plants to 1m high; leaf blades lanceolate to oblanceolate, 2-15cm long; involucre 1-2(4)cm long, the bracts lance-linear to oblong; rays 1.5-4.5cm long; disk corollas about 5-6mm long. Hills, slopes, and open woods. NW, SW, SE.

Helianthus L. Sunflower

Annual or perennial herbs with opposite, or the upper alternate, simple leaves; heads 1 to several, with ray and disk flowers, the rays yellow; involucral bracts herbaceous, subequal or imbricate; receptacle chaffy; pappus of deciduous awns and/or scales.

Reference: Heiser, C. B., Jr. et al. 1969. Mem. Torrey Club 22(3):1-218.

1. Plants annual
   2. Central bracts of receptacle with long, white, multicellular hairs at tip; involucral bracts lanceolate or lance-ovate, long-tapering to tip — _H. petiolaris_
   2. Central bracts of receptacle inconspicuously hairy; involucral bracts mostly ovate to ovate-oblong, abruptly contracted above the middle — _H. annuus_
1. Plants perennial
   3. Involucral bracts mostly ovate to lance-ovate, abruptly acute or obtuse; lobes of disk corollas red, purple, or yellow
      4. Lobes of disk corollas red or purple; NE Wyo. — _H. rigidus_
      4. Lobes of disk corollas yellow; SE Wyo. — _H. pumilus_

3. Involucral bracts mostly lanceolate or lance-linear, usually long-attenuate; lobes of disk corollas yellow

5. Leaves usually folded lengthwise at midrib, strictly pinnately veined; extreme E Wyo. _H. maximiliani_

5. Leaves usually not folded, some usually somewhat palmately 3 veined at base in addition to pinnately veined; widespread _H. nuttallii_

Helianthus annuus L., Sp. Pl. 904. 1753. Annual to 2m high; leaf blades ovate to subcordate, 2-30cm long, toothed to subentire; involucre 15-30mm long; rays 2-5cm long; disk corollas about 6-8mm long. Plains, hills, and disturbed areas. Statewide.

Helianthus maximiliani Schrad., Ind. Sem. Hort. Gött. 1835. Rhizomatous perennial to 2m high; leaf blades lanceolate or lance-linear, 2-20cm long, minutely toothed to entire; involucre 1-2cm long; rays 1.5-4cm long; disk corollas about 5-6mm long. Plains, hills, and disturbed areas. Crook and Laramie Cos.

Helianthus nuttallii T. & G., Fl. N. Am. 2:324. 1842. Rhizomatous or tuberous rooted perennial to 2m high; leaf blades mostly lanceolate or lance-linear, 2-16cm long, entire or toothed; involucre 1-2cm long; rays 1.5-3.5cm long; disk corollas about 5-7mm long. Moist places. Statewide.

Helianthus petiolaris Nutt., Journ. Acad. Phila. 2:115. 1821. Annual to 1m high; leaf blades deltoid, lanceolate, or ovate, 1-8cm long, entire or occasionally toothed; involucre 9-15mm long; rays 1.5-4cm long; disk corollas about 5mm long. Plains, hills, and disturbed areas. Statewide.

Helianthus pumilus Nutt., Trans. Am. Phil. Soc. II, 7:366. 1841. Taprooted perennial to 1m high; leaf blades mostly ovate or lanceolate, 2-12cm long, entire to serrate; involucre 8-13mm long; rays 1.5-3cm long; disk corollas about 5-6mm long. Plains, hills, and canyons. SE.

Helianthus rigidus (Cass.) Desf., Cat. Pl. 3:184. 1829. Rhizomatous perennial to 2m high; leaf blades ovate or lanceolate to oblong or oblanceolate, 1-14cm long, serrate to entire; involucre 7-13mm long; rays 1.5-3.5cm long; disk corollas about 6mm long. Plains and hills. NE. *H. laetiflorus* Pers. and *H. scaberrimus* Ell. of authors, *H. subrhomboideus* Rydb.

Heterotheca Cass. Golden Aster

Taprooted perennial herbs with alternate, mostly entire leaves; heads few to many, with ray and disk flowers, the rays yellow; involucral bracts imbricate; receptacle naked (rarely slightly chaffy); pappus of capillary bristles and usually an outer series of short scales or bristles also.

References: Harms, V. L. 1968. Rhodora 70:301-303.
1968. Wrightia 4:8-20.

Upper part of plant hirsute to sericeous-strigose, somewhat canescent, occasionally greenish, not glandular (rarely with sessile glands on leaves); upper leaves soft, the hairs mostly somewhat appressed; outer pappus often inconspicuous *H. villosa*

Upper part of plant less pubescent, greenish, or if as above, then either glandular or the upper leaves rigid or the hairs of leaves somewhat erect; outer pappus usually conspicuous

- Heads somewhat sessile, usually closely subtended by leaves of peduncle which usually do not grade into involucral bracts *H. fulcrata*
- Heads appearing peduncled, not subtended by leaves or else these grading into involucral bracts *H. horrida*

Heterotheca fulcrata (Greene) Shinners, Field & Lab. 19:71. 1951. Plants to 5dm high; leaf blades mostly oblanceolate to elliptic or oblong, 0.5-6cm long; involucre 6-12mm long; rays 5-15mm long; disk corollas about 5-6mm long. Plains, hills, and slopes. NE, SE. *Chrysopsis resinolens* A. Nels., *C. fulcrata* Greene.

Heterotheca horrida (Rydb.) Harms, Wrightia 4:17. 1968. Similar to H. fulcrata except the heads appearing peduncled rather than sessile, the upper leaves often grading into the involucral bracts, and the glands, if present, predominantly sessile rather than raised on hairs. Plains, hills, and slopes. NW, SW, SE. Chrysopsis arida A. Nels.

Heterotheca villosa (Pursh) Shinners, Field & Lab. 19:71. 1951. Plants to 5dm high; leaf blades oblanceolate to lanceolate, 0.5-5cm long; involucre 5-10mm long; rays 6-14mm long; disk corollas about 5-6mm long. Plains, hills, and slopes. NW, NE, SE. Chrysopsis villosa (Pursh) Nutt. ex DC., C. hispida (Hook.) DC., C. bakeri Greene, C. foliosa Nutt., C. mollis Nutt., C. depressa Rydb.

Hieracium L. Hawkweed

Fibrous rooted perennials from a very short rhizome, often with some stellate hairs at least above; leaves alternate or basal, entire to toothed (rarely slightly lobed); heads 1 to many, with all ray flowers; involucral bracts somewhat imbricate; receptacle naked; pappus of capillary bristles.

1. Basal and lowest stem leaves small and early deciduous, the middle leaves larger, the upper ones reduced; involucre with few or no long hairs *H. umbellatum*
1. Basal and lowest stem leaves larger than the progressively reduced middle and upper leaves, or stem leaves lacking; involucre often with many long hairs
   2. Rays white or ochroleucous; stellate hairs lacking *H. albiflorum*
   2. Rays yellow (sometimes drying whitish); stellate hairs usually present at least on involucre
      3. Leaves glabrous or sometimes very short hairy, mostly all basal and usually 10cm or less long *H. gracile*
      3. Leaves mostly somewhat long hairy, at least marginally, mostly with some reduced stem leaves, the lowest ones sometimes over 10cm long
         4. Plants somewhat long-hairy throughout, not glaucous *H. cynoglossoides*
         4. Plants long-hairy below, subglabrous or glandular-puberulent and often glaucous above *H. scouleri*

Hieracium albiflorum Hook., Fl. Bor. Am. 1:298. 1833. Plants to 12dm high; leaf blades oblanceolate to lanceolate, 2-15cm long, the margins entire or slightly toothed; heads several to many (rarely 1); involucre 6-15mm long; corollas white or ochroleucous, 6-15mm long. Woods and slopes. Statewide.

<u>Hieracium</u> <u>cynoglossoides</u> Arv.-Touv., Spicil. Hierac. 20. 1881. Plants to 13dm high; leaf blades oblanceolate to lanceolate or oblong, 2-25cm long, entire; heads few to many; involucre 7-12mm long; corollas yellow or sometimes drying whitish, 10-20mm long. Woods, meadows, hills, and slopes. Statewide.

<u>Hieracium</u> <u>gracile</u> Hook., Fl. Bor. Am. 1:298. 1833. Plants to 4dm high; leaf blades mostly oblanceolate or obovate, 1-10cm long, denticulate or entire; heads 1 to several; involucre 5-10mm long; corollas yellow, sometimes drying whitish, 5-12mm long. Meadows, woods, and slopes in the mountains. Statewide.

<u>Hieracium</u> <u>scouleri</u> Hook., Fl. Bor. Am. 1:298. 1833. Plants to 1m high; leaf blades oblanceolate or elliptic, 2-20cm long, mostly entire; heads few to many; involucre 6-11mm long; corollas yellow, 6-15mm long. Woods, hills, and slopes. Teton Co.

<u>Hieracium</u> <u>umbellatum</u> L., Sp. Pl. 804. 1753. Plants to 12dm high; leaf blades oblanceolate to lanceolate, 1-10cm long, irregularly toothed to entire or sometimes slightly lobed; heads several; involucre 7-13mm long; corollas yellow, 8-20mm long. Woods and thickets. NE, SE. <u>H</u>. <u>canadense</u> Michx.

Hulsea T. & G.

Hulsea algida Gray, Proc. Am. Acad. 6:547. 1865. Taprooted or somewhat rhizomatous, glandular-pubescent perennial herb to 4dm high; leaves alternate, the blades oblanceolate or obovate to oblong, 1-10cm long, crenate to shallowly pinnately lobed; heads solitary on each stem; involucre 13-20mm long, the bracts herbaceous and subequal in 2 or 3 series; receptacle naked; pappus of scales which are connate at base; rays yellow, 7-18mm long; disk corollas about 4-5mm long. Rocky places in the high mountains. Yellowstone Park and Park Co. H. carnosa Rydb.

Hymenopappus L'Her.

Taprooted biennial or perennial herbs with alternate and basal, mostly 1-2 times pinnately divided leaves with linear segments; heads few to many, lacking ray flowers; involucral bracts subequal, in 2-3 series, scarious on margins; receptacle naked; pappus of scales, these sometimes minute.

Reference: Turner, B. L. 1956. Rhodora 58:163-186, 208-242, 250-269, 295-308.

Plants biennial, the roots with a single crown; corollas white (sometimes drying yellowish); Crook Co. *H. tenuifolius*

Plants perennial, the roots with usually several crowns; corollas usually yellow; widespread

Plants of SW Wyo., or if elsewhere, the anthers about 2mm long, the ultimate leaf segments about 5mm or less long, rarely longer, and the pappus mostly 0.1-0.8mm long *H. filifolius*

Plants not in SW Wyo.; anthers mostly 2.5-3mm long; ultimate leaf segments often well over 5mm long; pappus 0.8-2mm long *H. polycephalus*

Hymenopappus filifolius Hook., Fl. Bor. Am. 1:317. 1833. Perennial to 5dm high; leaves 1-8cm long; involucre 5-10mm long; disk corollas yellow or rarely white, 2-5mm long. Plains and hills. Statewide. *H. luteus* Nutt.

Hymenopappus polycephalus Osterh., Torreya 18:90. 1918. Perennial to 6dm high; leaves 1-15cm long; involucre 5-9mm long; disk corollas yellow, 2.5-3.5mm long. Plains and hills. NW, NE, SE.

Hymenopappus tenuifolius Pursh, Fl. Am. Sept. 742. 1814. Biennial to 1m high; leaves 1-15cm long; involucre 5-8mm long; disk corollas white, sometimes drying yellowish, 2.5-3.5mm long. Plains and hills. Crook Co.

<u>Hymenoxys</u> Cass.

Perennial herbs with alternate or basal, entire to pinnately or ternately divided leaves; heads 1 to several, with usually both ray and disk flowers, the rays yellow and 3 lobed at tip; involucral bracts in 2 or 3 series, mostly herbaceous; receptacle naked; pappus of scales.

1. Leaves entire, all basal or nearly so
   2. Involucral bracts sparsely pubescent to glabrous at least toward tip, with thin scarious margins — *H. torreyana*
   2. Involucral bracts usually densely pubescent, the margins sometimes whitish but hardly thin and scarious — *H. acaulis*
1. Leaves, or some of them, ternately or pinnately divided, some on the stems
   3. Plants alpine or subalpine; involucral bracts similar — *H. grandiflora*
   3. Plants not alpine or subalpine; involucral bracts in 2 dissimilar series — *H. richardsonii*

Hymenoxys acaulis (Pursh) Parker, Madroño 10:159. 1950. Plants to 3dm high from an ultimate taproot; leaves all basal, the blades linear to oblanceolate, 0.5-8cm long, entire; heads solitary on each stem; involucre 4-9mm long, the bracts all similar; rays 5-18mm long, very rarely lacking; disk corollas about 4-5mm long. Plains, hills, and slopes. Statewide. Tetraneuris acaulis (Pursh) Greene, T. septentrionalis Rydb., T. simplex A. Nels., T. lanigera Dan., T. brevifolia Greene, T. incana A. Nels., T. eradiata A. Nels., T. epunctata A. Nels., Actinella acaulis (Pursh) Nutt., A. lanata Nutt., A. simplex (A. Nels.) A. Nels., A. incana (A. Nels.) A. Nels., A. eradiata (A. Nels.) A. Nels., A. epunctata (A. Nels.) A. Nels.

Hymenoxys grandiflora (T. & G. ex Gray) Parker, Madroño 10:159. 1950. Plants to 3dm high from a taproot; leaf blades mostly 1-2 times pinnately or ternately divided, 1-6cm long; heads 1 to few; involucre 9-20mm long, the bracts all similar; rays 12-30mm long; disk corollas about 5-6mm long. Alpine and subalpine. Statewide. Rydbergia grandiflora (T. & G. ex Gray) Greene.

Hymenoxys richardsonii (Hook.) Cockerell, Bull. Torrey Club 31:468. 1904. Plants to 2dm high from a taproot; leaf blades mostly ternate with linear segments, rarely simple and linear, 1-10cm long; heads mostly 1-3 per stem; involucre 4-8mm long, the bracts in 2 dissimilar series; rays 4-18mm long; disk corollas about 4mm long. Plains, hills, and slopes. SW, SE, NE. *H. floribunda* (Gray) Cockerell, *H. macrantha* (A. Nels.) Rydb., *H. pumila* (Greene) Rydb., *Hymenopappus ligulaeflorus* A. Nels., *Picradenia ligulaeflora* (A. Nels.) A. Nels., *P. macrantha* A. Nels.

Hymenoxys torreyana (Nutt.) Parker, Madroño 10:159. 1950. Plants to 2dm high from an ultimate taproot; leaves all basal, the blades linear to oblanceolate, 0.5-9cm long, entire; heads solitary on each stem; involucre 5-10mm long, the bracts all similar; rays 8-20mm long; disk corollas about 5mm long. Plains, hills, and slopes. Statewide. *Actinella torreyana* Nutt., *Tetraneuris torreyana* (Nutt.) Greene.

Iva L.

Annual or perennial herbs with opposite, or the upper alternate, simple leaves; heads several to many, lacking ray flowers; disk flowers unisexual, the pistillate few and marginal in each head; involucral bracts in 1 or 2 series, usually subequal; receptacle usually chaffy; pappus lacking; corolla sometimes lacking; anthers scarcely united.

Reference: Bassett, I. J. et al. 1962. Canad. Journ. Bot. 40:1243-1249.

Plants perennial; leaves entire; heads axillary *I. axillaris*
Plants annual; leaves toothed; heads in a panicle *I. xanthifolia*

Iva axillaris Pursh, Fl. Am. Sept. 743. 1814. Poverty Weed. Perennial from a creeping rootstock, to 6dm high; leaf blades obovate or oblanceolate to lanceolate, 0.5-4cm long, entire; involucre 2-5mm long, the 4-6 bracts connate at least at base; pistillate corollas about 1mm long, the staminate about 2.5-3mm long. Plains, hills, and disturbed areas. Statewide.

Iva xanthifolia Nutt., Gen. Pl. 2:185. 1818. Annual to 2m high; leaf blades ovate to deltoid, 1.5-20cm long, coarsely serrate; involucre 1-3mm long, with 5 outer bracts and 5 somewhat membranous inner ones; pistillate corollas lacking or nearly so, the staminate about 2mm long. Disturbed areas. Statewide.

Lactuca L. Lettuce

Annual to perennial herbs with alternate, entire to pinnatifid leaves; heads many, the flowers all ligulate, yellow, blue, or white; involucral bracts imbricate; receptacle naked; pappus of capillary bristles.

1. Achenes with a single conspicuous median nerve on each face, occasionally with an additional, less prominent pair; pappus white ... *L. ludoviciana*
1. Achenes with several evident nerves on each face; pappus white or brownish
  2. Pappus brownish; achene often beakless or nearly so ... *L. biennis*
  2. Pappus white; achene often with a well developed beak
    3. Fruiting involucres mostly 15-20mm long; upper leaves usually not clasping; perennial ... *L. oblongifolia*

3. Fruiting involucres 9-15mm long; upper leaves usually clasping the stem; annual or biennial _L_. _serriola_

Lactuca _biennis_ (Moench) Fern., Rhodora 42:300. 1940. Annual or biennial to 2m high; leaf blades lanceolate to oblanceolate, 0.5-4dm long, pinnatifid or some merely toothed; involucre 10-14mm long; corollas blue to white or sometimes yellow, 7-15mm long. Moist places. Sheridan and Albany Cos. _L_. _spicata_ (Lam.) Hitchc. of authors.

Lactuca _ludoviciana_ (Nutt.) Riddell, Syn. Fl. W. St. 51. 1835. Biennial or short lived perennial to 15dm high; leaf blades oblanceolate to lanceolate, 2-30cm long, toothed to pinnately lobed; involucre 10-22mm long; corollas yellow or sometimes blue, 7-15mm long. Plains, hills, and meadows. NE, SE.

Lactuca _oblongifolia_ Nutt. in Fraser, Cat. No. 47. 1813. Perennial from a deep creeping rootstock, to 1m high; leaf blades oblong to oblanceolate or lanceolate, entire above to pinnately lobed below, 2-18cm long; involucre 10-20mm long; corollas blue, 10-20mm long. Meadows, woods, thickets, and disturbed areas. Statewide. _L_. _pulchella_ (Pursh) DC., _L_. _sylvatica_ A. Nels.

Lactuca serriola L., Cent. Pl. 2:29. 1756. Biennial or winter annual to 15dm high; leaf blades oblanceolate or obovate to lanceolate, 1-25cm long, pinnately lobed or toothed or some rarely entire; involucre 8-15mm long; corollas yellow or drying blue, 6-10mm long. Disturbed areas. NW, NE, SE. *L. scariola* L., *L. integrata* (Gren. & Godr.) A. Nels.

Liatris Gaertn. ex Schreb. Blazing Star

Perennial herbs with alternate, entire or ciliate leaves; heads mostly in a spike-like or raceme-like inflorescence, ray flowers lacking; involucral bracts imbricate in several series; receptacle naked; pappus of barbellate or plumose capillary bristles; corollas pink-purple or white.

References: Gaiser, L. O. 1946. Rhodora 48:165-183, 216-263, 273-326, 331-382, 393-412.
Cruise, J. E. 1964. Canad. Journ. Bot. 42:1445-1455.

Pappus plumose; involucral bracts mostly cuspidate or acuminate at tip *L. punctata*

Pappus barbellate; involucral bracts rounded to rarely acute at tip

Heads 15 or fewer flowered; involucres 8-11mm long
L. lancifolia

Heads mostly 30 or more flowered; involucres 11-20mm long
L. ligulistylis

Liatris lancifolia (Greene) Kittell in Tidestrom & Kittell, Fl. Ariz. & N. M. 370. 1941. Plants to 12dm high; leaf blades punctate, linear, linear-lanceolate, or linear-oblanceolate, 1-25cm long; involucre 8-11mm long, the bracts rounded to rarely acute at tip; flowers mostly 9-15 per head, the corollas about 6-8mm long, the tube glabrous within; pappus barbellate. Plains, hills, and meadows. Goshen Co.

Liatris ligulistylis (A. Nels.) Schum., Just's Bot. Jahresb. 29(1):569. 1903. Plants to 6dm high; leaf blades lance-oblong to oblanceolate, 1-15cm long; involucre 11-20mm long, the bracts with rounded and erose tips; flowers 30 or more per head, the corollas 7-11mm long, the tube glabrous within; pappus barbellate. Plains, hills, and open woods. NE, SE. *Laciniaria ligulistylis* A. Nels.

Liatris punctata Hook., Fl. Bor. Am. 1:306. 1833. Plants to 6dm high; leaf blades punctate, linear or linear-oblanceolate, 1-16cm long; involucre 10-18mm long, the bracts mostly cuspidate or acuminate; flowers 3-9 per head, the corollas 8-12mm long, the tube hairy toward base within; pappus plumose. Plains and hills. NW, NE, SE.

Lygodesmia D. Don Skeletonweed

Perennial herbs with alternate, linear or lance-linear, entire leaves; heads 1 to few, the flowers all ligulate, pink or purple or rarely white; involucral bracts of 4-8 inner long ones and several outer short ones; receptacle naked; pappus of capillary bristles.

Involucres in flower (15)17-23mm long; some leaves usually 5cm or more long, mostly persistent *L. grandiflora*
Involucres mostly 9-17mm long in flower; leaves mostly all less than 5cm long, the larger deciduous *L. juncea*

Lygodesmia grandiflora (Nutt.) T. & G., Fl. N. Am. 2:485. 1843. Plants to 4dm high; leaves 1-15cm long; heads solitary at tip of stem or branches; corollas pinkish or purplish, 2-3.5cm long; involucre (15)17-23mm long. Plains and hills. SW, SE.

Lygodesmia juncea (Pursh) D. Don, Edinb. New Phil. Journ. 6:311. 1829. Plants to 6dm high, much branched; leaves 3-45mm long; heads mostly terminating branches, the corollas pink or very rarely white, 1-2.5cm long; involucre 9-17(18)mm long. Plains, hills, and slopes. Statewide.

Machaeranthera Nees

Annual to perennial herbs from a taproot and often branching caudex; leaves alternate, usually spinulose-tipped, entire to pinnately dissected; heads 1 to many; rays white, blue, pink, rose, or purplish, sometimes lacking; involucre of several series of bracts which are herbaceous or greenish at least at tip; receptacle naked; pappus of capillary bristles.

Reference: Cronquist, A. & D. D. Keck. 1957. Brittonia 9:231-239.

1. Leaves, at least the lower, once or twice pinnatifid _M. tanacetifolia_
1. Leaves entire or merely shallowly lobed or toothed
   2. Plants perennials from a stout woody taproot and usually branched caudex
      3. Rays lacking (very rarely present and white); leaves coarsely spinulose-toothed _M. grindelioides_

3. Rays present, pink, rose, or purplish; leaves entire or spinulose-toothed
   4. Leaves mostly entire; W Wyo. *M. commixta*
   4. Leaves spinulose-toothed; Carbon Co. *M. coloradoensis*

2. Plants annuals, biennials, or short-lived perennials from a slender taproot
   5. Leaves glabrous or nearly so except sometimes on margins, sometimes glandular; SE Wyo. *M. linearis*
   5. Leaves distinctly cinereous-puberulent at least beneath, sometimes also glandular; widespread
      6. Plants often perennial; involucral bracts mostly averaging 1-2mm wide; extreme W Wyo. *M. commixta*
      6. Plants rarely perennial; involucral bracts averaging about 1mm or less wide; widespread *M. canescens*

Machaeranthera canescens (Pursh) Gray, Pl. Wright. 1:89. 1852. Taprooted biennial or short-lived perennial to 6dm high; leaf blades linear, oblanceolate, or oblong, 0.5-7cm long, entire or toothed; heads several to many; involucre 5-10mm long; rays blue-purple or pinkish, 5-13mm long; disk corollas about 5-6mm long. Plains, hills, and slopes. Statewide. *Aster canescens* Pursh, *M. linearis* Rydb., *M. viscosa* (Nutt.) Greene, *M. pulverulenta* (Nutt.) Greene, *M. ramosa* A. Nels.

Machaeranthera coloradoensis (Gray) Osterh., Torreya 27:64. 1927. Perennial from a taproot and usually branched caudex, to 1dm high; leaf blades oblanceolate to oblong or sometimes obovate, 0.5-3cm long, spinulose-toothed; heads 1 per stem; involucre 7-9mm long; rays pink, rose, or purplish, 8-15mm long; disk corollas about 4-5mm long. Mountain slopes and rock outcrops. Carbon Co. *Aster coloradoensis* Gray.

Machaeranthera commixta Greene, Pittonia 4:71. 1899. Similar to *M. canescens* but often perennial with mostly wider involucral bracts and the leaves sometimes obovate. Hills, slopes, and woods. NW, SW. *M. superba* A. Nels. The distinction between *M. canescens* and *M. commixta* is not clear-cut and they might better be treated as conspecific. Also, the isotype (RM) of *M. subalpina* Greene appears very close to *M. commixta*. If the latter is maintained as a species and *M. subalpina* is conspecific with it, the valid name is *M. subalpina*.

Machaeranthera grindelioides (Nutt.) Shinners, Field & Lab. 18:40. 1950. Taprooted perennial with a branched caudex, to 3dm high; leaf blades oblanceolate to oblong, rarely obovate or elliptic, 0.5-5cm long, spinulose-toothed; heads 1 to several; involucre 5-11mm long; rays lacking or very rarely present and white; disk corollas 5-8mm long. Plains and hills. Statewide. *Haplopappus nuttallii* T. & G.

Machaeranthera linearis Greene, Bull. Torrey Club 24:511. 1897. Taprooted annual to 1m high; leaf blades linear, lanceolate, or oblanceolate, 0.5-8cm long, entire or toothed; heads several to many; involucre 6-11mm long; rays white, pink, or purplish, 7-15mm long; disk corollas about 5mm long. Plains and hills. SE. *Aster linearis* (Greene) Cory.

Machaeranthera tanacetifolia (H. B. K.) Nees, Gen. & Sp. Aster. 225. 1832. Taprooted annual to 3dm high; leaves, at least the lower, mostly once or twice pinnatifid, 1-7cm long; heads 1 to many; involucre 6-15mm long; rays blue or purplish, 10-20mm long; disk corollas about 4-5mm long. Plains and hills. NW, NE, SE. *Aster tanacetifolius* H. B. K., *M*. *coronopifolia* (Nutt.) A. Nels.

Madia Mol. Tarweed

Madia glomerata Hook., Fl. Bor. Am. 2:24. 1834. Annual to 8dm high, pubescent throughout, glandular above; leaves mostly linear, 1-7cm long, opposite below, alternate above, entire; heads several, mostly in small clusters; involucre 6-9mm long, the bracts mostly 4 in 1 series; receptacle often with bracts between ray and disk flowers; pappus lacking; rays mostly 1-3, lacking in some heads, about 2-4mm long, yellow; disk corollas about 3mm long. Plains, hills, slopes, and disturbed areas. Statewide.

Malacothrix DC. Desert Dandelion

Taprooted annuals with a basal rosette of toothed to pinnatifid leaves and a few alternate ones on the stem; heads 1 to many, the flowers all ligulate, yellow; involucral bracts somewhat imbricate, the outer series shorter than the inner; receptacle naked or with a few setae; pappus of capillary bristles.

Reference: Williams, E. W. 1957. Am. Midl. Nat. 58:494-512.

Pappus of a ring of inner deciduous bristles and 1-5 outer persistent bristles; plants usually glandular ... *M. torreyi*
Pappus bristles all alike, deciduous; plants usually not glandular ... *M. sonchoides*

Malacothrix sonchoides (Nutt.) T. & G., Fl. N. Am. 2:486. 1843. Plants to 25cm high; leaf blades 1-8cm long; involucre 6-13mm long; corollas 6-18mm long. Plains and hills. NW, SE.

Malacothrix torreyi Gray, Proc. Am. Acad. 9:213. 1874. Similar to *M. sonchoides* except with 1-5 persistent pappus bristles, usually glandular, and with mostly more slender involucral bracts. Plains and hills. Carbon Co. *M. runcinata* A. Nels.

Matricaria L.

Glabrous or glabrate annuals with alternate, pinnately dissected leaves; heads few to many, either lacking ray flowers or these present and white; involucral bracts somewhat imbricate, with scarious margins; receptacle naked; pappus a short crown or lacking.

Rays present ... *M. chamomilla*
Rays lacking ... *M. matricarioides*

Matricaria chamomilla L., Sp. Pl. 891. 1753. Plants to 4dm high; leaves 1-6cm long; involucre 2-4mm long; rays 5-12mm long; disk corollas 5 toothed, about 1.5mm long. Disturbed areas. Reported from Teton Co.

Matricaria matricarioides (Less.) Porter in Porter & Britt., Mem. Torrey Club 5:341. 1894. Pineapple Weed. Plants to 3dm high; leaves 0.5-5cm long; heads lacking ray flowers; involucre 3-5mm long; disk corollas 4 toothed, about 1mm long. Disturbed areas. NW, NE, SE.

Microseris D. Don

Reference: Chambers, K. L. 1957. Contr. Dudley Herb. 5:57-68.

Microseris nutans (Hook.) Schultz-Bip., Pollichia 22-24:309. 1866. Taprooted perennial herb to 5dm high; leaves mostly toward base, the blades linear to elliptic or oblanceolate, 2-20cm long, entire to runcinate-pinnatifid; heads 1 to few, the flowers all ligulate; involucre 1-2cm long, the outer bracts shorter than the inner; receptacle naked; pappus of short narrow scales tipped by a plumose bristle; corollas yellow, often drying purplish, 10-20mm long. Plains, hills, slopes, and meadows. Statewide. Ptilocalais nutans (Hook.) Greene.

<u>Nothocalais</u> (Gray) Greene

Perennial taprooted herbs with basal, entire or ciliate margined leaves; heads 1 per stem, the flowers all ligulate and yellow or drying purplish; involucral bracts somewhat imbricate or sometimes subequal; receptacle naked; pappus of capillary bristles or slender scales.

Reference: Chambers, K. L. 1957. Contr. Dudley Herb. 5:57-68.

Pappus of 40-80 members; involucral bracts lanceolate to linear-lanceolate ..... *N. cuspidata*
Pappus of 10-30 members; involucral bracts ovate to broadly lanceolate ..... *N. nigrescens*

Nothocalais cuspidata (Pursh) Greene, Bull. Calif. Acad. Sci. 2:55. 1886. Plants to 3dm high; leaf blades linear to oblong-lanceolate or oblong-elliptic, 3-25cm long, often crisp-margined; involucre 17-30mm long; pappus of capillary bristles and bristle-like scales; corollas 15-30mm long. Plains and hills. NE, SE. *Microseris cuspidata* (Pursh) Schultz-Bip.

Nothocalais nigrescens (Henderson) Heller, Muhlenbergia 1:8. 1900. Plants to 3dm high; leaf blades linear, oblong-elliptic, or oblong-oblanceolate, 2-20cm long, rarely a few reduced ones on stem; involucre 14-23mm long; pappus of bristle-like scales; corollas 15-30mm long. Plains, hills, slopes, and meadows. SW, NW, NE. *Microseris nigrescens* Henderson.

Onopordum L. Thistle

Reference: Dress, W. J. 1966. Baileya 14:74-86.

Onopordum acanthium L., Sp. Pl. 827. 1753. Spiny biennial to 3m high, the stems winged; leaves alternate, the blades lanceolate, elliptic, or oblanceolate, lobed or coarsely toothed, 0.3-4dm long; heads several, lacking ray flowers; involucre 18-30mm long, the bracts all spine-tipped; receptacle appearing scaly, the scales connate and forming a honeycomb pattern; pappus of capillary bristles; corollas 15-30mm long. Disturbed areas. Crook and Platte Cos.

Palafoxia Lag.

Reference: Baltzer, E. A. 1944. Ann. Mo. Bot. Gard. 31:249-278.

Palafoxia macrolepis (Rydb.) Cory, Rhodora 48:86. 1946. Pubescent annual to 4dm high, glandular above; leaves mostly alternate, the blades linear to lanceolate, 1-5cm long, entire; heads several; involucre 7-11mm long, the bracts in 2 or 3 series; receptacle naked; rays lacking; pappus of scales; disk corollas rose or pinkish, 6-13mm long, with long linear lobes. Plains and hills. Converse Co. *Othake texanum* (DC.) Bush var. *macrolepis* (Rydb.) Ammerman.

Parthenium L. Feverfew

Reference: Rollins, R. C. 1950. Contr. Gray Herb. 172:1-73.

Parthenium alpinum (Nutt.) T. & G., Fl. N. Am. 2:285. 1842. Acaulescent perennial herb to 5cm high; leaf blades linear to oblanceolate or obovate, 0.3-3cm long, entire, canescent; heads sessile or nearly so amongst leaves; involucre about 5-6mm long, the bracts in 2 or 3 series; receptacle somewhat chaffy; marginal flowers pistillate with achenes, without rays, the pappus of 2 teeth; disk flowers staminate, or perfect but without achenes, the corollas about 4-5mm long. Plains and hills. SE. *Bolophyta alpina* Nutt.

<u>Petasites</u> Mill. Butterbur

<u>Petasites sagittatus</u> (Banks ex Pursh) Gray, Bot. Calif. 1:407. 1876. Rhizomatous perennial to 1m high, somewhat tomentose; leaves basal (stems bracteate), the blades cordate or sagittate, 2-35cm long, dentate to subentire; heads several to many; involucre 6-10mm long, the bracts mostly subequal in 1 series, sometimes with a few much reduced ones at base; receptacle naked; pappus of capillary bristles; flowers usually unisexual or hermaphroditic without achenes; rays white or drying yellowish, 7-15mm long, sometimes lacking; disk corollas 6-10mm long. Wet places. Albany Co.

<u>Petradoria</u> Greene

Reference: Anderson, L. C. 1963. Trans. Kans. Acad. Sci. 66:632-684.

<u>Petradoria pumila</u> (Nutt.) Greene, Erythea 3:13. 1895. Taprooted perennial herb with a woody caudex, to 2dm high; leaves basal and alternate, the blades mostly linear to oblanceolate or oblong, 1-10cm long, the margins entire or scabrous; heads several to many; involucre 4-7mm long, the bracts mostly in vertical ranks; receptacle naked or bristly; pappus of capillary bristles; ray flowers 0-3 per head, pistillate, yellow, 4-9mm long; disk flowers 1-5 per head, bisexual but with abortive achenes, the corollas 4-6.5mm long. Plains and hills. Sweetwater and Uinta Cos.

Picradeniopsis Rydb. ex Britt.

Reference: Stuessy, T. F. et al. 1973. Brittonia 25:40-56.

Picradeniopsis oppositifolia (Nutt.) Rydb. ex Britt., Man. 1008. 1901. Perennial to 25cm high from a creeping rootstock; leaves opposite at least below, 1-6cm long, mostly divided into a few narrow divisions or toothed, the divisions sometimes again divided; heads several; receptacle naked; pappus of scales; involucre 4-8mm long, the bracts in 2 or 3 series; rays yellow, 2-7mm long. Plains and hills. Statewide. *Bahia oppositifolia* (Nutt.) DC.

Platyschkuhria (Gray) Rydb.

Reference: Ellison, W. L. 1971. Brittonia 23:269-279.

Platyschkuhria integrifolia (Gray) Rydb., Bull. Torrey Club 33:155. 1906. Perennial herb from a caudex and somewhat creeping rootstock, to 6dm high; leaves basal and alternate (stem leaves rarely lacking), the blades mostly elliptic to obovate, 1-8cm long, entire or very rarely trifid; heads 1 to several per stem; involucre 7-12mm long, the bracts mostly subequal in 2 series; receptacle naked; pappus of scales; rays yellow, 6-16mm long; disk corollas 3-7mm long. Plains and hills. NW, SW. *Bahia nudicaulis* Gray.

Prenanthes L. Rattlesnake-root

Prenanthes racemosa Michx., Fl. Bor. Am. 2:84. 1803. Perennial herb to 1m high; leaves alternate, the blades ovate above to obovate or oblanceolate below, 2-18cm long, entire to dentate; heads several to many, the flowers all ligulate; involucre 10-14mm long, the bracts mostly subequal in 1 or 2 series, often with a short outer series also; receptacle naked; pappus of capillary bristles; rays pink, rose, or purplish, 8-17mm long. Moist places. Albany Co.

Psilocarphus Nutt.

Reference: Cronquist, A. 1950. Res. Stud. St. Coll. Wash. 18:71-89.

Psilocarphus brevissimus Nutt., Trans. Am. Phil. Soc. II, 7:340. 1840. White-woolly annual to 10cm high; leaves mostly opposite, the blades mostly lance-oblong, lance-linear, or linear-oblong, 3-20mm long, entire; heads 1 to few, lacking ray flowers; involucre lacking, the heads subtended by foliage leaves; pappus lacking; marginal pistillate flowers each enclosed by a saccate, woolly, receptacular bract 2.5-4mm long which bears a hyaline appendage just below tip, only the style exserted; central flowers few, bractless, and functionally staminate. Drying shores of ponds. Campbell Co.

Ratibida Raf. Coneflower

Reference: Richards, E. L. 1968. Rhodora 70:348-393.

Ratibida columnifera (Nutt.) Woot. & Standl., Contr. U. S. Nat. Herb. 19:706. 1915. Taprooted perennial herb to 12dm high; leaves alternate, mostly pinnately divided, 1-15cm long; heads several; involucre 3-10mm long, the bracts mostly linear in 1 series; receptacle columnar, chaffy; pappus of 1 or 2 teeth or lacking; rays yellow or purple, 7-35mm long; disk corollas 1-3mm long. Plains, hills, and disturbed areas. Statewide. R. columnaris (Sims) D. Don.

Rudbeckia L. Coneflower

Biennial or perennial herbs with alternate, entire to pinnatifid or palmatifid leaves; heads 1 to few, with or without ray flowers; involucral bracts in 2 or 3 series, herbaceous, often reflexed; receptacle cylindrical, columnar, or hemispherical, chaffy; pappus a short crown or lacking.

References: Perdue, R. E., Jr. 1957. Rhodora 59:293-299.
Jones, G. N. 1957. Madroño 14:131-133.

Rays lacking — R. occidentalis

Rays present

  Leaves mostly laciniate-pinnatifid or palmatifid — R. laciniata

  Leaves all toothed or entire — R. hirta

Rudbeckia hirta L., Sp. Pl. 907. 1753. Hirsute biennial or short lived perennial to 1m high; leaf blades oblanceolate to lanceolate or oblong, 2-20cm long, entire or toothed; involucre 12-26mm long; rays yellow or rarely orange, 15-40mm long; pappus lacking; disk corollas about 3-4mm long. Plains, meadows, open woods, and disturbed areas. NE, SE. R. flava Moore.

Rudbeckia laciniata L., Sp. Pl. 906. 1753. Perennial to 2m high; leaf blades mostly laciniate-pinnatifid or palmatifid, 4-30cm long; involucre 8-25mm long; rays yellow, 2-6cm long; pappus a short crown or lacking; disk corollas about 3-5mm long. Moist places. NE, SE. R. ampla A. Nels.

Rudbeckia occidentalis Nutt., Trans. Am. Phil. Soc. II, 7:355. 1840. Perennial to 2m high; leaf blades mostly ovate, 3-25cm long, entire or toothed, very rarely pinnatifid; involucre 10-27mm long; ray flowers lacking; pappus a short crown; disk corollas about 3-4mm long. Moist places in or near the mountains. NW, SW.

Senecio L. Groundsel

Perennial herbs, or rarely annual, with alternate or basal, entire to bipinnatifid leaves; heads 1 to many, with or without ray flowers, the rays yellow to rarely orange or reddish; involucral bracts herbaceous, sometimes black-tipped, subequal and usually uniseriate, sometimes with much smaller bracteoles at base; receptacle naked; pappus of capillary bristles. There are undoubtedly several genera included here, but it seems best to maintain the traditional treatment until more comprehensive studies are undertaken.

References: Greenman, J. M. 1915. Ann. Mo. Bot. Gard. 2:573-626.
1916. Ann. Mo. Bot. Gard. 3:85-194.
1917. Ann. Mo. Bot. Gard. 4:15-36.
1918. Ann. Mo. Bot. Gard. 5:37-108.
Barkley, T. M. 1960. Leafl. West. Bot. 9:97-113.
1962. Trans. Kans. Acad. Sci. 65:318-408.
1963. Rhodora 65:65-67.
1968. Brittonia 20:267-284.
1968. Southw. Nat. 13:109-115.
Ediger, R. I. 1970. Sida 3:504-524.
Packer, J. G. 1972. Canad. Journ. Bot. 50:507-518.

1. Plants annual weeds with some pinnately lobed leaves which are little if at all reduced upward; rays lacking *S. vulgaris*

1. Plants native perennials, the leaves lobed or not, often much reduced upward; rays present or lacking

2. Plants with a woody base, the leaves mostly linear or divided into linear segments, the stems leafy with the leaves little if at all reduced upward; eastern plains

3. Leaves mostly pinnately divided into linear segments *S. riddellii*

3. Leaves mostly simple and linear, occasionally with a pair of lobes toward base *S. spartioides*

2. Plants not as above

4. Rays lacking or minute

5. Leaves, at least the upper ones, lobed about halfway or more to midrib

6. Stem leaves with mostly rounded lobes; inflorescence a corymbose cyme *S. debilis*

6. Stem leaves with mostly pointed lobes; inflorescence subumbellate

7. Leaves usually moderately lobed; heads usually 1-6(10); corolla lobes red or orange; Big Horn Mts. *S. pauciflorus*

7. Leaves usually lobed well over halfway to midrib; heads usually 6-20; corolla lobes yellow; Park Co. *S. indecorus*

5. Leaves entire to coarsely toothed

8. Heads 1-8 per stem, nodding, the involucres 10-18mm long ... *S. bigelovii*

8. Heads 10 or more per stem, mostly erect, the involucres 3-10mm long

9. Leaves sharply and irregularly toothed; stems usually less than 5mm wide at base ... *S. rapifolius*

9. Leaves entire or nearly so; stems usually over 5mm wide at base ... *S. hydrophilus*

4. Rays present

10. Stems leafy, the leaves little if at all reduced upward, basal tuft usually lacking

11. Leaf blades mostly triangular and on petioles, coarsely toothed ... *S. triangularis*

11. Leaf blades not triangular, petioled or not, toothed or lobed to subentire

12. Leaf blades mostly lobed halfway or more to midrib, usually acute at tip ... *S. eremophilus*

12. Leaf blades subentire or toothed or rarely lobed less than halfway to midrib, acute to rounded at tip

13. Leaf blades all 4.5cm or less long, often rounded at tip ... *S. fremontii*

13. Leaf blades, or some of them, well over 5cm long, usually acute at tip ... *S. serra*

10. Stems not leafy or the leaves usually reduced upward, basal tuft usually present

14. Leaves, at least the upper ones, lobed
  15. Plants somewhat woolly or tomentose nearly throughout at flowering time
    16. Leaves mostly all entire or subentire, only 1 or 2 upper ones lobed ... *S. canus*
    16. Leaves usually all toothed or lobed
      17. Leaves mostly all similarly lobed; Albany Co. ... *S. fendleri*
      17. Leaves usually not all similarly lobed, the lower usually merely toothed; NE Wyo. ... *S. plattensis*
  15. Plants glabrous or essentially so at flowering time, rarely with a few persisting patches of tomentum
    18. Leaves mostly all similarly lobed, rarely broader than lanceolate or oblanceolate; Albany Co. ... *S. fendleri*
    18. Leaves usually not all similarly lobed, the lower toothed or less deeply lobed, often obovate or deltoid-orbicular; widespread
      19. Leaves mostly similar throughout, the blades elliptic to obovate, 4.5cm or less long ... *S. fremontii*
      19. Leaves mostly dimorphic, the lower different in size or shape from the upper, the blades variously shaped, sometimes over 4.5cm long

20. Heads 1 or 2, rarely 3; plants with slender scaly rhizomes; rays sometimes orange; northern and western mountains *S. cymbalarioides*

20. Heads only occasionally as few as 3; plants with a caudex or sometimes a stout rhizome; rays mostly yellow; widespread

21. Cauline leaves mostly lobed about halfway or less to midrib, usually clasping; peduncles of heads rarely over 3cm long; plants mostly alpine and subalpine or high montane *S. dimorphophyllus*

21. Cauline leaves usually more deeply lobed or not clasping; peduncles often over 3cm long; plants mostly lower in elevation

22. Basal leaf blades predominantly truncate or subcordate at base, mostly toothed

23. Rays usually orange-red; S Wyo. *S. crocatus*

23. Rays usually yellow; widespread *S. pseudaureus*

22. Basal leaf blades predominantly tapering at base, sometimes lobed

24. Plants taprooted, on dry plains and hills in SW Wyo.; basal leaves often lobed *S. multilobatus*

24. Plants usually from a rhizome or caudex and without a taproot, often in moist places, widespread; basal leaves usually not lobed

25. Basal leaves subentire to coarsely dentate especially above middle, often long-tapering to base; plants somewhat taprooted; plains, hills, and lower slopes in E Wyo. *S. tridenticulatus*

25. Basal leaves various; plants mostly not taprooted; mostly in the mountains, rarely on the plains; widespread

26. Basal leaves thickish and subsucculent at least when fresh *S. streptanthifolius*

26. Basal leaves thin, not subsucculent *S. pauperculus*

14. Leaves entire or toothed

27. Plants somewhat woolly or tomentose at flowering time, at least in inflorescence or toward base

28. Largest mature heads 1.5-2cm wide, the involucre 12mm or more long; Yellowstone Park *S. megacephalus*

28. Largest mature heads rarely over 1.5cm wide, the involucre 10(12)mm or less long; widespread

29. Leaves greenish, all basal or nearly so, the blades 5cm or less long and 14mm or less wide, rarely with a few much reduced leaves on stem about 1mm wide at their middle; plants 25cm or less high, usually in the mountains
S. *werneriifolius*

29. Leaves basal and on the stem, or if only basal, some usually over 5cm long or over 14mm wide or else the plants over 25cm high or the leaves densely white or silver pubescent; habitat various

30. Plants alpine or subalpine, mostly completely covered with loose cobwebby hairs which often cover coarse multicellular hairs; disk corollas usually orange; Park Co. S. *fuscatus*

30. Plants without the above combination of characters

31. Heads mostly 30 or more per stem; mountains of Lincoln Co. S. *atratus*

31. Heads rarely as many as 30 per stem; habitat various; widespread

32. Pubescence rather loose and often sparse; plants from fibrous roots, lacking a rhizome or caudex or nearly so
S. *integerrimus*

32. Pubescence often close and dense; plants with a well developed caudex or rhizome

33. Stems usually several to many from a branched caudex; leaves usually white or silvery pubescent; involucral bracts not black-tipped _S. canus_

33. Stems usually 1 from a simple caudex or rhizome; leaves often greenish; involucral bracts usually black-tipped

34. Pubescence of leaves usually conspicuous without magnification; involucral bracts commonly about 21, rarely as few as 13 _S. sphaerocephalus_

34. Pubescence of leaves usually not conspicuous without magnification; involucral bracts commonly about 13, rarely as many as 21 _S. lugens_

27. Plants glabrous or nearly so at flowering time

35. Heads mostly (15)25 or more per stem _S. hydrophilus_

35. Heads less than 15 per stem

36. Leaves mostly 4.5cm or less long and 2cm or less wide; stems usually many from a caudex; heads upright _S. fremontii_

36. Leaves often larger; stems often solitary; heads sometimes nodding
    37. Heads mostly 1-3(4) per stem, usually nodding; involucral bracts not black-tipped ..... *S. amplectens*
    37. Heads mostly 1-12 per stem, not nodding; involucral bracts usually black-tipped ..... *S. crassulus*

Senecio amplectens Gray in Parry, Am. Journ. Sci. II, 33:240. 1862. Plants to 4dm high from a rhizome or caudex and fibrous roots, glabrous or slightly floccose when young; basal leaf blades oblanceolate or obovate to elliptic, 2-15cm long, dentate to entire; cauline leaves reduced or lacking; heads mostly 1-4, nodding; involucre 9-17mm long; rays 1-2.5cm long. High mountain slopes. NW. *S. holmii* Greene, *Ligularia amplectens* (Gray) Weber, *L. holmii* (Greene) Weber.

Senecio atratus Greene, Pittonia 3:105. 1896. Plants to 8dm high with stout rootstocks, floccose-tomentose; basal leaf blades elliptic, oblanceolate, or obovate, 4-15cm long, entire to dentate; heads many; involucre 5-8mm long; rays 5-12mm long. Rocky slopes. Lincoln Co.

Senecio bigelovii Gray in Torrey, Pac. R. R. Rep. 4:111. 1857. Plants to 1m high from fibrous roots, glabrate to pilose; lower leaf blades oblanceolate to lance-elliptic, 6-20cm long, dentate to subentire; cauline leaves reduced; heads 1 to several; involucre 10-18mm long; rays lacking. Mountain slopes and woods. Albany and Carbon Cos. Ligularia bigelovii (Gray) Weber.

Senecio canus Hook., Fl. Bor. Am. 1:333. 1834. Plants to 4dm high from a branched caudex or taproot, usually tomentose; lower leaf blades oblanceolate, elliptic, or obovate, 1-7cm long, entire to toothed; heads several; involucre 5-11mm long; rays 6-15mm long. Plains, hills, and slopes. Statewide. S. hallii Britt., S. purshianus Nutt., S. laramiensis A. Nels.

Senecio crassulus Gray, Proc. Am. Acad. 19:54. 1883. Plants to 7dm high with fibrous roots from a short caudex or rhizome, glabrous or nearly so; lower leaf blades oblanceolate or obovate to elliptic, 1.5-15cm long, entire to dentate; cauline leaves reduced and becoming lanceolate; heads several or rarely 1; involucre 5-13mm long; rays 6-20mm long. Slopes, meadows, and woods in the mountains. NW, SW, SE. S. semiplexicaulis Rydb.

<u>Senecio crocatus</u> Rydb., Bull. Torrey Club 24:299. 1897. Plants to 7dm high from a caudex, glabrous or nearly so except in axils of bracts; lower leaf blades oblanceolate to subcordate, 2-8cm long, entire to crenate-dentate; upper leaves more elongate and lobed; heads few to many; involucre 6-8mm long; rays 6-13mm long, orange-red to deep yellow. Mountain meadows and slopes. Albany and Sweetwater Cos.

<u>Senecio cymbalarioides</u> Buek, Gen., Sp. & Syn. Cand. 2:VI. 1840. Plants to 4dm high with a slender rhizome, glabrous; basal leaf blades mostly obovate or suborbicular, 0.5-3cm long, mostly crenate; cauline leaves more elongate, usually lobed; heads 1 or sometimes 2 or 3; involucre 5-9mm long; rays 7-17mm long. Alpine and subalpine or occasionally lower. SW, NW, NE. <u>S. subnudus</u> DC. of authors. Packer puts <u>S. ovinus</u> Greene in synonymy here.

<u>Senecio debilis</u> Nutt., Trans. Am. Phil. Soc. II, 7:408. 1841. Plants to 5dm high with fibrous roots from a short caudex, glabrous or slightly floccose-tomentose when young; lower leaf blades elliptic to deltoid-ovate, 2-8cm long, entire to lobed; upper leaves reduced and lobed; heads several to many; involucre 5-9mm long; rays lacking; disk corollas about 5-6mm long. Moist meadows. NW, SW, SE. <u>S. discoideus</u> (Hook.) Britt. of authors.

Senecio dimorphophyllus Greene, Pittonia 4:109. 1900. Plants to 4dm high from a caudex, glabrous or glabrate; basal leaf blades cordate or ovate to obovate, 1-4cm long, entire or crenulate; upper leaves more elongate and lobed; heads few to several; involucre 5-9mm long; rays 6-15mm long. Meadows, slopes, and woods in the high mountains. NW, SE.

Senecio eremophilus Richards. in Frankl., Narr. 1st Journ. 2:31. 1824. Plants to 12dm high with a taproot or short caudex, glabrous or glabrate; leaf blades elliptic or oblanceolate, 2-15cm long, laciniate-pinnatifid or sometimes a few merely toothed; heads several to many; involucre 5-10mm long; rays 6-20mm long. Woods and slopes in or near the mountains. NW, NE, SE. *S. ambrosioides* Rydb.

Senecio fendleri Gray, Pl Fendl. 108. 1849. Plants to 6dm high from a rhizome or caudex, tomentose to rarely glabrate; leaf blades lanceolate to oblanceolate or oblong, 1-9cm long, mostly pinnatifid or sometimes a few toothed; heads few to many; involucre 4-7mm long; rays 5-12mm long. Hills and slopes. Albany Co. *S. nelsonii* Rydb.

Senecio fremontii T. & G., Fl. N. Am. 2:445. 1843. Plants to 4dm high with a taproot and branched caudex, glabrous or glabrate, the stems often decumbent or trailing; leaf blades mostly elliptic or oblanceolate to obovate, 0.5-4.5cm long, toothed or shallowly lobed; heads several; involucre 7-12mm long; rays 6-18mm long. Rocky places in the mountains. Statewide.

Senecio fuscatus Hayek, Allg. Bot. Zeit. 23:4. 1917. Plants to 2dm high with a short caudex, usually densely and loosely tomentose; basal leaf blades elliptic, oblanceolate, or obovate, 1-7cm long, entire or irregularly toothed; cauline leaves reduced, becoming lanceolate; heads few to several; involucre 5-10mm long; rays 8-20mm long. Alpine and subalpine. Park Co. S. bivestitus Cronq.

Senecio hydrophilus Nutt., Trans. Am. Phil. Soc. II, 7:411. 1841. Plants to 1.5m high with fibrous roots and a hollow stem, glabrous and often glaucous; basal leaf blades lanceolate to oblanceolate or oblong, 1-3dm long, entire or nearly so; cauline leaves reduced, becoming lanceolate or oblong; heads many; involucre 5-10mm long; rays 4-10mm long or lacking. Wet places. NW, SW; SE.

Senecio indecorus Greene, Fl. Francis. 470. 1897. Plants to 8dm high with a short caudex and fibrous roots, glabrous or slightly floccose-tomentose when young; basal leaf blades elliptic to deltoid-ovate, 1-9cm long, toothed to lobed; cauline leaves more elongate and incised-pinnatifid; heads several to many; involucre 6-10mm long; rays lacking or very short. Wet meadows, woods, and bogs. Park Co.

Senecio integerrimus Nutt., Gen. Pl. 2:165. 1818. Plants to 7dm high with fibrous roots, pubescent or glabrate; basal leaf blades elliptic to obovate or rarely linear, 1-20cm long, entire to irregularly dentate; cauline leaves more elongate; heads few to several; involucre 5-10(12)mm long; rays 6-15mm long. Plains, hills, open woods, and slopes. Statewide. *S. perplexus* A. Nels., *S. dispar* A. Nels.

Senecio lugens Richards. in Frankl., Narr. 1st Journ. 747. 1823. Plants to 5dm high with a short rhizome or caudex and fibrous roots, tomentulose or glabrate; basal leaf blades elliptic to oblanceolate or obovate, 2-20cm long, denticulate; cauline leaves much reduced; heads few to several; involucre 5-9mm long; rays 7-20mm long. Woods, meadows, and slopes in the mountains. Statewide. *S. glaucescens* Rydb.

Senecio megacephalus Nutt., Trans. Am. Phil. Soc. II, 7:410. 1841. Plants to 6dm high with a stout caudex or short rhizome, tomentose to glabrate; lower leaf blades oblanceolate or elliptic, 2-20cm long, entire or denticulate; upper leaves reduced; heads 1-5; involucre 12-16mm long; rays 15-25mm long. Rocky places in the mountains. Yellowstone Park. S. solitarius Rydb. The type locality remains uncertain. Nuttall cites "On the plains of the Platte, towards the Rocky Mountains." On the label of the type is "R. Mts & Platte." The species is not found on the plains and has not been collected south of Yellowstone Park, well north of Nuttall's route of 1834. The type is in early fruit, a condition not expected until late July or August. It may have been collected in or near the Pioneer Mountains of central Idaho about August 15.

Senecio multilobatus T. & G. ex Gray, Pl. Fendl. 109. 1849. Plants to 4dm high from a taproot, glabrous or glabrate; basal leaf blades oblanceolate or obovate, 1-7cm long, pinnatifid or sometimes dentate; cauline leaves more elongate and pinnatifid; heads few to many; involucre 5-9mm long; rays 5-13mm long. Plains and hills. SW. S. uintahensis (A. Nels.) Greenm.

Senecio pauciflorus Pursh, Fl. Am. Sept. 529. 1814. Plants to 5dm high with a short caudex and fibrous roots, glabrous or floccose-tomentose when young; basal leaf blades ovate-elliptic to suborbicular or cordate, 1-5cm long, subentire, toothed, or rarely lobed; cauline leaves more elongate and toothed to pinnatifid; heads few to several; involucre 6-10mm long; rays lacking or very short. Meadows, slopes, and cliffs in the mountains. Big Horn and Sheridan Cos. S. discoideus (Hook.) Britt.

Senecio pauperculus Michx., Fl. Bor. Am. 2:120. 1803. Plants to 6dm high with a short caudex or rhizome, slightly floccose-tomentose to glabrate; basal leaf blades oblanceolate or elliptic to suborbicular, 0.7-5cm long, crenate, serrate, or subentire; cauline leaves more elongate and pinnatifid; heads few to several; involucre 5-9mm long; rays 5-15mm long, rarely lacking. Moist places. Statewide. S. tweedyi Rydb., S. balsamitae Muhl. ex Willd., S. flavovirens Rydb.

Senecio plattensis Nutt., Trans. Am. Phil. Soc. II, 7:413. 1841. Plants to 6dm high from a short caudex and fibrous roots, floccose-tomentose; basal leaf blades deltoid-ovate to elliptic, 1-6cm long, toothed or sometimes lobed; cauline leaves more elongate and lobed; heads several; involucre 5-8mm long; rays 8-15mm long. Plains and hills. NE.

Senecio pseudaureus Rydb., Bull. Torrey Club 24:298. 1897. Plants to 7dm high with a short rhizome or caudex and fibrous roots, glabrous or glabrate; basal leaf blades mostly ovate, deltoid, or cordate, 0.5-8cm long, crenate-serrate; cauline leaves more elongate, lobed at least at base; heads few to many; involucre 5-8mm long; rays 6-14mm long. Moist places. NW, NE, SE. S. flavulus Greene.

Senecio rapifolius Nutt., Trans. Am. Phil. Soc. II, 7:409. 1841. Plants to 5dm high with fibrous roots from a short caudex, glabrous or glabrate and often glaucous; basal leaf blades oblanceolate or obovate to deltoid-ovate, 2-13cm long, mostly irregularly toothed; cauline leaves becoming lanceolate or oblong and usually reduced; heads several to many; involucre 3-7mm long; rays lacking. Hills, slopes, and rocky places. NE, SE.

Senecio riddellii T. & G., Fl. N. Am. 2:444. 1843. Plants to 1m high from a woody base, glabrous; leaves 1-10cm long, mostly pinnately divided into linear segments; heads few to many; involucre 7-12mm long; rays 1-2cm long. Sandy plains. NE, SE.

Senecio serra Hook., Fl. Bor. Am. 1:333. 1834. Plants to 2m high with fibrous roots from a caudex, often rhizomatous also, glabrous or puberulent; leaf blades oblanceolate to lanceolate, 3-20cm long, serrate to subentire; heads many; involucre 5-9mm long; rays 5-20mm long. Moist places. Statewide. *S. admirabilis* Greene.

Senecio spartioides T. & G., Fl. N. Am. 2:438. 1843. Plants to 8dm high, somewhat woody at base, glabrous or puberulent; leaf blades mostly linear, 1-12cm long, entire or sometimes few toothed or lobed toward base; heads many; involucre 7-12mm long; rays 8-20mm long. Plains and hills. SE.

Senecio sphaerocephalus Greene, Pittonia 3:106. 1896. Plants to 8dm high with a thick rhizome, tomentose to nearly glabrate; basal leaf blades oblanceolate or obovate to elliptic, 2-20cm long, entire or denticulate; cauline leaves much reduced; heads few to many; involucre 5-10mm long; rays 6-15mm long. Moist places in the mountains. SW, NW, NE. *S. altus* Rydb.

Senecio streptanthifolius Greene, Erythea 3:23. 1895. Plants to 6dm high from a caudex or short rhizome, glabrous or slightly floccose-tomentose when young; lower leaf blades elliptic to suborbicular, 1-6cm long, toothed to entire; upper leaves more elongate and usually lobed; heads few to many; involucre 5-10mm long; rays 6-13mm long. Woods, slopes, and meadows in or near the mountains. Statewide. *S. acutidens* Rydb., *S. cymbalarioides* Nutt., *S. longipetiolatus* Rydb., *S. rubricaulis* Greene, *S. rydbergii* A. Nels.

Senecio triangularis Hook., Fl. Bor. Am. 1:332. 1834. Plants to 15dm high with fibrous roots, glabrous or puberulent; leaf blades mostly triangular or nearly so, 2-18cm long, 1-8cm wide, coarsely toothed; heads few to many; involucre 7-10mm long; rays 7-20mm long. Moist places in or near the mountains. Statewide.

Senecio tridenticulatus Rydb., Bull. Torrey Club 27:175. 1900. Plants to 3dm high from a taproot, glabrous to slightly floccose-tomentulose; lower leaf blades mostly oblanceolate, 1-8cm long, entire to coarsely toothed; heads few to many; involucre 6-10mm long; rays 6-15mm long. Open woods, plains, and slopes. NE, SE.

Senecio vulgaris L., Sp. Pl. 867. 1753. Annual to 4dm high with a taproot or fibrous roots, hairy or glabrate; leaf blades oblanceolate or obovate to elliptic, 1-10cm long, coarsely toothed to bipinnatifid; heads few to many; involucre 5-8mm long; rays lacking; disk corollas about 5mm long. Disturbed areas. Albany and Crook Cos.

Senecio werneriifolius (Gray) Gray, Proc. Am. Acad. 19:54. 1883. Plants to 25cm high with a branched caudex, tomentulose to glabrate; leaves mostly basal, the blades oblanceolate or elliptic to obovate, 0.5-5cm long, entire or nearly so; heads 1 to few; involucre 5-10mm long; rays 5-15mm long. Rocky places in the mountains. NW, SE. S. saxosus Klatt, S. petrocallis Greene, S. perennans A. Nels.

<u>Shinnersoseris</u> Tomb

Reference: Tomb, A. S. 1973. Sida 5:183-189.

<u>Shinnersoseris</u> <u>rostrata</u> (Gray) Tomb, Sida 5:186. 1973. Taprooted annual to 8dm high; lower leaves opposite and deciduous, the upper alternate, the blades linear, 1.5-13cm long, entire; heads several to many, the flowers all ligulate; involucre 12-20mm long, the bracts linear or lance-linear, the inner series long, the outer series very short; receptacle naked; pappus of capillary bristles; corollas pink or rose, 7-17mm long. Sandy plains and hills. Goshen Co. <u>Lygodesmia</u> <u>rostrata</u> (Gray) Gray.

<u>Solidago</u> L. Goldenrod

Perennial herbs with a rhizome or caudex and alternate, entire or toothed leaves; heads few to many; involucral bracts usually somewhat imbricate in several series; receptacle naked; pappus of capillary bristles; rays yellow, rather short.

References: Kapoor, B. M. & J. R. Beaudry. 1966. Canad. Journ. Genet. Cytol. 8:422-443.

Croat, T. 1972. Brittonia 24:317-326.

1. Plants with well developed, slender, creeping rhizomes; basal leaves not well developed in most species
   2. Leaves punctate, sometimes obscurely so, mostly less than 1cm wide, without a basal cluster; rays mostly 15-30 per head; inflorescence corymbose
      3. Inflorescence usually compact; involucral bracts broadly lanceolate and somewhat blunt at tip ... *S. graminifolia*
      3. Inflorescence often interrupted; involucral bracts narrowly lanceolate and somewhat pointed at tip ... *S. occidentalis*
   2. Leaves not punctate (rarely so but then with a basal cluster of leaves), some often over 1cm wide; rays mostly 13 or fewer per head, rarely as many as 17; inflorescence mostly racemose, paniculiform, or cymose
      4. Stems glabrous below the inflorescence (rarely pubescent and with a basal cluster of leaves); leaf surfaces usually glabrous
         5. Rays mostly 5-8mm long; stems pubescent below inflorescence ... *S. multiradiata*
         5. Rays mostly 3-5mm long; stems only rarely pubescent below inflorescence
            6. Stems puberulent in the inflorescence, often glaucous; plants mostly 5-20dm high; rays mostly about 13 per head ... *S. gigantea*

6. Stems often glabrous throughout or nearly so, not glaucous; plants mostly 2-5(9)dm high; rays usually about 8 per head *S. missouriensis*

4. Stems pubescent at least between middle and inflorescence; basal cluster of leaves usually lacking, the surfaces pubescent to subglabrous

7. Rays usually about 8 or fewer per head, 3-6mm long; middle stem leaves often 4 times or less as long as wide

8. Involucral bracts mostly broadest near middle and obtuse at tip *S. mollis*

8. Involucral bracts mostly broadest at base and acute at tip *S. sparsiflora*

7. Rays usually about 13 per head, 1-4mm long; middle stem leaves mostly over 4 times as long as wide *S. canadensis*

1. Plants with usually a short, stout rhizome or a caudex, rarely with slender rhizomes; basal leaves usually well developed

9. Leaves glabrous although sometimes with ciliate margins

10. Lower leaves with ciliate-margined petioles; rays mostly about 13 per head *S. multiradiata*

10. Lower leaves without the petioles ciliate-margined; rays mostly about 8 per head *S. spathulata*

9. Leaves pubescent with short spreading hairs or puberulent

11. Involucral bracts somewhat longitudinally striate; achenes glabrous or nearly so; basal leaves mostly 2-8cm wide *S. rigida*

11. Involucral bracts not striate; achenes pubescent throughout; basal leaves mostly 0.3-2cm wide

 12. Disk flowers mostly 5-9 per head, the rays about as many or more; inflorescence of somewhat secund branches or elongate and often nodding at tip *S. nemoralis*

 12. Disk flowers mostly 8-16 per head, the rays usually fewer; inflorescence usually relatively broad, usually not nodding nor with secund branches *S. nana*

Solidago canadensis L., Sp. Pl. 878. 1753. Plants to 1m high with creeping rhizomes; basal leaves lacking or reduced; cauline leaf blades lanceolate to oblanceolate, 1-15cm long, serrate to entire; inflorescence somewhat triangular; involucre 2-5mm long; rays 1-4mm long. Open, often moist places or occasionally in woods. Statewide. *S. elongata* Nutt.

Solidago gigantea Ait., Hort. Kew. 3:211. 1789. Similar to *S. canadensis* except to 2m high, the stems often glaucous and glabrous below inflorescence, the leaf surface glabrous rather than pubescent, and the involucral bracts mostly abruptly tapered to tip rather than long-tapering. Open moist places. Statewide. *S. serotina* Ait.

<u>Solidago</u> <u>graminifolia</u> (L.) Salisb., Prodr. 199. 1796. Plants to 1m high with creeping rhizomes; basal leaves lacking or reduced; cauline leaf blades mostly lance-linear or oblong, 2-12cm long, the margins mostly entire and scabrous; inflorescence somewhat flat-topped; involucre 3-5mm long; rays 1.5-3.5mm long. Open moist places. Albany and Platte Cos. <u>S</u>. <u>camporum</u> (Greene) A. Nels., <u>Euthamia</u> <u>graminifolia</u> (L.) Nutt. ex Cass.

<u>Solidago</u> <u>missouriensis</u> Nutt., Journ. Acad. Phila. 7:32. 1834. Plants to 5(9)dm high with creeping rhizomes and often also a caudex; basal leaf blades oblanceolate, 2-20cm long, entire and ciliolate or sometimes toothed; cauline leaves reduced, becoming lanceolate or linear; inflorescence somewhat triangular or oblong; involucre 3-5mm long; rays 2-4mm long. Plains, hills, slopes, and open woods. Statewide. <u>S</u>. <u>concinna</u> A. Nels.

<u>Solidago</u> <u>mollis</u> Bartl., Ind. Sem. Hort. Gött. 1836:5. 1836. Plants to 6dm high with creeping rhizomes; basal leaves lacking; cauline leaf blades oblanceolate or obovate to lance-ovate, 1-9cm long, entire or toothed; inflorescence somewhat oval or triangular; involucre 3-6mm long; rays 3-5mm long. Plains, hills, and open woods. Statewide.

<u>Solidago</u> <u>multiradiata</u> Ait., Hort. Kew. 3:218. 1789. Plants to 5dm high with a rhizome or branched caudex; lower leaf blades mostly oblanceolate, 1-12cm long, entire or toothed; cauline leaves reduced; inflorescence somewhat oval or elongate; involucre 4-7mm long; rays 5-8mm long. Woods and slopes in the mountains. Statewide. <u>S</u>. <u>dilatata</u> A. Nels., <u>S</u>. <u>corymbosa</u> Nutt.

<u>Solidago</u> <u>nana</u> Nutt., Trans. Am. Phil. Soc. II, 7:327. 1840. Plants to 4dm high with a stout rhizome or branched caudex; basal leaf blades mostly oblanceolate or obovate, 1-8cm long, entire or toothed; cauline leaves reduced; inflorescence somewhat hemispheric, fan shaped, or elongate; involucre 3-6mm long; rays 3-6mm long. Plains, hills, slopes, and open woods. SW, NW, NE.

<u>Solidago</u> <u>nemoralis</u> Ait., Hort. Kew. 3:213. 1789. Plants to 6dm high with a branched caudex; basal leaf blades mostly oblanceolate, 1-12cm long, entire or toothed; cauline leaves reduced; inflorescence somewhat elongate; involucre 3-6mm long; rays 3-6mm long. Plains, hills, and open woods. NE, SE. <u>S</u>. <u>diffusa</u> A. Nels., <u>S</u>. <u>pulcherrima</u> A. Nels.

Solidago occidentalis (Nutt.) T. & G., Fl. N. Am. 2:226. 1842. Plants to 2m high with creeping rhizomes; basal leaves lacking; cauline leaf blades mostly linear or oblong, 1-9cm long, the margins entire and scabrous; inflorescence often scattered along stem or compact and flat-topped; involucre 3-5mm long; rays 2-5mm long. Open moist places. Carbon Co. *Euthamia occidentalis* Nutt.

Solidago rigida L., Sp. Pl. 880. 1753. Plants to 8dm high with a simple or branched caudex; lower leaf blades obovate to elliptic or ovate, 3-15cm long, entire or slightly toothed; upper leaves reduced, mostly ovate or lance-ovate; inflorescence somewhat flat-topped and broad or oval; involucre 5-7mm long; rays 3-8mm long. Plains, hills, bottomlands, and disturbed areas. NE, SE. *Oligoneuron rigidum* (L.) Small.

Solidago sparsiflora Gray, Proc. Am. Acad. 12:58. 1876. Plants to 8dm high with rhizomes; basal leaf blades oblanceolate, 2-7cm long, entire or slightly serrate; cauline leaves reduced; inflorescence triangular or elongate; involucre 3-5mm long; rays 3-6mm long. Rocky slopes and canyons. Statewide. *S. trinervata* Greene.

Solidago _spathulata_ DC., Prodr. 5:339. 1836. Plants to 8dm high, with a stout rhizome or caudex; basal leaf blades oblanceolate or obovate, 1-12cm long, toothed to subentire; cauline leaves reduced; inflorescence somewhat elongate or capitate; involucre 4-6mm long; rays 4-8mm long. Plains, hills, slopes, and woods mostly in or near the mountains. Statewide. _S_. _decumbens_ Greene.

Sonchus L. Sow Thistle

Annual or perennial herbs with alternate, toothed to pinnatifid, usually prickly-margined and auriculate leaves; heads several; flowers all ligulate, yellow; involucral bracts usually imbricate in several series; receptacle naked; pappus of capillary bristles.

References: Hsieh, T. et al. 1972. Am. Journ. Bot. 59:789-796.
Boulos, P. L. 1973. Bot. Notiser 126:155-196.

1. Plants annual; heads mostly 1.5-2.5cm wide in flower including rays
   2. Auricles at base of leaves acute; mature achenes transversely tuberculate-rugulose — _S_. _oleraceus_
   2. Auricles at base of leaves rounded; mature achenes not rugulose — _S_. _asper_

1. Plants perennial with deep, horizontal, rhizome-like roots; heads mostly 2.5-5cm wide in flower including rays

3. Involucres and peduncles with coarse gland-tipped hairs *S. arvensis*

3. Involucres and peduncles glabrous or slightly tomentose *S. uliginosus*

Sonchus arvensis L., Sp. Pl. 793. 1753. Perennial with deep rhizome-like roots, to 2m high; leaf blades oblanceolate or oblong to elliptic, 3-30cm long, the lower and middle ones pinnately lobed or pinnatifid; involucre 15-24mm long, glandular-pubescent; corollas 1-3cm long. Disturbed areas. Sweetwater Co.

Sonchus asper (L.) Hill, Herb. Brit. 1:47. 1769. Taprooted annual to 1m high; leaf blades ovate to obovate, 2-20cm long, toothed to occasionally pinnatifid; involucre 8-15mm long; corollas 3-15mm long. Disturbed areas. SW, SE, NE.

Sonchus oleraceus L., Sp. Pl. 794. 1753. Similar to *S. asper* except the leaf auricles acute rather than rounded, the lower leaves definitely petioled, and the achenes transversely tuberculate-rugulose when mature. Disturbed areas. NE, SE.

Sonchus uliginosus Bieb., Fl. Taur.-Cauc. 2:238. 1808. Similar to *S. arvensis* except lacking glandular hairs on the involucre and peduncles and averaging slightly smaller throughout. Disturbed areas. NW, SW, SE.

Stephanomeria Nutt. Skeletonweed

Annual to perennial herbs with alternate, entire to bipinnatifid leaves; heads several to many; flowers all ligulate, pink or lavender or sometimes white; involucral bracts in 1 subequal series but also with some much shorter outer bracts; receptacle naked; pappus of capillary bristles, these plumose at least above.

Reference: Gottlieb, L. D. 1972. Madroño 21:463-481.

1. Plants annual or biennial; pappus plumose only on the upper 1/3 to 2/3 — *S. exigua*
1. Plants perennial; pappus plumose to base or nearly so
   2. Pappus bristles, at least the main axis, brownish; Platte Co. — *S. pauciflora*
   2. Pappus bristles white; widespread
      3. Leaves, at least the lower, mostly runcinate-pinnatifid; achenes often pitted or rugose — *S. runcinata*
      3. Leaves entire or toothed (rarely a few lobed); achenes smooth or nearly so — *S. tenuifolia*

Stephanomeria exigua Nutt., Trans. Am. Phil. Soc. II, 7:428. 1841. Annual or biennial to 6dm high; leaf blades lanceolate to oblanceolate or linear, the larger 1-5cm long, the lower toothed to bipinnatifid, the upper scale-like; involucre 5-10mm long; corollas 7-14mm long; pappus not plumose to base. Plains and hills. The type was apparently collected in the Green River Basin by Nuttall.

Stephanomeria pauciflora (Torr.) A. Nels. in Coult. & Nels., New Man. Rocky Mts. 588, 610. 1909. Perennial to 6dm high usually with a woody caudex; leaf blades linear to elliptic or oblanceolate, the larger 1-7cm long, entire, toothed, or runcinate-pinnatifid, the upper often scale-like; involucre 9-13mm long; corollas 8-15mm long; pappus plumose to within about 1mm of base. Plains and hills. Platte Co. (Brittonia 26:210. 1974).

Stephanomeria runcinata Nutt., Trans. Am. Phil. Soc. II, 7:428. 1841. Taprooted perennial to 3dm high; leaf blades linear, oblong, or oblanceolate, the larger 1-7cm long, runcinate-pinnatifid to entire, the upper scale-like; involucre 9-13mm long; corollas 9-16mm long; pappus plumose to base or nearly so. Plains, hills, and slopes. Statewide.

Stephanomeria tenuifolia (Torr.) Hall, Univ. Calif. Publ. Bot. 3:256. 1907. Perennial to 7dm high with a taproot, caudex, or creeping rhizome; leaf blades linear to oblanceolate, the larger 1-8cm long, entire or toothed or rarely a few lobed, the upper often scale-like; involucre 7-11mm long; corollas 8-15mm long; pappus plumose to base or nearly so. Plains, hills, and slopes, often where rocky. NW, SE.

Tanacetum L. Tansy

Perennial herbs with alternate and often basal, pinnately dissected to ternate or toothed or rarely entire leaves; heads in a corymb-like or capitate inflorescence or solitary; ray flowers small or lacking; outer flowers pistillate; involucral bracts mostly imbricate, dry and usually scarious-margined; receptacle naked or hairy; pappus a short crown or lacking.

1. Leaves pinnately compound, mostly well over 4cm long T. vulgare
1. Leaves merely 3-5 lobed or cleft from tip or else entire, 4cm or less long
  2. Heads solitary on each stem; Albany Co. T. simplex
  2. Heads usually several per stem, sometimes in a dense globose cluster and appearing like 1; widespread
    3. Heads sessile in a dense globose head; lobes of leaves mostly well over 3mm long T. capitatum
    3. Heads peduncled or subsessile in a corymb or loose head; lobes of leaves usually 3mm or less long T. nuttallii

Tanacetum capitatum (Nutt.) T. & G., Fl. N. Am. 2:415. 1843. Plants to 2dm high, slightly woody at base; leaves clustered toward base, 5-40mm long, mostly deeply 3-5 cleft, the lobes sometimes again cleft; inflorescence of heads a dense globose head; involucre 3-5mm long; corollas about 2.5-3mm long, the rays poorly if at all developed. Plains and hills. NW, SW, SE.

Tanacetum nuttallii T. & G., Fl. N. Am. 2:415. 1843. Plants to 2dm high, somewhat woody at base; leaves clustered toward base, the blades cuneate, 3-15mm long, 3 toothed or shallowly lobed at tip, becoming entire above; inflorescence of heads a corymb or loose head; involucre 3-6mm long; corollas about 2mm long, the rays poorly if at all developed. Plains and hills. SW. Sphaeromeria argentea Nutt.

Tanacetum simplex A. Nels., Bull. Torrey Club 26:484. 1899. Plants to 12cm high; leaves clustered toward base, the blades linear and entire or some 2 or 3 toothed or lobed toward tip, 3-40mm long; heads solitary on each stem; involucre 3-7mm long; corollas about 2.5-3mm long, the rays usually lacking. Plains and hills. Endemic in Albany Co.

Tanacetum vulgare L., Sp. Pl. 844. 1753. Plants to 15dm high with a stout rhizome; leaves 1-2 times pinnatifid, 3-20cm long, punctate; ray flowers lacking; involucre 4-6mm long; corollas about 1.5-2mm long. Disturbed areas. SW, NW, NE.

__Taraxacum__ Wiggers Dandelion

Taprooted perennial herbs with basal, entire to pinnatifid leaves; heads solitary on each stem, with all ray flowers, these yellow or drying purplish or pinkish; involucral bracts biseriate, the outer usually shorter and often reflexed; receptacle naked; pappus of capillary bristles.

Reference: Fernald, M. L. 1933. Rhodora 35:369-386.

1. Plants introduced and weedy; leaves mostly lobed more than halfway to midrib; rarely alpine
   2. Achenes red, purple, or reddish-brown at maturity, the beak mostly 0.5-3 times as long as body; leaves usually deeply cut throughout, without an enlarged terminal segment; inner involucral bracts often corniculate __T. laevigatum__
   2. Achenes olive to brown at maturity, the beak mostly 2.5-4 times as long as body; leaves mostly moderately cut, sometimes with an enlarged terminal segment; inner involucral bracts usually not corniculate __T. officinale__
1. Plants native, mostly in the mountains; either the plants alpine or the leaves mostly lobed less than halfway to midrib
   3. Achenes red or reddish-brown at maturity, usually sharply quadrangular; inner involucral bracts rarely corniculate __T. eriophorum__

3. Achenes olive to brown or black at maturity, at least below the uppermost part, mostly obscurely quadrangular; inner involucral bracts sometimes corniculate
   4. Achenes olive to light brown; inner involucral bracts often corniculate ... *T. ceratophorum*
   4. Achenes blackish or dark brown; inner involucral bracts rarely corniculate ... *T. lyratum*

Taraxacum ceratophorum (Ledeb.) DC., Prodr. 7:146. 1838. Plants to 4dm high; leaf blades oblanceolate or elliptic, 2-20cm long, toothed or slightly lobed; involucre 10-25mm long, the inner bracts often corniculate, the outer appressed to spreading and usually wider; corollas 5-20mm long; body of achene 3-5mm long, gray-brown to olive, the beak 2-4 times as long as body. Moist places in the mountains. Statewide. *T. dumetorum* Greene.

Taraxacum eriophorum Rydb., Mem. N. Y. Bot. Gard. 1:454. 1900. Similar to *T. ceratophorum* except the involucral bracts rarely corniculate and the mature achenes red or reddish-brown and often sharply quadrangular. Moist places mostly in the mountains. Statewide. *T. ammophilum* A. Nels. ex Greene, *T. angustifolium* Greene.

Taraxacum laevigatum (Willd.) DC., Cat. Hort. Monspel. 149. 1813. Plants to 3dm high; leaf blades elliptic to oblanceolate, 3-20cm long, pinnately lobed nearly to midrib; involucre 1-2cm long, the inner bracts usually somewhat corniculate, the outer appressed to reflexed and about as wide as inner; corollas 8-20mm long; body of achene 2.5-4mm long, red, reddish-brown, or reddish-purple, the beak 0.5-3 times as long as body. Disturbed areas. NE, SE.

Taraxacum lyratum (Ledeb.) DC., Prodr. 7:148. 1838. Plants to 10cm high; leaf blades elliptic or oblanceolate, 1-6cm long, entire to pinnatifid; involucre 6-16mm long, the bracts often blackish, rarely corniculate, the outer appressed to reflexed; corollas 3-12mm long; body of achene 2-5mm long, dark brown to blackish (or reddish at tip), the beak about as long as body. Alpine and subalpine. Statewide. *T. scopulorum* (Gray) Rydb.

Taraxacum officinale Weber in Wiggers, Prim. Fl. Holsat. 56. 1780. Common Dandelion. Plants to 5dm high; leaf blades oblanceolate or elliptic, 3-30cm long, mostly runcinate-pinnatifid; involucre 12-30mm long, the inner bracts usually not corniculate, the outer reflexed and slightly if at all wider; corollas 8-25mm long; body of achene 3-4mm long, gray-brown to olive-brown, the beak 2.5-4 times as long as body. Disturbed areas. Statewide. *T. mexicanum* DC.

Tetradymia DC. Horsebrush

Low, somewhat canescent or tomentose shrubs with alternate, often fascicled, entire leaves; heads few to several, lacking ray flowers; involucre of 4-6 subequal bracts; receptacle naked; pappus of capillary bristles.

Reference: Strother, J. L. 1974. Brittonia 26:177-202.

Plants spiny
  Flowers 5-9 per head; branches tomentose; heads solitary in upper axils ... _T. spinosa_
  Flowers 4 per head; branches glabrate in age; heads clustered at ends of branches ... _T. nuttallii_
Plants not spiny ... _T. canescens_

Tetradymia canescens DC., Prodr. 6:440. 1838. Plants not spiny, to 7dm high; primary leaves linear to oblanceolate, 0.5-3cm long; involucre 6-10mm long, with usually 4 bracts; flowers 4 per head; corollas 7-15mm long. Plains, hills, and slopes. Statewide. _T. inermis_ Nutt., _T. multicaulis_ A. Nels.

Tetradymia nuttallii T. & G., Fl. N. Am. 2:447. 1843. Plants spiny, to 1m high; leaves mostly fascicled in axils of spines, oblanceolate or linear-oblanceolate, 0.5-2cm long; involucre 4-10mm long, with 4 or 5 bracts; flowers 4 per head; corollas 8-14mm long. Plains and hills. SW, SE.

<u>Tetradymia</u> <u>spinosa</u> H. & A., Bot. Beechey Voy. 360. 1839. Plants spiny, to 13dm high; leaves mostly fascicled in axils of spines, linear, subterete, 2-12mm long; involucre 8-13mm long, with 5 or 6 bracts; flowers 5-9 per head; corollas 8-15mm long. Plains and hills. NW, SW, SE.

<u>Thelesperma</u> Less. Greenthread

Annual to perennial herbs with opposite (or alternate above or rarely all basal), mostly pinnately divided leaves; heads 1 to several, long-peduncled; involucral bracts biseriate and dimorphic, the inner united 1/3 to 2/3 their length; receptacle chaffy; rays yellow or lacking; pappus of 2 often retrorsely barbed awns or a few teeth or scales.

References: Alexander, E. J. 1955. N. Am. Fl. II, 2:65-69.
Shinners, L. H. 1966. Sida 2:348.

Rays conspicuous — <u>T</u>. <u>filifolium</u>

Rays inconspicuous or lacking
- Leaves mostly toward base; pappus mostly less than 1.5mm long; NW Wyo. — <u>T</u>. <u>marginatum</u>
- Leaves scattered along stem; pappus mostly over 1.5mm long; SE Wyo. — <u>T</u>. <u>megapotamicum</u>

<u>Thelesperma</u> <u>filifolium</u> (Hook.) Gray, Journ. Bot. & Kew Misc. 1:252. 1849. Annual to short lived perennial to 45cm high; leaf blades 1-7cm long, once or twice pinnately divided into linear segments; inner involucral bracts 7-10mm long; rays 10-16mm long; disk corollas 4-6mm long. Plains and hills. Laramie Co. <u>T</u>. <u>trifidum</u> (Poir.) Britt. and <u>T</u>. <u>ambiguum</u> Gray of authors, <u>T</u>. <u>intermedium</u> Rydb.

<u>Thelesperma</u> <u>marginatum</u> Rydb., Mem. N. Y. Bot. Gard. 1:421. 1900. Perennial to 2dm high; leaf blades 1-7cm long, once or twice pinnately divided into linear segments; inner involucral bracts 7-10mm long; rays lacking; disk corollas 4-7mm long. Plains, hills, and slopes. Park and Fremont Cos.

<u>Thelesperma</u> <u>megapotamicum</u> (Spreng.) Kuntze, Rev. Gen. 3(3):182. 1898. Perennial to 7dm high; leaf blades 1-12cm long, once or twice pinnately divided into linear segments; inner involucral bracts 5-10mm long; rays lacking or rarely poorly developed; disk corollas 5-8.5mm long. Plains and hills. SE. <u>T</u>. <u>gracile</u> (Torr.) Gray.

Townsendia Hook. Daisy

Taprooted annual to perennial herbs with alternate or all basal, entire leaves; heads several or solitary, sometimes sessile amongst basal leaves; involucral bracts somewhat imbricate, the middle usually green, the margins usually fringed or erose; receptacle naked or slightly honeycombed with minute bristly scales; rays white, pink, blue, or violet; pappus of disk flowers of barbellate, bristle-like scales, of the ray flowers similar or reduced and chaffy.

Reference: Beaman, J. H. 1957. Contr. Gray Herb. 183:1-151.

1. Involucral bracts with rigid acuminate tips and broad hyaline margins, the central green portion only about 1/3 the bract width; usually caulescent biennial; plains and hills of SE Wyo. _T. grandiflora_
1. Involucral bracts usually without rigid acuminate tips, occasionally with broad hyaline margins; caulescent or not, annual to perennial; widespread
   2. Plants with mostly erect stems usually over 5cm high (except some depauperate alpine specimens); rays blue or purple, rarely whitish; involucral bracts acuminate; heads mostly 1.5-3.5cm wide; montane to alpine _T. parryi_
   2. Plants not as above

3. Plants densely caespitose, to 4cm high, with long, tangled, loose, woolly hairs; heads 16mm or less wide; pappus deciduous on mature achenes; plains and hills *T. spathulata*

3. Plants not as above

4. Involucral bracts in 5-7 series, linear to narrowly lanceolate or rarely wider, the tips mostly acuminate to acute

5. Pappus of disk flowers usually over 8mm long, those of rays similar; midveins of leaves usually conspicuous *T. exscapa*

5. Pappus of disk flowers usually less than 8mm long, the ray pappus sometimes shorter than disk pappus; midveins of leaves often not conspicuous

6. Lowest leaves usually obovate or broadly spatulate; largest heads over 17mm wide; plants well up in the mountains *T. condensata*

6. Lowest leaves usually all linear or narrowly oblanceolate; heads mostly less than 17mm wide; plants mostly on the plains or foothills

7. Involucral bracts long-tapering at tip, the tip with a tuft of long cilia or rarely just scarious; common *T. hookeri*

7. Involucral bracts, at least the outer, not long-tapering at tip, acute or blunt, with or without cilia; very rare *T. leptotes*

4. Involucral bracts in 2-5 series, broadly lanceolate to ovate or elliptic, the tips obtuse to acute

8. Leaves or achenes (at least above) or both glabrate *T. alpigena*

8. Leaves and achenes conspicuously pubescent

9. Stems densely pubescent, white or grayish-white, the stem surface obscured *T. incana*

9. Stems less densely pubescent, usually green or purple, the stem surface usually readily visible

10. Hairs of achenes glochidiate (appearing like a small ball at tip of hair under low magnification); plants of the southern plains and deserts *T. strigosa*

10. Hairs of achenes not glochidiate; plants of the NW mountains or rarely the western plains

11. Pappus deciduous near maturity of achenes; plants mostly villous with crinkled loose hairs *T. condensata*

11. Pappus persistent; plants mostly strigose with straight stiff hairs *T. florifer*

<u>Townsendia</u> <u>alpigena</u> Piper, Bull. Torrey Club 27:394. 1900. Caespitose perennial to 8cm high, strigose or glabrate in age; leaves oblanceolate to obovate, 1-4cm long; heads sessile or on scapes to 5cm long; involucre 6-12mm long; rays white, pink, blue, or violet, 6-15mm long; disk corollas 3-5.5mm long. Slopes and ridges in the mountains. NW, SW. <u>T</u>. <u>dejecta</u> A. Nels. The name <u>T</u>. <u>montana</u> was not validly published by Jones (Zoe 4:262. 1893) because he stated, "Other forms that may eventually prove to be <u>T</u>. <u>scapigera</u> I have given the provisional name of <u>T</u>. <u>montana</u>." See Article 34 of International Code.

<u>Townsendia</u> <u>condensata</u> Parry ex Gray, Proc. Am. Acad. 16:83. 1880. Caespitose perennial to 6cm high, strigose to villous or woolly; leaves mostly basal, oblanceolate or obovate or broadly spatulate, 5-30mm long; heads sessile or solitary or corymbose at tips of short stems; involucre 7-18mm long; rays white, pink, or lavender, 8-16mm long; disk corollas 4-6.5mm long. Mountain slopes. Park Co. This name was not validly published by Eaton (Am. Nat. 8:213. 1874) since he thought the plant might be <u>T</u>. <u>incana</u> Nutt. and stated, "If not, it may properly bear the name which Dr. Parry has proposed." <u>T</u>. <u>anomala</u> Heiser.

Townsendia exscapa (Richards.) Porter in Porter & Britt., Mem. Torrey Club 5:321. 1894. Caespitose perennial to 5cm high, usually strigose; leaves basal, linear-oblanceolate, 1-6cm long; heads sessile or nearly so amongst leaves; involucre 10-22mm long; rays white or pink, 12-25mm long; disk corollas 6-12.5mm long. Plains and hills. Crook and Laramie Cos.

Townsendia florifer (Hook.) Gray, Proc. Am. Acad. 16:84. 1880. Caulescent taprooted annual to short-lived perennial to 2dm high, strigose to subglabrate; basal leaves oblanceolate to obovate, 1-6cm long; cauline leaves similar or reduced; heads few to several near tips of stems; involucre 6.5-13mm long; rays white to pinkish or lavender, 8-18mm long; disk corollas 3-6.5mm long; ray pappus usually reduced. Plains, hills, and lower mountain slopes. Teton Co. T. watsonii Gray.

Townsendia grandiflora Nutt., Trans. Am. Phil. Soc. II, 7:306. 1840. Usually caulescent taprooted biennial to 2dm high, strigose; leaf blades linear or oblanceolate, 1-9cm long; heads mostly near ends of stems; involucre 10-18mm long; rays white or with a broad pink or purple stripe, 12-23mm long; disk corollas 4-6mm long; ray pappus reduced. Plains and hills. SE.

<u>Townsendia</u> <u>hookeri</u> Beaman, Contr. Gray Herb. 183:95. 1957. Caespitose perennial to 5cm high, strigose-sericeous; leaves basal, linear to oblanceolate, 1-4.5cm long; heads sessile or rarely on peduncles to 13mm long; involucre 9-15mm long; rays white, cream, or pinkish, 8-15mm long; disk corollas 4-6.5mm long. Plains, hills, and slopes. NW, NE, SE.

<u>Townsendia</u> <u>incana</u> Nutt., Trans. Am. Phil. Soc. II, 7:305. 1840. Caespitose or suberect perennial or rarely biennial to 6(10)cm high, strigose-canescent; leaves oblanceolate, 1-4cm long; heads subsessile or mostly at tips of stems; involucre 7-14mm long; rays white or pinkish, 7-15mm long; disk corollas 3.5-6.5mm long; ray pappus somewhat reduced. Plains and hills. NW, SW, SE.

<u>Townsendia</u> <u>leptotes</u> (Gray) Osterh., Muhlenbergia 4:69. 1908. Caespitose perennial to 5cm high, strigose to subglabrate; leaves basal, linear to oblanceolate, 0.5-5cm long; heads sessile or rarely on peduncles to 15mm long; involucre 5-15mm long; rays white, cream, pink, or blue, 7-15mm long; disk corollas 3-7mm long. Slopes and ridges in the mountains. NW.

Townsendia parryi Eaton in Parry, Am. Nat. 8:212. 1874. Biennial or short lived perennial to 3dm high, strigose to subglabrate; basal leaves oblanceolate to obovate, 1-7cm long; cauline leaves similar or reduced; heads usually 1 per stem; involucre 9-18mm long; rays blue or purplish, rarely white or pink, 1-2.5cm long; disk corollas 4-7mm long. Mostly open areas in the mountains. SW, NW, NE. T. alpina (Gray) Rydb.

Townsendia spathulata Nutt., Trans. Am. Phil. Soc. II, 7:305. 1840. Caespitose biennial or perennial to 3cm high, woolly-villous; leaves oblanceolate or obovate, 3-20mm long; heads sessile or nearly so; involucre 6-10mm long; rays pink to purple or bronze, 6-12mm long; disk corollas 3.5-6mm long. Plains and hills. Endemic in Sweetwater, Fremont, and Natrona Cos.

Townsendia strigosa Nutt., Trans. Am. Phil. Soc. II, 7:306. 1840. Caulescent taprooted biennial to 2dm high, strigose-pilose; basal leaves oblanceolate, 1-4cm long; cauline leaves usually reduced; heads 1 to few; involucre 5-10mm long; rays white to pink, 5-15mm long; disk corollas 3-5mm long. Plains and hills. SW.

Tragopogon L. Goatsbeard

Annual or biennial taprooted herbs with alternate, linear or lance-linear, entire, somewhat clasping leaves; heads solitary at ends of stems or branches; flowers all ligulate, yellow or purple; involucral bracts uniseriate and subequal; receptacle naked although roughened; pappus of plumose bristles.

Rays purple — _T. porrifolius_

Rays yellow

- Outer ray flowers usually exceeding involucral bracts; achenes mostly 15-25mm long including beak — _T. pratensis_
- Outer ray flowers exceeded by involucral bracts; achenes mostly 25-36mm long including beak — _T. dubius_

Tragopogon dubius Scop., Fl. Carn. 2nd ed. 2:95. 1772. Plants to 1m high; leaves 4-30cm long; peduncles enlarged below heads; involucre 25-40mm long at anthesis, to 7cm long in fruit, surpassing the yellow rays which are 15-30mm long; achenes mostly 25-36mm long including beak. Disturbed areas. Statewide.

Tragopogon _porrifolius_ L., Sp. Pl. 789. 1753. Plants to 1m high; leaves 3-30cm long; peduncles enlarged below heads; involucre 25-40mm long at anthesis, to 7cm long in fruit, equaling or surpassing the purple rays which are 15-35mm long; achenes 25-40mm long including beak. Disturbed areas. NW, NE, SE.

Tragopogon _pratensis_ L., Sp. Pl. 789. 1753. Plants to 8dm high; leaves 3-30cm long; peduncles not enlarged below heads or very slightly so in fruit; involucre (18)20-30mm long at anthesis, to 40mm in fruit, equaling or shorter than the yellow rays which are 20-35mm long; achenes 15-25mm long including beak. Disturbed areas. Albany Co.

Verbesina L. Crownbeard

Reference: Coleman, J. R. 1966. Am. Midl. Nat. 76:475-481.

Verbesina _encelioides_ (Cav.) Benth. & Hook. ex Gray, Bot. Calif. 1:350. 1876. Annual to 1m high; leaves opposite below, alternate above, the blades lanceolate to ovate-deltoid, 2-12cm long, coarsely serrate; heads few to many; involucre 6-20mm long, the bracts herbaceous, usually in 3 series; receptacle chaffy; rays yellow or rarely orange, (12)15-27mm long; pappus of 2 awns or rarely lacking; disk corollas about 4-5mm long. Plains and hills. Campbell Co. _Ximenesia_ _encelioides_ Cav.

Viguiera H. B. K. Goldeneye

Reference: Blake, S. F. 1918. Contr. Gray Herb. 54:1-205.

Viguiera multiflora (Nutt.) Blake, Contr. Gray Herb. 54:108. 1918. Perennial herb to 13dm high; leaves opposite except above, the blades lance-ovate to lance-linear, 1-8cm long, entire or slightly toothed; heads few to many, with ray and disk flowers; involucre 5-12mm long, the bracts herbaceous, in 2 to several series; receptacle chaffy; pappus lacking; rays yellow, 7-19mm long; disk corollas about 3-4mm long. Hills and slopes mostly in or near the mountains. NW, SW, SE. *Gymnolomia multiflora* (Nutt.) Benth. & Hook. ex Gray.

Wyethia Nutt. Mule's Ears

Taprooted perennial herbs with alternate and basal leaves; heads 1 to several, with ray and disk flowers, the rays yellow, white, or cream; involucral bracts herbaceous, in several series; receptacle chaffy, the bracts clasping the achenes; pappus a short, toothed or fringed crown.

Reference: Weber, W. A. 1946. Am. Midl. Nat. 35:400-452.

Rays white or cream colored, sometimes drying yellowish; plants pubescent at least above; some leaves usually over 2.5cm wide — W. helianthoides

Rays yellow; plants glabrous or pubescent, if pubescent, the leaves usually all less than 2cm wide

Plants glabrous and resinous — W. amplexicaulis

Plants hispid or scabrous, not resinous — W. scabra

Wyethia amplexicaulis (Nutt.) Nutt., Trans. Am. Phil. Soc. II, 7:352. 1840. Plants to 8dm high; basal leaf blades elliptic or oblanceolate, 1-5dm long, entire or sometimes toothed, glabrous; cauline leaves reduced, becoming ovate or lance-ovate; involucre 15-33mm long; rays yellow, 2.5-6cm long; disk corollas about 1cm long. Meadows and slopes. Statewide.

Wyethia helianthoides Nutt., Journ. Acad. Phila. 7:40. 1834. Plants to 8dm high; basal leaf blades mostly elliptic or elliptic-oblanceolate, 9-35cm long, entire or nearly so and cilate, hirsute or villous-hirsute; cauline leaves similar but reduced and becoming lanceolate; involucre 14-30mm long; rays white or cream colored or drying yellowish, 2.5-5cm long; disk corollas about 1cm long. Moist meadows in the mountains. Yellowstone Park and Teton Co.

Wyethia scabra Hook., Lond. Journ. Bot. 6:245. 1847. Plants to 5dm high; basal leaf blades not enlarged, the cauline mostly oblong, oblong-lanceolate, or oblong-oblanceolate, 3-15cm long, entire, scabrous; involucre 15-35mm long; rays yellow, 2.5-5cm long; disk corollas about 7-8mm long. Plains and hills. NW, SW, SE.

Xanthium L. Cocklebur

Reference: Löve, D. & P. Dansereau. 1959. Canad. Journ. Bot. 37:173-208.

Xanthium strumarium L., Sp. Pl. 987. 1753. Annual to 2m high; leaves alternate, the blades ovate to cordate or reniform, 2-12cm long, toothed and sometimes shallowly lobed; heads unisexual, the staminate above the pistillate; involucral bracts in 1-3 series, the pistillate enclosing the 2 flowers and forming a bur with hooked prickles; bur 1-3.5cm long; corolla lacking in pistillate flowers, about 3mm long in staminate; anthers free, filaments united; receptacle chaffy; pappus lacking. Disturbed areas. SW, SE, NE. *X. pennsylvanicum* Wallr., *X. echinatum* Murr.

Xylorhiza Nutt. Woody Aster

Xylorhiza glabriuscula Nutt., Trans. Am. Phil. Soc. II, 7:297. 1840. Perennial from a taproot and branched caudex, to 25cm high; leaves alternate, the blades oblong, oblanceolate, or linear-oblanceolate, 1-8cm long, entire; heads 1 to few per stem; involucre (7)8-15mm long, of several series of bracts; rays white to pinkish or brownish in age, 10-25mm long; disk corollas about 5mm long; receptacle naked; pappus of capillary bristles. Plains and hills. Statewide. *Aster glabriusculus* (Nutt.) T. & G., *Machaeranthera glabriuscula* (Nutt.) Cronq. & Keck, *X. parryi* Gray.

Herbs with alternate, simple leaves or parasites without green leaves; flowers bisexual, regular, solitary or paired or in cymose clusters, mostly axillary; sepals usually 5, separate or united; petals usually 5, united, sometimes to the tip; stamens 5, attached to the corolla tube and alternate with the lobes; ovary 1, superior; carpels 2; styles 1 or 2; locules 2 or incompletely 2; ovules 1 or 2 per locule; placentation axile; fruit a capsule.

1. Plants parasitic, the leaves reduced to scales or lacking, twining on above-ground part of host plant — Cuscuta
1. Plants not as above
  2. Corolla about 5-6mm long; styles 2, each deeply 2 cleft — Evolvulus
  2. Corolla 1.5-10cm long; style 1
    3. Stems not twining or trailing, the plants bushy; leaf blades linear or oblong to narrowly lanceolate or elliptic — Ipomoea
    3. Stems twining or trailing; leaf blades sagittate or hastate
      4. Calyx enclosed by 2 bracts, the bracts cordate or ovate — Calystegia
      4. Calyx not enclosed by bracts, the bracts linear and borne much below the calyx — Convolvulus

Calystegia R. Br. Hedge Bindweed

References: Lewis, W. H. & R. L. Oliver. 1965. Ann. Mo. Bot. Gard. 52:217-222.

Brummitt, R. K. 1965. Ann. Mo. Bot. Gard. 52:214-216.

Calystegia sepium (L.) R. Br., Prodr. Fl. Nov. Holl. 483. 1810. Perennial with twining or trailing stems to 3m long; leaf blades hastate or sagittate, 3-12cm long; flowers solitary or paired on axillary peduncles; peduncles with cordate-ovate bracts enclosing the calyx; calyx 12-20mm long; corolla funnelform, white to pink, 3.5-7cm long; locule 1 or incompletely 2. Disturbed areas. NE, SE. _Convolvulus sepium_ L., _Convolvulus repens_ L.

Convolvulus L. Field Bindweed; Morning Glory

Convolvulus arvensis L., Sp. Pl. 153. 1753. Perennial with twining or trailing stems to 2m long; leaf blades hastate or sagittate, 1-7cm long; flowers solitary or paired on axillary peduncles; peduncles mostly with 2 linear bracts about midlength or above on single flowered peduncles or on 1 of pedicels of paired flowers; calyx 4-5mm long; corolla funnelform, white to pinkish-purple, 15-25mm long; locules 2. Disturbed areas. NW, NE, SE.

Cuscuta L. Dodder

Plants parasitic on above-ground portion of other plants, the stems twining about the host; leaves reduced to alternate minute scales or lacking; flowers in cymose or capitate clusters; corolla campanulate to urceolate, mostly white or yellowish, often with toothed or fringed appendages within; locules 2, each with 2 ovules; styles 2. Grammica Lour.

References: Yuncker, T. G. 1965. N. Am. Fl. II, 4:1-51.
Hadač, E. & J. Chrtek. 1970. Folia Geobot. Phytotax. Praha 5:443-445.

1. Stigmas long and narrow
  2. Calyx lobes as long as wide or shorter than wide, the tips obtuse, turgid, and slightly recurved — C. approximata
  2. Calyx lobes longer than wide, the tips somewhat acute, not turgid or recurved — C. epithymum
1. Stigmas capitate
  3. Appendages lacking on corolla within; corolla lobes sharply acute or acuminate — C. occidentalis
  3. Appendages present on corolla within; corolla lobes not sharply acute or acuminate
    4. Capsules somewhat beaked; sepals broadly rounded at tip and overlapping — C. umbrosa
    4. Capsules often with a terminal depression, sometimes thickened around the styles but not beaked; sepals mostly acute at tip, overlapping or not

5. Capsule thickened around base of styles; flowers minutely papillate ..... *C. indecora*
5. Capsule not thickened around base of styles; flowers not papillate ..... *C. plattensis*

Cuscuta approximata Bab., Ann. & Mag. Nat. Hist. 13:253. 1844. Flowers 2.5-4mm long; calyx enclosing the corolla tube, the lobes of calyx mostly wider than long; corolla lobes nearly equaling the tube; stigmas filiform. On a variety of forbs, especially legumes. NE, SE. *C. anthemi* A. Nels., *C. gracilis* Rydb.

Cuscuta epithymum Murray, Syst. Veg. 13:140. 1774. Flowers 2.5-4mm long; calyx as long as or shorter than corolla tube, the lobes somewhat triangular; corolla lobes shorter than the tube; stigmas filiform. On a variety of forbs. Sublette and Carbon Cos.

Cuscuta indecora Choisy, Mém. Soc. Phys. Nat. Genève 9:278. 1841. Flowers 2.5-4mm long; calyx mostly much shorter than corolla tube, the lobes triangular-ovate; corolla lobes shorter than the tube; stigmas capitate. On a variety of forbs and shrubs. Sheridan and Platte Cos.

Cuscuta _occidentalis_ Millsp. in Millsp. & Nutt., Field Mus. Publ. Bot. 5:204. 1923. Flowers 2-3.5mm long; calyx about as long as corolla tube, the lobes mostly acute; corolla lobes about as long as tube, sharply acute or acuminate; stigmas capitate. On a variety of forbs. Sweetwater Co.

Cuscuta _plattensis_ A. Nels., Bull. Torrey Club 26:131. 1899. Flowers mostly 3-5mm long; calyx mostly shorter than corolla tube, the lobes triangular; corolla lobes about equaling or shorter than the tube; stigmas capitate. On a variety of forbs and shrubs. Platte and Converse Cos.

Cuscuta _umbrosa_ Beyrich ex Hook., Fl. Bor. Am. 2:78. 1837. Flowers about 2.5-4mm long; calyx shorter than corolla tube, the lobes broadly ovate to rotund; corolla lobes less than half as long as tube; stigmas capitate; capsules somewhat beaked. On a variety of forbs and shrubs. Platte Co. _C._ _megalocarpa_ Rydb.

Evolvulus L.

Reference: Van Ooststroom, S. J. 1934. Meded. Bot. Mus. Herb. Rijksuniv. Utrecht 14:1-267.

Evolvulus nuttallianus Schult. in R. & S., Syst. Veg. 6:198. 1820. Perennial herb to 25cm high, densely hairy; leaf blades mostly elliptic, 5-20mm long; flowers solitary and axillary; calyx 4-5mm long; corolla funnelform, rose or purplish, 5-6mm long; locules 2; styles 2, each deeply 2 cleft. Plains and hills. NE, SE. *E. pilosus* Nutt.

Ipomoea L. Morning Glory

Ipomoea leptophylla Torr., 1st Rep. Frem. Exp. 90. 1843. Perennial herb with large deep roots; stems to 12dm long; leaf blades linear or oblong to narrowly lanceolate or elliptic, 3-13cm long; flowers solitary or few in leaf axils; calyx 5-12mm long; corolla funnelform, rose to pink-purple, 3.5-10cm long; locules 2; style 1. Plains, hills, and disturbed areas. SE.

Shrubs or herbs with simple, entire, opposite or apparently whorled leaves; flowers usually bisexual, regular, in cymes or a flower-like involucrate head; sepals 4, united; petals 4, separate; stamens 4; ovary 1, inferior; carpels 2-4; style 1; locules 2; ovule 1 per locule, pendulous; fruit a drupe.

Reference: Rickett, H. W. 1945. N. Am. Fl. 28B:299-311.

Cornus L. Dogwood

Plants herbaceous or woody only at base, less than 20cm high
C. canadensis

Plants shrubby throughout, mostly well over 20cm high
C. stolonifera

Cornus canadensis L., Sp. Pl. 118. 1753. Bunchberry. Rhizomatous herb or subshrub to 20cm high; true leaves 4-7 in a terminal whorl, elliptic or obovate-elliptic, 2-8cm long; flowers in a semicapitate cyme subtended by 4 whitish bracts 1-2cm long, these often confused for petals; petals 1-1.5mm long, greenish-white or purplish; fruit red. Moist woods. NW, NE.

Cornus stolonifera Michx., Fl. Bor. Am. 1:92. 1803. Red-osier Dogwood. Shrub to 4m high with the year-old stems red or reddish-purple; leaves opposite, the blades ovate or elliptic, (1)3-12cm long; flowers in cymes terminating the branches; petals white, 2-4mm long; fruit white or bluish. Streambanks and moist areas. Statewide. *C. baileyi* Coult. & Evans.

CRASSULACEAE Stonecrop Family

Herbs, usually succulent, with simple, entire or rarely toothed leaves; flowers mostly bisexual but sometimes unisexual, regular, usually in cymes or solitary in leaf axils; sepals mostly 4 or 5, separate or united at base; petals mostly 4 or 5, separate or rarely united at base; stamens mostly 10 but rarely 4, 5, or 8; ovaries solitary but apparently 4 or 5, superior; carpels mostly 4 or 5, separate or sometimes united at base; style 1 per carpel; locules 1 per carpel; ovules few to many; placentation parietal; fruit a follicle or capsule.

Plants perennial; flowers usually 5-merous *Sedum*
Plants annual; flowers usually 4-merous *Tillaea*

Sedum L. Stonecrop

Perennial herbs with alternate or opposite leaves; flowers usually in cymes; sepals mostly 5; petals mostly 5; stamens mostly 10 but rarely 5 or 8.

References: Fernald, M. L. 1947. Rhodora 49:79-81.
Clausen, R. T. 1948. Cactus & Succ. Journ. 20:143-146.

1. Petals greenish-white, pink, or purple; most leaves on the flowering stems
   2. Petals greenish-white to pink, 7-11mm long; flowers bisexual — S. rhodanthum
   2. Petals usually purple but sometimes pink, less than 5mm long; flowers usually unisexual — S. rosea
1. Petals yellow, occasionally drying pink; most leaves basal along the creeping stems or on sterile shoots
   3. Leaves of flowering stems mostly opposite, oval to obovate or spatulate — S. debile
   3. Leaves (or leaf scars) of flowering stems alternate, linear to lanceolate
      4. Leaves mostly keeled and acuminate; upper stem leaves usually with bulbil-like flowers in their axils; carpels divergent in fruit — S. stenopetalum
      4. Leaves not keeled or acuminate; upper stem leaves rarely with bulbil-like flowers in their axils; carpels mostly erect in fruit — S. lanceolatum

Sedum debile Wats., Bot. King Exp. 102. 1871. Plants to 12cm high; leaves opposite or subopposite, oval to obovate or spatulate, 3-8mm long, 2-6mm wide; sepals 3-4mm long; petals yellow, 6-8mm long, often connate at base. Rocky areas in the mountains. NW, SW.

Sedum lanceolatum Torrey, Ann. Lyc. N. Y. 2:205. 1828. Plants to 2dm high; leaves alternate, linear to lanceolate or rarely ovate, 3-15mm long, 1-4mm wide, the upper ones mostly deciduous before flowering; sepals 2-4mm long; petals yellow, 5-8mm long, separate. Gravelly or rocky hills and slopes. Statewide. S. stenopetalum Pursh of early authors.

Sedum rhodanthum Gray, Am. Journ. Sci. II, 33:405. 1862. Plants to 3.5dm high; leaves alternate, oblong-lanceolate to elliptic-oblanceolate, 6-28mm long, 2-7mm wide; sepals 4-7mm long; petals pink, white, or greenish, 7-11mm long, separate. Wet areas in the mountains. Statewide.

Sedum rosea (L.) Scop., Fl. Carn. 2nd ed. 1:326. 1771. Plants to 18cm high; leaves oblanceolate to obovate or elliptic, 5-20mm long, 2-10mm wide; flowers mostly unisexual; sepals 1-3mm long; petals purple or sometimes pink, 2-4mm long, separate. Moist areas in the high mountains. Statewide. S. integrifolium (Raf.) A. Nels.

Sedum stenopetalum Pursh, Fl. Am. Sept. 324. 1814. Plants to 2dm high; leaves alternate, linear to lanceolate, keeled or nerved dorsally, 5-20mm long, 1-3mm wide, the middle ones usually deciduous by flowering; sepals 2-3mm long; petals yellow, 5-9mm long, separate. Plains, hills, slopes, and open woods. Teton Co. S. douglasii Hook.

Tillaea L. Pigmy Weed

Tillaea aquatica L., Sp. Pl. 128. 1753. Glabrous annual with stems to 10cm long; leaves opposite, linear or linear-oblanceolate, 2-7mm long, entire, with connate-sheathing bases; flowers solitary in axils; calyx 4 lobed, 0.5-1mm long; petals membranous, whitish, about 1-1.5mm long; stamens 4; fruit a follicle. Shallow water and shores. Yellowstone Park.

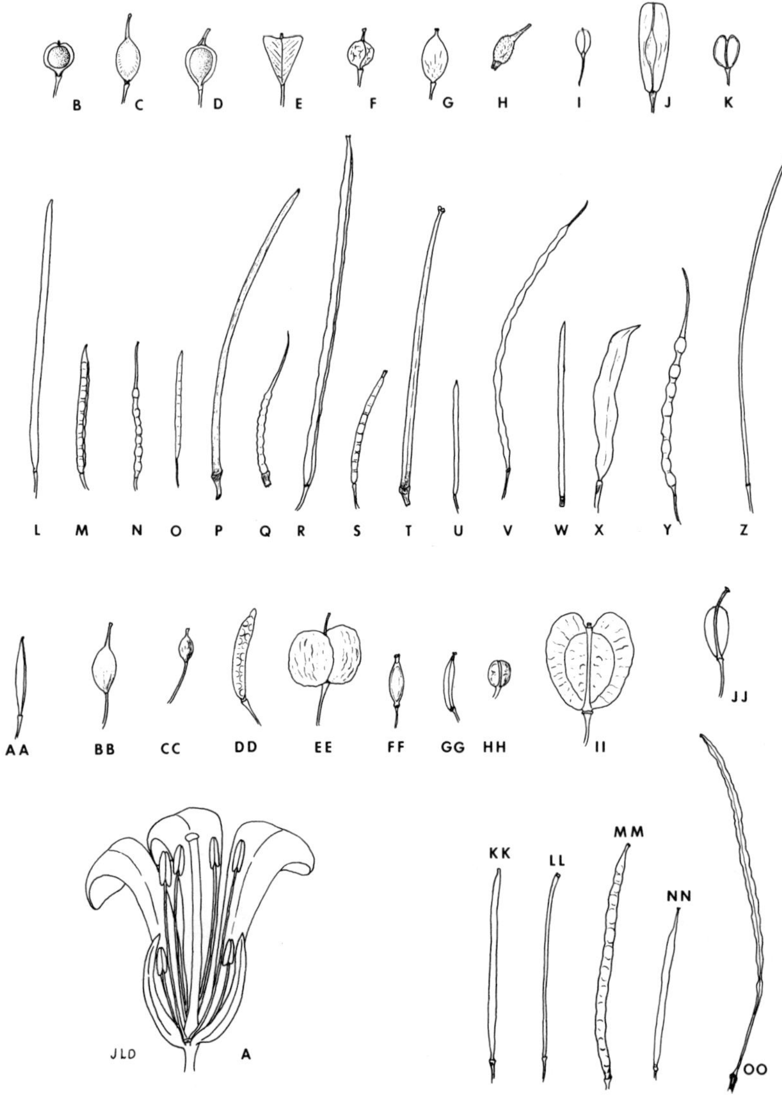
B C D E F G H I J K
L M N O P Q R S T U V W X Y Z
AA BB CC DD EE FF GG HH II JJ
JLD A
KK LL MM NN OO

Herbs with simple or compound, alternate or basal leaves; flowers bisexual, regular, in simple or compound racemes; sepals 4, separate; petals 4 and separate or rarely lacking; stamens usually 6 (rarely 2 or 4), 2 usually shorter than the other 4; ovary 1, superior; carpels 2 or 4; style 1 or lacking; locules 2 (rarely 1); ovules 1 to many; placentation usually parietal; fruit a silique or silicle.

The following key is based primarily on fruit characters. A key based primarily on flower and vegetative characters follows this key. The first key is most reliable.

Figure 7. Cruciferae. A. Generalized flower with one petal and sepal removed (X 10). B-K. Fruits (X 1.3): B. Alyssum; C. Berteroa; D. Camelina; E. Capsella; F. Cardaria; G. Draba; H. Euclidium; I. Hutchinsia; J. Isatis; K. Lepidium. L-Z. Fruits (X 0.7): L. Arabis; M. Barbarea; N. Brassica; O. Cardamine; P. Caulanthus; Q. Chorispora; R. Conringia; S. Erucastrum; T. Erysimum; U. Halimolobos; V. Hesperis; W. Malcolmia; X. Parrya; Y. Raphanus; Z. Sisymbrium. AA-JJ. Fruits (X 1.3): AA. Descurainia; BB. Lesquerella; CC. Lobularia; DD. Nasturtium; EE. Physaria; FF. Rorippa; GG. Smelowskia; HH. Subularia; II. Thlaspi arvense; JJ. Thlaspi montanum. KK-OO. Fruits (X 0.7): KK. Thelypodium; LL. Thelypodiopsis; MM. Streptanthus; NN. Streptanthella; OO. Stanleya.

1. Mature fruit less than 4 times as long as wide (exclude style)
   2. Plants aquatic, the leaves all basal or nearly so and grass-like — Subularia
   2. Plants not as above
      3. Fruits flattened except sometimes where extended by the seeds — GROUP I
      3. Fruits terete, subterete, or quadrangular in cross section — GROUP II
1. Mature fruit over 4 times as long as wide
   4. Fruit stalked at base, the stalk usually extending 7mm or more beyond the receptacle (do not confuse this stalk with the pedicel); petals yellow — Stanleya
   4. Fruit not stalked at base, the valves extending all the way to base, or, the stalk less than 3mm long; petals yellow or not
      5. Plants with glandular-tipped hairs; fruits with a stout, prominent beak
         6. Plants alpine or subalpine; mature fruits mostly 4mm or more wide — Parrya
         6. Plants below subalpine; mature fruits 2mm or less wide — Chorispora
      5. Plants not with both glandular-tipped hairs (rarely papillose) and prominently beaked fruits
         7. Pubescence, at least in part, of stellate or branched hairs — GROUP III
         7. Pubescence all of simple hairs or the plants glabrous (at least check base of plant for hairs)
            8. Plants glandular-pubescent; alpine or subalpine — Parrya
            8. Plants not glandular-pubescent; alpine or not

9. Fruits usually conspicuously constricted between the seeds, the constrictions about half as wide as the fruit, also with a prominent beak at tip and often a short, narrow seedless portion at base of fruit, the margin of the valves inconspicuous at base and tip

10. Stem leaves mostly auriculate *Raphanus*

10. Stem leaves not auriculate *Thelypodium*

9. Fruits only rarely constricted between the seeds, if beaked, usually lacking a narrow seedless portion at base and the margins of valves conspicuous at base and tip

11. Fruits usually broader than linear, rarely as much as 8 times as long as wide GROUP IV

11. Fruits linear, often over 8 times longer than wide

12. Fruits definitely flattened GROUP V

12. Fruits terete or 4 angled, not flattened GROUP VI

GROUP I

1. Fruits oblong to club-shaped, 1 seeded, the septum lacking, 1cm or more long when mature *Isatis*

1. Fruits various shaped but 2 or more seeded, the septum present, often much less than 1cm long

2. Fruits flattened at a right angle to the septum

3. Seeds solitary in each chamber of the fruit

4. Plants rhizomatous, with simple, entire to dentate leaves; fruits not notched at tip, the style usually 0.7mm or more long — Cardaria

4. Plants without the above combination of characters — Lepidium

3. Seeds 2 or more in each chamber of the fruit

5. Stem leaves not auriculate nor clasping the stem

6. Styles 1.5mm or more long — Lesquerella

6. Styles lacking or nearly so — Hutchinsia

5. Stem leaves auriculate or clasping the stem

7. Fruit triangular, widest at top; plants hairy at least below — Capsella

7. Fruit not triangular, or if so, widest below the top; plants glabrous or hairy

8. Plants glabrous — Thlaspi

8. Plants hairy at least at base — Cardaria

2. Fruits flattened parallel to the septum

9. Seeds 1 or 2 in each chamber of the fruit; leaves mostly along the stem — Alyssum

9. Seeds more than 2 in each chamber of the fruit, or if rarely 1 or 2, the leaves all basal or nearly so

10. Leaves mostly compound or deeply divided — Smelowskia

10. Leaves mostly simple and not deeply divided

11. Seeds often minutely winged; plants annual, erect, mostly over 3dm high; petals white; plants weedy; styles 1.5mm or more long — Berteroa

11. Seeds not winged; plants often perennial, often less than 2dm high and matted; petals yellow or white; plants usually not weedy; styles usually much less than 1.5mm long — Draba

GROUP II

1. Hairs simple or lacking
   2. Fruits subglobose or oval in outline; leaves entire or toothed, the upper ones auriculate; plants rhizomatous — *Cardaria*
   2. Fruits not subglobose or oval, distinctly longer than wide, or if oval, the leaves either lobed or divided or not auriculate or the plants not rhizomatous — *Rorippa*
1. Hairs, or some of them, branched or stellate (at least check base of plant for hairs)
   3. Fruits notched at tip, often appearing like 2 parts grown together — *Physaria*
   3. Fruits rounded or pointed at tip
      4. Fruits on stout stalks (pedicels) 1-2mm long — *Euclidium*
      4. Fruits on slender pedicels usually longer than the fruit
         5. Seeds 1 in each chamber of the fruit
            6. Pubescence stellate — *Lesquerella*
            6. Pubescence not stellate — *Lobularia*
         5. Seeds 2 or more in each chamber of the fruit
            7. Stem leaves compound or deeply divided
               8. Plants annual or biennial with a taproot, mostly below subalpine — *Descurainia*
               8. Plants perennial with a branched caudex, alpine or subalpine — *Smelowskia*
            7. Stem leaves simple and entire or merely toothed
               9. Stem leaves numerous and auriculate or clasping the stem; basal leaves few or lacking — *Camelina*
               9. Stem leaves few and not auriculate or clasping the stem; basal leaves numerous — *Lesquerella*

GROUP III

1. Fruits conspicuously constricted between the seeds, the constrictions about half as wide as the fruit — Hesperis

1. Fruits not conspicuously constricted between the seeds

  2. Hairs 2 branched from near the base, the branches appressed to the plant surface or nearly so (sometimes with appressed 3 branched hairs also); flowers yellow or rarely reddish or purplish — Erysimum

  2. Hairs more than 2 branched or 2 branched from above and the branches not appressed to the plant surface (rarely 2 branched and appressed in the nearly glabrous *Arabis drummondii*); flowers various

    3. Leaves compound or nearly so

      4. Plants perennial with a branching caudex, alpine or subalpine — Smelowskia

      4. Plants annual or biennial with a taproot, mostly below subalpine — Descurainia

    3. Leaves simple

      5. Fruits usually not linear (sometimes broadly so), usually strongly flattened, mostly less than 8 times as long as wide — Draba

      5. Fruits usually about as wide as deep, or if flattened, then narrowly linear, usually over 8 times longer than wide

6. Fruits distinctly flattened — Arabis
6. Fruits terete or subterete
7. Seeds in 2 rows in each chamber, at least in part — Halimolobos
7. Seeds in 1 row in each chamber
8. Plants annual, 0.5-4dm high; petals 6-11mm long — Malcolmia
8. Plants perennial, mostly 5-13dm high; petals 16-25mm long — Hesperis

GROUP IV

1. Fruits 1 seeded, the septum lacking — Isatis
1. Fruits usually at least 2 seeded, the septum present
2. Fruits flattened at a right angle to the septum — Thlaspi
2. Fruits flattened parallel to the septum or terete
3. Fruits terete or nearly so
4. Mature fruits 5mm or more wide; annual or biennial with usually purple or white petals — Raphanus
4. Mature fruits much less than 5mm wide; if annual or biennial, then with yellow petals
5. Petals white; leaves, or some of them, pinnately compound; plants glabrous, aquatic, usually rooting at the nodes — Nasturtium

5. Petals yellow; leaves often not compound; plants glabrous or not, aquatic or not, usually not rooting at nodes _Rorippa_

3. Fruits flattened

6. Petals white to purple; plants glabrous; some mature fruits 2cm or more long _Parrya_

6. Petals yellow or fading to white; plants pubescent; fruits much less than 2cm long _Draba_

GROUP V

1. Petals yellow

2. Stem leaves auriculate or cordate-clasping _Streptanthus_

2. Stem leaves not auriculate or cordate-clasping _Streptanthella_

1. Petals white, pink, or purple

3. Fruits with a beak 3-10mm long at tip, pendant _Streptanthella_

3. Fruits not beaked or the beak less than 3mm long, only occasionally pendant

4. Leaves, or some of them, pinnately compound _Cardamine_

4. Leaves all simple, entire to pinnatifid

5. Plants glabrous; middle and upper leaves cordate-clasping, some over 1cm wide; sepals 6mm or more long at anthesis; in dry areas of SW Wyoming _Streptanthus_

5. Plants not as above

6. Valves of fruit nerved; seeds often narrowly winged; leaves not cordate
   7. Mature fruits 4mm or more wide; some leaves usually over 6mm wide; plants glabrous, alpine or subalpine ..... Parrya
   7. Mature fruits 3mm or less wide; leaves usually all 6mm or less wide and the plants pubescent if alpine ..... Arabis
6. Valves of fruit not nerved; seeds wingless; leaves mostly cordate ..... Cardamine

GROUP VI

1. Styles beak-like, usually sharply differentiated from body of fruit, 1.5mm or more long on at least some fruits
   2. Stem leaves mostly auriculate; petals not yellow ..... Thelypodium
   2. Stem leaves not auriculate, or if so, the petals yellow
      3. Valves nerveless ..... Cardamine
      3. Valves with 1 or more somewhat raised nerves
         4. Beaks usually over 3mm long, if shorter, some or all leaves simple and not deeply lobed or divided

5. Petals yellow, often fading to white; leaves usually toothed or else auriculate

6. Sepals 3-5mm long; plants annual — Brassica

6. Sepals 2-3mm long; plants biennial or perennial — Barbarea

5. Petals white to rose or purple; leaves entire or nearly so, not auriculate — Thelypodium

4. Beaks mostly 3mm or less long, the leaves all compound or else deeply lobed or divided — Erucastrum

1. Styles very short or lacking, not beak-like (rarely to 1.5mm)

7. Petals yellow or rarely cream colored (sometimes fading to white on drying)

8. Leaves entire or nearly so

9. Plants with taproots; leaves 5-35mm wide — Conringia

9. Plants with creeping rhizomes; leaves 1-9mm wide — Sisymbrium

8. Leaves dentate to pinnatifid or pinnate

10. Fruits mostly 12mm long or less, spreading to ascending — Rorippa

10. Fruits mostly over 15mm long when mature, mostly ascending to erect

11. Stem leaves auriculate or cordate-clasping — Barbarea

11. Stem leaves not auriculate or cordate-clasping — Sisymbrium

7. Petals white, pink, or purple

12. Fruits conspicuously constricted between the seeds, or with a stipe at base, or both — Thelypodium

12. Fruits not conspicuously constricted between the seeds, without a stipe at base
 13. Leaves entire or nearly so, often auriculate at least above
  14. Styles in fruit usually 0.6mm or more wide, the stigmas slightly bilobed — Thelypodiopsis
  14. Styles in fruit usually 0.4mm or less wide, the stigmas usually not bilobed — Thelypodium
 13. Leaves, at least the basal ones, dentate or sinuate to pinnatifid or pinnate, usually not auriculate
  15. Seeds in 2 rows in each chamber, at least in part; fruits 15mm or less long — Nasturtium
  15. Seeds in 1 row in each chamber; fruits often over 15mm long
   16. Pedicels and sepals conspicuously hairy, the pedicels usually 5mm or less long, the sepals 8-12mm long; stems stout and hollow — Caulanthus
   16. Pedicels and sepals mostly glabrous or nearly so, the pedicels often over 5mm long, the sepals 6mm or less long; stems mostly solid
    17. Styles in fruit 0.5mm or less wide; plants mostly of wet areas — Cardamine
    17. Styles in fruit 0.6mm or more wide or lacking; plants mostly of dry areas but occasionally in moist areas
     18. Stem leaves auriculate, entire to slightly toothed — Thelypodiopsis
     18. Stem leaves not auriculate, mostly pinnately compound or pinnatifid — Sisymbrium

Key based primarily on flower and vegetative characters.

1. Hairs branched at least in part
   2. Hairs 2 branched, the branches appressed to the plant surface (rarely with a few 3 branched hairs, these also appressed) (see also <u>Arabis</u> <u>drummondii</u>)
      3. Petals yellow or sometimes reddish or purplish — <u>Erysimum</u>
      3. Petals white or bluish tinged — <u>Lobularia</u>
   2. Hairs not as above
      4. Stem leaves, at least the upper ones, auriculate or clasping the stem
         5. Plants annual
            6. Basal leaves rosulate and lobed — <u>Capsella</u>
            6. Basal leaves not rosulate, entire or minutely toothed — <u>Camelina</u>
         5. Plants biennial or perennial
            7. Petals (3)4mm or more long; widespread — <u>Arabis</u>
            7. Petals 2-4mm long; known only from Albany Co. — <u>Halimolobos</u>
      4. Stem leaves not auriculate or clasping the stem
         8. Leaves pinnately compound or nearly so
            9. Plants perennial with a branching caudex, alpine or subalpine — <u>Smelowskia</u>
            9. Plants annual or biennial with a taproot, mostly below subalpine — <u>Descurainia</u>
         8. Leaves simple and entire or toothed or rarely a few pinnatifid
            10. Petals yellow, rarely tinged with red or purple or drying white

11. Plants annual or biennial, the leaves mostly along the stem; petals 2.5-4mm long

12. Leaves linear to oblanceolate, 1-6mm wide, entire; sepals 1.5-3.5mm long *Alyssum*

12. Leaves ovate to obovate, 1-15mm wide, entire or toothed; sepals 1-2mm long *Draba nemorosa*

11. Plants perennial, or if not, the leaves mostly basal and rosulate and the petals usually longer

13. Styles of young fruits averaging 1.5mm or less long *Draba*

13. Styles of young fruits averaging over 1.5mm long (also see *Draba streptocarpa*)

14. Young fruits notched at tip *Physaria*

14. Young fruits pointed or round at tip *Lesquerella*

10. Petals white, purple, pink, or rose

15. Petals about 1mm long; plants annual

16. Plants of dry areas with some leaves over 7mm wide *Euclidium*

16. Plants of moist areas with the leaves 1-7mm wide *Hutchinsia*

15. Petals 2mm or more long; plants annual to perennial

17. Plants perennial; petals mostly 16-25mm long

18. Leaves 1-3mm wide *Arabis pulchra*

18. Leaves 10mm or more wide *Hesperis*

17. Plants not both perennial and with petals 16-25mm long

19. Plants annual, the petals pink or purplish and 6-11mm long — Malcolmia

19. Plants biennial or perennial if with pink or purplish petals or the petals shorter

20. Plants annual or biennial, the leaves mostly along the stem and 1-6mm wide, entire; petals 2.5-4mm long — Alyssum

20. Plants not as above

21. Plants annual

22. Petals 4-6mm long, often notched at tip; sepals 2-3mm long — Berteroa

22. Petals 2-4(5)mm long, not notched; sepals 1-2mm long — Draba

21. Plants biennial or perennial

23. Young fruits linear; petals and sepals of various lengths — Arabis

23. Young fruits not linear; petals 2-5mm long; sepals 1-3mm long — Draba

1. Hairs all simple or lacking

24. Plants aquatic, the leaves all basal or nearly so and grass-like — Subularia

24. Plants not as above

25. Plants glandular-hairy

26. Plants alpine or subalpine — Parrya

26. Plants below subalpine — Chorispora

25. Plants not glandular-hairy

27. Petals lacking or less than 0.5mm long — Lepidium

27. Petals present, 0.5mm or more long

28. Plants glabrous and glaucous except the sepals and pedicels densely pubescent, the stem stout and hollow; petals purple or drying brownish, 10-17mm long; plants of dry areas in SW Wyoming — Caulanthus

28. Plants not as above

29. Petals about 1mm long, white; plants annual

30. Stamens 2 or 4, of if 6, the plants over 15cm high and often in dry areas — Lepidium

30. Stamens 6; plants 15cm or less high, in moist areas — Hutchinsia

29. Petals, if white, (1.5)2mm or more long; plants annual to perennial

31. Petals yellow (sometimes drying white)

32. Upper leaves auriculate or clasping the stem

33. Leaves all entire or merely toothed

34. Petals 0.5-4mm long

35. Ovules solitary in ovaries, the ovary 1 celled — Isatis

35. Ovules several to many per ovary, the ovary 2 celled — Rorippa

34. Petals 5mm or more long

36. Plants annual

37. Sepals 3-5mm long *Brassica*

37. Sepals 6-8mm long *Conringia*

36. Plants biennial or perennial

38. Sepals 12-20mm long *Stanleya*

38. Sepals 6-11mm long *Streptanthus*

33. Leaves, or some of them, lobed to compound

39. Petals 0.5-5mm long

40. Leaves dimorphic, the lower much divided into linear segments, the upper mostly entire and strongly cordate-clasping or perfoliate; petals about 1.5mm long *Lepidium*

40. Leaves not as above; petals of various lengths

41. Petals 0.5-3mm long; sepals 0.5-2(2.5) mm long *Rorippa*

41. Petals 3-5mm long; sepals 2-4mm long

42. Plants rhizomatous perennials *Rorippa*

42. Plants biennial, without rhizomes *Barbarea*

39. Petals 6-20mm long

43. Plants annual *Brassica*

43. Plants biennial or perennial (see also *Barbarea*) *Stanleya*

32. Upper leaves not auriculate or clasping the stem

44. Sepals 8-16mm long; ovary and young fruit on a stalk usually 5mm or more long *Stanleya*

44. Sepals less than 8mm long, or if not, the ovary and young fruit without a stalk

45. Petals 0.5-5mm long

46. Leaves all entire or merely toothed

47. Leaves mostly basal, the plants often alpine or subalpine *Draba*

47. Leaves mostly along the stem, the plants rarely alpine *Rorippa*

46. Leaves, or some of them, lobed to compound

48. Petals 4-5mm long; sepals 2-4mm long

49. Plants rhizomatous perennials *Rorippa*

49. Plants annual or biennial, not rhizomatous *Erucastrum*

48. Petals 0.5-4mm long; sepals 0.5-2.5mm long

50. Plants perennial, in dry areas of southern Wyoming *Lepidium*

50. Plants mostly annual or biennial, mostly in moist areas *Rorippa*

45. Petals 5mm or more long

51. Plants annual or biennial

52. Petals 10-20mm long; sepals 5-10mm long *Raphanus*

52. Petals or sepals or both shorter than above
53. Leaves all entire or merely toothed
54. Leaves 1-12mm wide; petals 5-7mm long *Streptanthella*
54. Leaves 10-70mm wide; petals 6-15mm long *Brassica*
53. Leaves, or some of them, lobed to compound
55. Styles of young fruits lacking or nearly so; petals 5-8mm long *Sisymbrium*
55. Styles of young fruits usually apparent; petals 4-15mm long
56. Leaves all compound or deeply lobed or divided; petals 4-7mm long *Erucastrum*
56. Leaves, or some of them, often simple and not deeply lobed or divided; petals 6-15mm long *Brassica*
51. Plants perennial
57. Sepals 3.5-5.5mm long *Sisymbrium linifolium*
57. Sepals 2-3mm long *Draba crassa*
31. Petals white, purple, pink, or rose

58. Upper leaves auriculate or clasping the stem
  59. Petals 2-7mm long; sepals 1-5mm long
    60. Sepals 1-2mm long
      61. Plants pubescent
        62. Plants annual — Lepidium
        62. Plants perennial — Cardaria
      61. Plants glabrous — Thlaspi
    60. Sepals 2-5mm long
      63. Plants glabrous; sepals 1-3mm long — Thlaspi
      63. Plants pubescent, or if glabrous, the sepals 3-5mm long
        64. Petals about equally wide their entire length or slightly narrowed at very base, rarely over 1.5mm wide — Arabis
        64. Petals long clawed, the upper expanded portion usually well over 1.5mm wide — Thelypodium
  59. Petals 7-25mm long; sepals 3-11mm long
    65. Sepals 6-11mm long
      66. Plants annual — Conringia
      66. Plants biennial or perennial
        67. Stem leaves mostly cordate; plants perennial, glabrous — Streptanthus
        67. Stem leaves mostly lanceolate; plants biennial or short lived perennials, glabrous or not — Thelypodium
    65. Sepals 2.5-6mm long

68. Plants alpine and subalpine Arabis lyallii

68. Plants below subalpine

69. Stigmas slightly bilobed especially on young fruits; inflorescence and ovaries often pubescent Thelypodiopsis

69. Stigmas not bilobed; inflorescence and ovaries glabrous Thelypodium

58. Upper leaves not auriculate or clasping the stem

70. Petals 10-25mm long; sepals 5-10mm long

71. Plants perennial, alpine or subalpine Parrya

71. Plants annual or biennial, below subalpine Raphanus

70. Petals 0.5-10(12)mm long; sepals 0.5-5(6)mm long

72. Plants glabrous perennials, usually rooting at the nodes, in wet areas, the leaves mostly pinnately compound; petals 3-5mm long; sepals 2-3mm long (see also Cardamine breweri) Nasturtium

72. Plants not as above

73. Petals 4mm long or less; sepals 0.5-2mm long

74. Young fruits linear; petals at least 2mm long; stamens 6 Cardamine

74. Young fruits not linear; petals often less than 2mm long; stamens sometimes 2 or 4 Lepidium

73. Petals 4mm or more long; sepals (1)3-6mm long

75. Leaves all entire or merely toothed

76. Leaves mostly cordate; plants rhizomatous — <u>Cardamine</u>

76. Leaves not cordate; plants rhizomatous or not

77. Plants glabrous annuals or biennials

78. Plants annual; flowers few in simple racemes — <u>Streptanthella</u>

78. Plants biennial; flowers many in compound racemes — <u>Thelypodium</u>

77. Plants either perennial or pubescent or both — <u>Arabis</u>

75. Leaves, or some of them, lobed to compound

79. Plants rhizomatous perennials — <u>Cardamine breweri</u>

79. Plants annual or biennial, not rhizomatous

80. Petals 2-4mm long — <u>Cardamine</u>

80. Petals 6-8mm long — <u>Sisymbrium</u>

Alyssum L.

Annual or biennial herbs with simple, entire leaves; flowers in a raceme, the petals yellow to white; fruits orbicular to ovate-elliptic, flattened parallel to the septum; seeds 1 or 2 per chamber.

References: Dudley, T. R. 1964. Journ. Arn. Arbor. 45:358-373.
1965. Journ. Arn. Arbor. 46:181-217.

Fruits glabrous; styles 0.6-1mm long ..... *A. desertorum*
Fruits pubescent with stellate hairs; styles 0.5mm long or less ..... *A. alyssoides*

Alyssum alyssoides (L.) L., Syst. Nat. 10th ed. 1130. 1759. Stems simple to freely branched, to 3dm high; leaves linear to oblanceolate, 3-25mm long, 1-6mm wide; sepals 1.5-3.5mm long, persisting until the fruit nearly matures; petals 2.5-4mm long; fruits 3-4mm long, with stellate hairs; styles 0.5mm long or less. Disturbed areas. NE, SE.

Alyssum desertorum Stapf, Denkschr. Akad. Wien 51(2):302. 1886. Similar to above species except the sepals early deciduous, the fruits glabrous, and the styles 0.6-1mm long. Disturbed areas. Statewide.

Arabis L. Rock Cress

Biennial or perennial or rarely annual herbs with petioled, usually pubescent and rosulate, basal leaves and usually sessile, often auriculate, cauline leaves; flowers in a raceme; petals white to pink or purple; fruits linear, usually glabrous but sometimes pubescent, flattened parallel to the partition; seeds usually numerous, often winged.

1. Mature fruiting pedicels erect to ascending
  2. Basal leaves obovate to broadly oblanceolate, usually obtuse or rounded at tip, often forming a flat rosette; seeds wingless or rarely winged and the wing less than 0.3mm wide (This is sometimes a difficult separation. It may be desirable to try both choices.)
    3. Seeds in 2 rows in each locule, at least below *A. glabra*
    3. Seeds in 1 row in each locule
      4. Stem leaves auriculate, sometimes obscurely so; pedicels mostly erect or nearly so *A. hirsuta*
      4. Stem leaves not auriculate; pedicels mostly ascending at about a 45° angle *A. nuttallii*
  2. Basal leaves mostly linear to narrowly oblanceolate (rarely lacking), acute or rarely obtuse at tip, ascending and not forming a flat rosette; seeds winged or wingless

5. Fruit less than 1.5mm wide; stems either with simple, spreading hairs at base or else the basal leaves densely fine-hairy with the hairs obscuring the leaf surface or nearly so

6. Basal leaf blades densely hairy; stems with appressed hairs near base; pedicels often hairy; plants of deserts ... A. crandallii

6. Basal leaf blades usually moderately hairy; lower part of stems usually with spreading hairs; pedicels usually glabrous; plants montane ... A. microphylla

5. Fruit 1.5-3.5mm wide; stems usually glabrous or glabrate or with mostly appressed, branched hairs at base and the basal leaf surfaces not obscured by hairs

7. Seeds in 2 rows in each locule ... A. drummondii

7. Seeds in 1 row in each locule or rarely with 2 imperfect rows

8. Plants less than 3dm high; stems several from a branched caudex; mostly alpine and subalpine

9. Pedicels 1-4 or rarely 6mm long; petals 4.5-7mm long; basal leaves pubescent, usually apparent without magnification ... A. lemmonii

9. Pedicels usually 6mm or more long; petals 6-10mm long; basal leaves glabrous to sparsely pubescent, the hairs usually not apparent without magnification ... A. lyallii

8. Plants mostly 3dm or more high; stems solitary or rarely more from a simple caudex; mostly below subalpine

10. Fruits nerved only at the very base or nerveless; hairs minute, 0.1mm or less long — _A. fructicosa_

10. Fruits nerved at least to about the middle; some hairs over 0.1mm long

11. Pubescence of lower stems and leaves usually somewhat sparse, the hairs predominantly simple to 3 branched; petioles often ciliate with simple hairs; widespread — _A. divaricarpa_

11. Pubescence of lower stems and leaves usually somewhat dense, the hairs predominantly 3 or more branched; petioles rarely ciliate with simple hairs; western Wyo. — _A. sparsiflora_

1. Mature fruiting pedicels diverging at right angles to stem to strictly reflexed

12. Basal leaves hirsute with mostly simple or forked hairs, also ciliate with simple hairs (see also _A. divaricarpa_)

13. Seeds in 2 rows in each locule — _A. fendleri_

13. Seeds in 1 row in each locule — _A. demissa_

12. Basal leaves pubescent with mostly dendritic or at least 3 branched hairs, usually not ciliate (basal leaves sometimes early deciduous)

14. Plants usually with a woody base; leaves linear to linear-oblanceolate; petals 8-20mm long; fruits mostly pendulous, pubescent; seeds biseriate *A. pulchra*

14. Plants without the above combination of characters

15. Basal leaves linear or linear-oblanceolate; seed wing nearly 0.5mm wide; stem leaves usually gray-pubescent *A. cobrensis*

15. Basal leaves mostly spatulate to oblanceolate (sometimes lacking); seed wing less than 0.3mm wide; stem leaves greenish

16. Pedicels 1-4 or rarely 6mm long; stem leaves mostly ovate or obovate, usually glabrous or glabrate; mostly alpine or subalpine *A. lemmonii*

16. Pedicels 4-20mm long; stem leaves oblong to lanceolate, mostly pubescent; alpine or not

17. Mature fruiting pedicels mostly spreading at right angles to the rachis, straight or arched downward; fruits spreading at right angles or arcuate

18. Plants 1-3dm high, loosely caespitose; stem leaves often few and small and remote; basal leaves often very numerous *A. microphylla*

18. Plants mostly 3-9dm high, rarely caespitose; stem leaves usually numerous and well developed, often crowded and overlapping near base of stem; basal leaves few to many

19. Basal leaves usually entire; stems densely appressed-hairy at least below, the hairs minute and mostly with more than 3 branches *A. lignifera*

19. Basal leaves, at least the outer ones, often dentate; stems often with spreading, coarse hairs below, the hairs sometimes only 2 or 3 branched or simple

20. Pubescence of lower stems and leaves usually somewhat sparce, the hairs predominantly simple to 3 branched; petioles often ciliate with simple hairs; widespread *A. divaricarpa*

20. Pubescence of lower stems and leaves usually somewhat dense, the hairs predominantly 3 or more branched; petioles rarely ciliate with simple hairs; western and southern Wyo.

21. Petals 8-12mm long, 2-4mm wide; pedicels mostly pubescent; western Wyo. *A. sparsiflora*

21. Petals 6-9mm long, 1.5-2.5mm wide; pedicels mostly glabrous; southern Wyo. *A. perennans*

17. Mature fruiting pedicels descending to reflexed, straight or nearly so; fruits pendulous to strictly appressed to rachis

22. Pedicels appressed to rachis or nearly so *A. holboellii*

22. Pedicels descending but not appressed to rachis

23. Stems usually with spreading hairs below; fruits straight or nearly so — *A. holboellii*

23. Stems finely appressed-hairy below; fruits often somewhat curved — *A. lignifera*

Arabis cobrensis Jones, Contr. West. Bot. 12:1. 1908. Stems to 4dm high; basal leaf blades linear or linear-oblanceolate, mostly 1-4cm long, 1-4mm wide, entire; cauline leaves narrowly lanceolate; sepals 2.5-3mm long; petals 4-6mm long, white to pink; fruits spreading to pendulous; seeds uniseriate. Sagebrush. NW, SW. *A. canescens* Nutt.

Arabis crandallii Robins., Bot. Gaz. 28:135. 1899. Stems to 4dm high; basal leaf blades oblanceolate to spatulate, 5-20mm long, 2-4mm wide, entire or minutely dentate; cauline leaves oblong to lanceolate; sepals 3-4mm long; petals 5-7mm long, white to pinkish; fruits erect to ascending; seeds uniseriate. Desert areas. Carbon Co.

Arabis demissa Greene, Pl. Baker. 3:8. 1901. Stems to 3dm high; basal leaf blades linear to oblanceolate, 5-30mm long, 1-4mm wide, entire; cauline leaves mostly oblong; sepals 2-3.5mm long; petals 4-6.5mm long, white to pink; fruits pendulous; seeds uniseriate. Plains and hills. Albany Co. *A. aprica* Osterh. ex A. Nels.

Arabis divaricarpa A. Nels., Bot. Gaz. 30:193. 1900. Stems to 8dm high; basal leaf blades oblanceolate, 1-5cm long, 2-7mm wide, entire or remotely toothed; cauline leaves oblong to lanceolate; sepals 3-5mm long; petals 6-10mm long, pink to reddish-purple; fruits mostly erect to spreading; seeds mostly uniseriate. Woods and slopes. Statewide.

Arabis drummondii Gray, Proc. Am. Acad. 6:187. 1864. Stems to 8dm high; basal leaf blades oblanceolate to elliptic, 1-6cm long, 2-8mm wide, entire or rarely dentate; cauline leaves somewhat lanceolate; sepals 3-4.5mm long; petals 7-12mm long, white to pink; fruits erect; seeds biseriate. Woods and slopes. Statewide.

Arabis fendleri (Wats.) Greene, Pittonia 3:156. 1897. Stems to 6dm high; basal leaf blades oblanceolate to linear-oblanceolate, 1-4cm long, 1-7mm wide, entire to dentate; cauline leaves oblong to lanceolate; sepals 2.5-4mm long; petals 4-7mm long, white to pink; fruits pendulous; seeds biseriate. Plains and hills. Albany and Uinta Cos. *Sisymbrium pauciflorum* Nutt.

Arabis fructicosa A. Nels., Bot. Gaz. 30:190. 1900. Stems to 8dm high; basal leaf blades oblanceolate to spatulate, 1-3cm long, 3-9mm wide, dentate or rarely entire; cauline leaves ovate to oblong; sepals 2-3mm long; petals 5-7mm long, white to purplish; fruits mostly spreading; seeds uniseriate. Dry areas. Known only from the type collection from Yellowstone Park.

Arabis glabra (L.) Bernh., Syst. Verz. Erfurt 195. 1800. Annual to short lived perennial to 15dm high; basal leaf blades oblanceolate or obovate, mostly 3-14cm long, 0.5-3cm wide, entire to toothed; cauline leaves ovate-lanceolate to lanceolate, often over 4cm long; sepals 3-5mm long; petals 4-6mm long, cream colored; fruits erect; seeds biseriate at least below. Streambanks and meadows. Statewide.

Arabis hirsuta (L.) Scop., Fl. Carn. 2(2):30. 1772. Annual to short lived perennial to 1m high; basal leaf blades oblanceolate to obovate, 1-6cm long, 5-25mm wide, entire to toothed; cauline leaves oblong-lanceolate to elliptic; sepals 2.5-5mm long; petals 3.5-6mm long, white to pinkish; fruits erect; seeds uniseriate. Moist to dry areas. NW, SW, SE.

Arabis holboellii Hornem., Fl. Dan. 11(32):5, pl. 1879. 1827. Stems to 1m high; basal leaf blades oblanceolate, 0.5-4cm long, 2-9mm wide, entire to toothed; cauline leaves mostly lanceolate; sepals 2.5-5mm long; petals 4-10mm long, white, pink, or purplish; fruits mostly pendant; seeds uniseriate to irregular. Sagebrush and mountain slopes. Statewide. A. pendulocarpa A. Nels., A. exilis A. Nels., A. caduca A. Nels., A. lignipes A. Nels.

Arabis lemmonii Wats., Proc. Am. Acad. 22:467. 1887. Stems to 3dm high; basal leaf blades oblanceolate, 3-25mm long, 2-6mm wide, entire or remotely toothed; cauline leaves mostly ovate to obovate; sepals 2.5-3.5mm long; petals 4.5-7mm long, rose-purple; fruits ascending to slightly reflexed; seeds uniseriate. Alpine and subalpine. NW, SW.

Arabis lignifera A. Nels., Bull. Torrey Club 26:123. 1899. Stems to 6dm high; basal leaf blades linear-oblanceolate to oblanceolate, 0.5-3cm long, 1-5mm wide, entire or rarely remotely toothed; cauline leaves lanceolate; sepals 2-4mm long; petals 3-9mm long, pink to purple; fruits spreading to drooping; seeds mostly uniseriate. Sagebrush and desert. NW, SW.

Arabis lyallii Wats., Proc. Am. Acad. 11:122. 1876. Stems to 3dm high; basal leaf blades elliptic-oblanceolate, 5-25mm long, 1-6mm wide, entire; cauline leaves oblong to ovate; sepals 3-5mm long; petals 6-10mm long, purple; fruits erect to spreading; seeds uniseriate. Alpine and subalpine. NW, SW. *A. oreophila* Rydb.

Arabis microphylla Nutt. in T. & G., Fl. N. Am. 1:82. 1838. Stems to 3dm high; basal leaf blades oblanceolate, 5-25mm long, 1-4mm wide, entire to toothed; cauline leaves oblong to lanceolate; sepals 2-4mm long; petals 4-8mm long, pink to purple; fruits erect to slightly reflexed; seeds uniseriate. Woods and slopes. SW, NW. *A. densicaulis* A. Nels.

Arabis nuttallii Robins. in Gray, Syn. Fl. 1(1):160. 1895. Stems to 3dm high; basal leaf blades obovate to oblanceolate, 0.5-4cm long, 2-20mm wide, usually entire; sepals 3-4mm long; petals 6-9mm long, white or purplish; fruits erect to ascending; seeds uniseriate. Meadows, slopes, and thickets. Statewide.

Arabis perennans Wats., Proc. Am. Acad. 22:467. 1887. Stems to 6dm high; basal leaf blades oblanceolate to obovate, 1-5cm long, 2-10mm wide, usually dentate; cauline leaves lanceolate or oblong, often dentate; sepals 3.5-4.5mm long; petals 6-9mm long, purple or pink; fruits spreading to pendulous; seeds uniseriate. Plains and hills. Albany Co.

Arabis pulchra Jones ex Wats., Proc. Am. Acad. 22:468. 1887. Stems to 6dm high; basal leaf blades linear or linear-oblanceolate, 1-6cm long, 1-3mm wide, entire or rarely dentate; cauline leaves similar; sepals 5-8mm long; petals 8-20mm long, purple or reddish to white; fruits reflexed to pendulous; seeds biseriate. Plains and hills. Carbon Co.

Arabis sparsiflora Nutt. in T. & G., Fl. N. Am. 1:81. 1838. Stems to 1m high; basal leaf blades oblanceolate, 1-6cm long, 2-10mm wide, entire to remotely toothed; cauline leaves lanceolate to oblanceolate; sepals 3-6mm long; petals 8-12mm long, white to purple; fruits ascending to drooping; seeds uniseriate. Sagebrush and slopes. SW, NW. *A. elegans* A. Nels., *A. perelegans* A. Nels.

Barbarea R. Br. Winter Cress

Glabrous to hirsute biennial or perennial herbs with simple or pinnately compound leaves, the upper mostly auriculate; flowers in racemes or panicles; petals yellow; fruits linear, terete or nearly so, erect to spreading-ascending; seeds uniseriate.

Styles mostly less than 1mm long; fruits 2-3.5cm long; petals 3-5mm long B. orthoceras

Styles mostly 1mm long or more; fruits 1-2.5cm long; petals 4-8mm long B. vulgaris

Barbarea orthoceras Ledeb., Hort. Dorp. 2. 1824. Plants to 6dm high; leaf blades oblong to obovate or some cordate, pinnate to lyrate-pinnatifid or some occasionally crenate, the terminal leaflet usually much larger than the others, 1-10cm long, 4-30mm wide, reduced upward; sepals 2-3mm long; petals 3-5mm long. Meadows, streambanks, and woods. NW, SE. B. americana Rydb.

Barbarea vulgaris R. Br. in Ait., Hort. Kew. II, 4:109. 1812. Plants to 6dm high; leaf blades oblong to obovate or cordate, simple and entire to lyrate-pinnatifid or the lower sometimes pinnate, the terminal leaflet usually much larger than the others, 1-8cm long, 4-25mm wide, reduced upward; sepals 2-3mm long; petals 4-8mm long. Disturbed areas, usually where moist. NW, SE.

Berteroa DC.

Berteroa incana (L.) DC., Regni Veg. 2:291. 1821. Pubescent annual to 1m high, the hairs simple and branched; leaves short-petioled or sessile, the blades oblanceolate to elliptic, 1-6cm long, 2-15mm wide, entire; flowers in a raceme or panicle; sepals 2-3mm long; petals 4-6mm long, white, some often notched at tip; fruit oblong-elliptic, 5-7mm long, about half as wide, stellate-pubescent, very slightly inflated; seeds biseriate, 3-7 per locule. Disturbed areas. Sublette Co.

Brassica L. Mustard

Glabrous to hirsute annual herbs with often lyrate-pinnatifid leaves below and simple and entire or irregularly toothed leaves above; flowers in racemes, the petals yellow, often fading to white; fruits linear, terete or quadrangular, prominently beaked; seeds uniseriate.

Stem leaves sessile and strongly auriculate-clasping *B. rapa*
Stem leaves petioled or sessile but not auriculate or clasping
  Beak of fruit usually with a single seed at base, the beak and valves with usually 3 raised nerves *B. kaber*
  Beak of fruit usually lacking a seed at base, usually 1 nerved, the valves with 1 raised nerve *B. juncea*

Brassica juncea (L.) Czern., Consp. Pl. Charc. 8. 1859. Plants glabrous to hirsute-hispid, to 15dm high; leaf blades lyrate-pinnatifid to sinuate-dentate, 2-12cm long, 1-7cm wide; leaves not auriculate; sepals 3-5mm long; petals 6-15mm long; fruits ascending to erect, the valves 1 nerved. Disturbed areas. NW, NE, SE.

Brassica kaber (DC.) Wheeler, Rhodora 40:306. 1938. Plants hirsute-hispid at least below, to 1m high; leaf blades sinuate-dentate to lyrate-pinnatifid, 2-12cm long, 1-6cm wide; leaves not auriculate; sepals 4-5mm long; petals 8-14mm long; fruits mostly spreading to ascending, the valves 3 nerved. Disturbed areas. NW, NE, SE. *B. arvensis* (L.) B. S. P. and *Sinapis arvensis* L. of authors.

Brassica rapa L., Sp. Pl. 666. 1753. Plants glabrous or sparsely hirsute, to 1m high; basal leaf blades usually irregularly lyrate-pinnatifid, 5-20cm long, 3-8cm wide; cauline leaves becoming auriculate and entire upward, ovate to oblong or narrowly cordate; sepals 3-5mm long; petals 6-10mm long; fruits ascending to spreading, the valves 1 nerved at base. Disturbed areas. Sublette Co. and Yellowstone Park. *B. campestris* L.

Camelina Crantz False Flax

Stems somewhat densely hairy below; fruits mostly 5-7mm long; pedicels rarely over 17mm long *C. microcarpa*

Stems glabrous or sparsely hairy below; fruits mostly 7-9mm long; pedicels often over 17mm long *C. sativa*

Camelina microcarpa Andrz. ex DC., Regni Veg. 2:517. 1821. Annual to 1m high, with simple and forked or stellate hairs at base; leaves mostly cauline, sessile or short-petioled, the blades lanceolate to oblanceolate, 0.5-10cm long, 2-25mm wide, entire or minutely toothed, auriculate above; sepals about 3mm long; petals 4-5mm long, white or pale yellow; fruits obovate, mostly 5-7mm long, usually inflated; seeds several per locule, biseriate. Disturbed areas. Statewide.

Camelina sativa (L.) Crantz, Stirp. Austr. 1:17. 1762. Similar to above species but nearly glabrous and the fruits averaging slightly longer and often on longer pedicels. Disturbed areas. Sheridan Co. and Yellowstone Park.

Capsella Medic. Shepherd's Purse

Capsella bursa-pastoris (L.) Medic., Pflanzengatt. 1:85. 1792. Hirsute and stellate annual to 5dm high; basal leaves rosulate, the blades oblanceolate, 1-8cm long, 5-35mm wide, nearly entire to lyrate-pinnatifid; cauline leaves becoming sessile and clasping, lanceolate to oblanceolate; flowers in racemes; sepals 1.5-2mm long; petals 2-3mm long, white; fruit triangular-obcordate, 4-8mm long, not quite as wide, flattened contrary to partition; seeds numerous. Disturbed areas. Statewide.

Cardamine L. Bitter Cress

Annual to perennial, glabrous or hirsute herbs with simple to pinnate leaves; flowers in racemes, the petals white to pink or rose; fruits linear, at least slightly flattened; seeds uniseriate.

1. Leaves simple and mostly cordate *C. cordifolia*
1. Leaves compound at least in part, usually not cordate (rarely all simple but not cordate)
   2. Plants perennial with rhizomes, these sometimes short and stout; petals 3-7mm long; leaves often over 2cm wide *C. breweri*
   2. Plants annual or biennial with a taproot; petals 2-4mm long; leaves mostly 2cm or less wide
      3. Fruits mostly 1.3-1.5mm wide, 15-24 seeded *C. oligosperma*
      3. Fruits mostly 0.6-1mm wide, 20-40 seeded *C. pensylvanica*

Cardamine breweri Wats., Proc. Am. Acad. 10:339. 1875. Erect to prostrate, rhizomatous perennial to 6dm high, glabrous or sparsely pubescent near base; leaves mostly cauline, some usually simple and some compound, the blades usually cordate-reniform or ovate, 1-10cm long, 1-6cm wide; sepals 1.5-2.5mm long; petals 3-7mm long, white; fruits mostly erect or ascending. Wet areas. NW, SW, SE.

Cardamine _cordifolia_ Gray, Pl. Fendl. 8. 1849. Erect, rhizomatous perennial to 6dm high, glabrous to densely pubescent with simple hairs; leaves mostly cauline, the blades simple and mostly cordate, 1.5-8cm long, not quite as wide, somewhat crenate; sepals 3-4.5mm long; petals 7-12mm long, white; fruits ascending to erect. Meadows and streambanks in the mountains. Albany and Carbon Cos.

Cardamine _oligosperma_ Nutt. in T. & G., Fl. N. Am. 1:85. 1838. Taprooted annual or biennial to 4dm high, glabrous or slightly hirsute; leaves, or some of them, pinnately compound, the blades elliptic to obovate or ovate in outline, 3-35mm long, 2-20mm wide; sepals 1-2mm long; petals 2-4mm long, usually white; fruits erect or ascending. Wet areas. SW, NW, NE. _C_. _unijuga_ Rydb.

Cardamine _pensylvanica_ Muhl. ex Willd., Sp. Pl. 3:486. 1800. Similar to above species but with narrower fruits and usually more seeds. Moist areas. Carbon Co.

Cardaria Desv. White-top

Rhizomatous herbs with simple hairs and simple, entire to dentate leaves which are usually auriculate above; flowers in corymbose racemes; petals white; fruit orbicular or ovoid to cordate, inflated or slightly flattened; seeds 1 or 2 per locule.

Reference: Mulligan, G. A. and C. Frankton. 1962. Canad. Journ. Bot. 40:1411-1425.

Fruits, sepals, and pedicels hairy *C. pubescens*
Fruits, sepals, and pedicels glabrous
  Mature fruits cordate at base or nearly so, often indented at septum *C. draba*
  Mature fruits not cordate at base, rarely indented at septum *C. chalepensis*

Cardaria chalepensis (L.) Hand.-Mazz., Ann. Nat. Hofm. Wien 27:55. 1913. Plants to 5dm high; leaf blades ovate-oblong to oblanceolate, 2-10cm long, 1-3.5cm wide; sepals about 2mm long; petals 3-4mm long; fruits somewhat ovoid, 2-5mm long, inflated or somewhat flattened, glabrous. Disturbed areas. SW, SE.

Cardaria draba (L.) Desv., Journ. Bot. 3:163. 1814. Similar to above species except the fruits cordate at base or nearly so and often indented at the septum especially toward base. Disturbed areas. Albany and Park Cos. *Lepidium draba* L.

Cardaria pubescens (Meyer) Jarmol. in Keller et al., Weeds of USSR 3:29. 1934. Plants to 5dm high; leaf blades ovate-oblong to oblanceolate, 1-10cm long, 3-35mm wide; sepals 1-2mm long; petals 2-3mm long; fruits orbicular, inflated, pubescent, 2-4mm long. Disturbed areas. NW, SW, SE.

Caulanthus Wats.

Caulanthus crassicaulis (Torr.) Wats., Bot. King Exp. 27. 1871. Biennial or perennial with a hollow stem to 1m high, glaucous, glabrous except in inflorescence; leaves mostly basal, the blades ovate-lanceolate to oblanceolate, 1-7cm long, 2-30mm wide, entire to lyrate-pinnatifid; cauline leaves greatly reduced and usually linear; sepals 8-12mm long, usually hirsute; petals 10-17mm long, purplish, brownish on drying; fruits linear, mostly terete and ascending. Plains and deserts. SW.

Chorispora R. Br. ex DC.

Chorispora _tenella_ (Pall.) DC., Regni Veg. 2:435. 1821. Glandular and often hirsute-pilose annual to 5dm high; leaf blades lanceolate to oblanceolate, 1-8cm long, 3-20mm wide, sinuate-dentate; flowers in racemes; sepals 5-8mm long; petals 9-13mm long, pinkish-purple; fruits linear, usually torulose, with a prominent beak usually well over 5mm long; seeds uniseriate. Dry areas. Statewide.

Conringia Adans. Hare's-ear Mustard

Conringia _orientalis_ (L.) Dumort., Fl. Belg. 123. 1827. Glabrous and glaucous annual to 7dm high; basal leaves obovate to oblanceolate, 1-9cm long, 5-35mm wide, entire or nearly so; cauline leaves mostly oblong-ovate or lanceolate, clasping; flowers in racemes; sepals 6-8mm long; petals 7-12mm long, yellow or sometimes cream colored; fruits linear, quadrangular; seeds uniseriate. Disturbed areas. NE, SE.

Descurainia Webb & Berth. Tansy Mustard

Annual or biennial, stellate-pubescent (sometimes with simple, glandular hairs also) herbs with leaves 1-3 times pinnately compound; flowers in racemes; petals yellow or cream; fruit linear or clavate to fusiform, terete or slightly quadrangular; seeds uniseriate or biseriate.

1. Leaves, at least the lower ones, 2 or 3 times compound; fruits narrowly linear, mostly 1mm wide or less, some usually 15mm or more long; seeds mostly 20-40, uniseriate; valves 1-3 nerved D. sophia
1. Leaves mostly once compound, or if not, the fruits not as above
   2. Fruits fusiform, widest near middle; pedicels 3-7mm long; style usually over 0.3mm long; seeds 1-3 per locule D. californica
   2. Fruits linear, or clavate and widest near top, or if fusiform, then without the other characters above
      3. Fruits clavate or subclavate; seeds often biseriate at least near middle D. pinnata
      3. Fruits linear or at least not clavate; seeds uniseriate D. richardsonii

Descurainia californica (Gray) Schulz, Pflanzenr. IV, 105, Heft 86:330. 1924. Annual or biennial to 8dm high, usually glabrate; leaf blades obovate to oblanceolate or ovate, once compound, 1-6cm long, 0.5-4cm wide; sepals 0.5-1.5mm long; petals 1-2mm long, yellow; fruits somewhat fusiform; seeds uniseriate, 1-3 per locule. Hills, slopes, and open woods. NW, SW.

Descurainia pinnata (Walt.) Britt., Mem. Torrey Club 5:173. 1894. Annual to 7dm high; leaves mostly cauline, the blades lanceolate to oblanceolate, once or twice pinnately compound, 1-10cm long, 0.5-4cm wide; sepals 1-2mm long; petals 1-3.5mm long, cream to yellow (rarely purplish); seeds mostly biseriate at least near middle, 3-20 per locule. Disturbed areas, streambanks, plains, and hills. Statewide. *Sophia filipes* (Gray) Heller, *S. pinnata* (Walt.) Howell, *S. hartwegiana* (Fourn.) Greene, *S. nelsonii* Rydb.

Descurainia richardsonii (Sweet) Schulz, Pflanzenr. IV, 105, Heft 86:318. 1924. Annual or biennial to 1m high; leaves mostly cauline, the blades ovate to oblanceolate, mostly once to nearly twice pinnately compound, 1-10cm long, 0.5-4cm wide; sepals 1-2.5mm long; petals 1-3.5mm long, yellow or drying whitish; seeds uniseriate, 4-14 per locule. Woods, thickets, and open areas. NW, SW, SE. *Sophia incisa* (Engelm.) Greene, *S. leptophylla* Rydb., *S. viscosa* Rydb., *S. ramosa* Rydb.

Descurainia sophia (L.) Webb ex Prantl in Engl. & Prantl, Nat. Pflanzenf. 3(2):192. 1889. Annual or biennial to 1m high; leaves mostly cauline, the blades oblong-ovate to oblanceolate, 1-3 times compound, 1-10cm long, 0.5-4cm wide, usually appearing lace-like; sepals 2-2.5mm long; petals 1-2mm long, yellowish, or white when dry; fruits somewhat torulose; seeds uniseriate, mostly 10-20 per locule. Disturbed areas and streambanks. Statewide.

Draba L.

Annual to perennial, usually pubescent herbs with entire to dentate leaves; flowers in racemes; petals yellow or white; fruits linear or oblong to ovate or obovate, usually flattened, glabrous or pubescent; seeds biseriate.

References: Rollins, R. C. 1953. Rhodora 55:229-235.
Mulligan, G. A. 1971. Canad. Journ. Bot. 49:89-93.
1972. Canad. Journ. Bot. 50:1763-1766.

1. Plants annual
   2. Upper stem and usually the pedicels pubescent ... *D. praealta*
   2. Upper stem and pedicels glabrous or nearly so

3. Leaves usually all basal in a rosette, rarely 1 or 2 on the stem, usually glabrous on upper surface D. crassifolia

3. Leaves usually not all basal, often pubescent on the upper surface

4. Leaves mostly entire, usually densely pubescent; fruits linear, mostly 1-2mm wide; plants of the plains and foothills D. reptans

4. Leaves, or some of them, usually toothed, pubescent or glabrous; fruits either at least 2mm wide or the plants montane or sometimes weedy

5. Pedicels 1-5 times as long as the elliptic to oblong-oblanceolate fruits, the fruits mostly 2-3mm wide; basal leaves rosulate or not, usually pubescent D. nemorosa

5. Pedicels rarely as much as 1.5 times as long as the narrowly oblong fruits, the fruits 1.5-2.3mm wide; basal leaves usually rosulate, often glabrate D. stenoloba

1. Plants biennial or perennial

6. Flowering stems with leaves, the plants usually not caespitose or matted

7. Style usually less than 0.2mm long

8. Leaves densely to moderately pubescent on both sides; petals white to cream; fruits pubescent D. praealta

8. Leaves usually sparsely pubescent to glabrous; petals white or yellow; fruits glabrous or pubescent

9. Pedicels mostly subequal to or shorter than the fruits, the fruits 1.5-3.5mm wide; stem leaves 1 or 2, rarely more; mostly alpine and subalpine

10. Petals usually yellow; hairs predominantly simple — *D. crassifolia*

10. Petals white; hairs predominantly branched — *D. glabella*

9. Pedicels mostly subequal to or longer than the fruits, the fruits 1.5-2.3mm wide; stem leaves usually several; mostly below subalpine — *D. stenoloba*

7. Style at least 0.2mm long

11. Petals white

12. Stem leaves 1 or 2, reduced; plants 8cm or less high; petals not notched at tip — *D. nivalis*

12. Stem leaves usually 2 to several, not much reduced; plants 2-30cm high; petals usually notched at tip, sometimes minutely so

13. Hairs of leaves, at least a few of them, pectinately branched (use high magnification); fruits mostly glabrous — *D. glabella*

13. Hairs of leaves branched but not pectinately so; fruits pubescent

14. Lower stem with all branched hairs or nearly so; plants mostly alpine and subalpine — *D. cana*

14. Lower stem with many simple hairs mixed with branched hairs; plants usually lower than subalpine — *D. praëalta*

11. Petals yellow (sometimes drying white)
 15. Stem leaves 2-6; basal leaves somewhat fleshy, mostly glabrous or sometimes ciliate; fruits usually glabrous ... D. crassa
 15. Stem leaves usually more than 6; basal leaves not fleshy, usually pubescent on both sides; fruits often pubescent
  16. Styles in fruit less than 1.5mm long; fruits usually pubescent all over; petals 4-6mm long D. aurea
  16. Styles on some fruits often over 1.5mm long; fruits pubescent only on margin or glabrous; petals mostly 6-8mm long ... D. streptocarpa

6. Flowering stems mostly without leaves, the plants usually caespitose and matted
 17. Pubescence of leaves, especially beneath, wholly or largely of 2 rayed doubly pectinate hairs
  18. Leaves averaging less than 1.5mm wide; hairs of leaves usually sessile
   19. Petals white or yellow; fruits oblong-oblanceolate or lanceolate, with doubly pectinate hairs; pedicels and scapes pubescent ... D. pectinipila
   19. Petals yellow (fading to white); fruits ovate to obovate, with simple hairs; pedicels and scapes glabrous ... D. oligosperma
  18. Leaves averaging about 2mm or more wide; hairs of leaves short-stalked ... D. incerta

17. Pubescence of leaves simple to stellate, not doubly pectinate

20. Style lacking or nearly so

21. Petals white at anthesis; leaves 0.5-3.5mm wide

22. Fruits 3-6mm long

23. Rosettes usually rather loose, the leaves often over 10mm long — *D. fladnizensis*

23. Rosettes usually very compact, the leaves mostly 2-6(10)mm long — *D. nivalis*

22. Fruits 7-20mm long

24. Fruits 4-7(10)mm long; funiculus shorter than the seed length — *D. nivalis*

24. Fruits (7)10-20mm long; funiculus usually equaling or longer than the seed length — *D. lonchocarpa*

21. Petals yellow at anthesis; leaves 1-5mm wide — *D. crassifolia*

20. Style usually at least 0.2mm long

25. Leaves glabrous on upper surface except sometimes for ciliate margins

26. Fruits usually glabrous; styles 0.5mm or less long — *D. apiculata*

26. Fruits usually pubescent; styles often over 0.5mm long — *D. densifolia*

25. Leaves pubescent on upper surface

27. Petals white; styles 0.5mm or less long
see separation 24 above

27. Petals yellow; styles usually over 0.5mm long

28. Leaves averaging less than 2mm wide, individual leaves rarely exceeding 2mm wide *D. paysonii*

28. Leaves averaging at least 2mm wide *D. ventosa*

Draba apiculata C. L. Hitchc., Univ. Wash. Publ. Biol. 11:72. 1941. Cushion forming, scapose perennial to 5cm high, usually with only simple hairs; leaves lanceolate to oblanceolate, 3-10mm long, 1-2mm wide, imbricate and fleshy, the margins with short, stiff, simple hairs, otherwise glabrous; sepals 1.5-3mm long; petals about 4mm long, pale yellow; fruits ovate to oblong-elliptic, usually glabrous. Alpine meadows and talus. NW.

Draba aurea Vahl ex Hornem., Fors. Dansk Oekon. Pl. 2:599. 1806. Short-lived perennial to 5dm high with simple and branched hairs; basal leaves rosulate, the blades oblanceolate, 0.5-4cm long, 2-12mm wide, mostly entire; cauline leaves ovate to oblanceolate, entire to dentate; sepals 2-3.5mm long; petals 4-6mm long, yellow; fruits lanceolate, pubescent or rarely glabrous, often twisted. Woods and meadows in the mountains. NW, SW, SE. *D. surculifera* A. Nels., *D. luteola* Greene, *D. uber* A. Nels.

Draba cana Rydb., Bull. Torrey Club 29:241. 1902. Perennial to 25cm high with mostly branched hairs; basal leaves rosulate, linear to oblanceolate, 2-20mm long, 1-4mm wide, entire to dentate; cauline leaves lanceolate to oblanceolate; sepals 1-2mm long; petals 2-4mm long, white; fruits elliptic-lanceolate to oblong, pubescent. Mostly alpine and subalpine. SW, NW, NE. D. lanceolata Royle of authors.

Draba crassa Rydb., Mem. N. Y. Bot. Gard. 1:182. 1900. Perennial to 15cm high, with simple and sometimes branched hairs, somewhat fleshy; basal leaves rosulate, the blades oblanceolate, 1-6cm long, 2-10mm wide, entire; cauline leaves ovate to obovate; sepals 2-3mm long; petals 3.5-7mm long, yellow; fruits elliptic-lanceolate, glabrous or not. Ridges and talus in the high mountains. Teton and Fremont Cos. D. chrysantha Wats. in part.

Draba crassifolia Grah., Edinb. New Phil. Journ. 7:182. 1829. Annual to perennial to 20cm high, with simple and sometimes branched hairs; leaves mostly basal and rosulate, linear-spatulate to oblanceolate, 2-25mm long, 1-5mm wide; cauline leaves usually lacking, occasionally 1 or 2, rarely more; sepals 1-2mm long; petals 2-3mm long, yellow or fading to white; fruits elliptic or lanceolate, glabrous, the style usually obsolete. Usually alpine and subalpine. NW, SW, SE.

Draba densifolia Nutt. in T. & G., Fl. N. Am. 1:104. 1838. Mat forming, scapose perennial to 15cm high, with simple and branched hairs; leaves numerous and imbricate, linear to oblanceolate, 2-8mm long, mostly 0.5-1.5mm wide, with stiff, simple hairs on the margins; sepals 2-3mm long; petals 2-5mm long, yellow; fruits ovate to elliptic, usually pubescent. Rocky ridges in the high mountains. Yellowstone Park and Sublette Co. D. globosa Payson.

Draba fladnizensis Wulfen in Jacq., Misc. Austr. Bot. 1:147. 1779. Perennial to 10cm high, with simple and branched hairs; leaves usually all basal, the blades mostly oblanceolate, 3-16mm long, 1-3mm wide; sepals 1-2mm long; petals 2-3mm long, white; fruits mostly oblong-ovate, usually glabrous. Alpine and subalpine. Yellowstone Park and Albany Co.

Draba glabella Pursh, Fl. Am. Sept. 434. 1814. Perennial to 3dm high, with branched and some simple hairs; leaves mostly basal and rosulate, the blades oblanceolate to elliptic, 0.4-2cm long, 1-8mm wide, entire or denticulate; sepals 2-3mm long; petals 3-5mm long, white; fruits linear-lanceolate or lanceolate, usually glabrous. Alpine and subalpine. Park Co.

Draba incerta Payson, Am. Journ. Bot. 4:261. 1917. Caespitose perennial to 20cm high, with simple and branched hairs; leaves linear-oblanceolate to oblanceolate, 3-15mm long, 1-3.5mm wide; scapes leafless or nearly so; sepals 2.5-3.5mm long; petals 4-5mm long, yellow or fading to white; fruits ovate to elliptic, usually pubescent. Alpine and subalpine. NW, SW.

Draba lonchocarpa Rydb., Mem. N. Y. Bot. Gard. 1:181. 1900. Caespitose, scapose perennial to 12cm high, with branched and some simple hairs; leaves linear to oblanceolate, 3-15mm long, 1-3.5mm wide; sepals 1-2mm long; petals 2.5-4mm long, white; fruits linear to elliptic or oblanceolate, glabrous or not. Alpine and subalpine where rocky, rarely lower. NW, SW.

Draba nemorosa L., Sp. Pl. 643. 1753. Annual to 3dm high, with simple and branched hairs; leaves ovate to obovate, 2-30mm long, 1-15mm wide, entire to dentate; sepals 1-2mm long; petals 2-4mm long, yellow to whitish; fruits elliptic to oblong-oblanceolate, glabrous or not. Plains, hills, and slopes. Statewide.

Draba nivalis Liljeb., Vet. Acad. Handl. 208. 1793. Strongly caespitose perennial to 8cm high, with branched and a few simple hairs; leaves mostly basal, the blades linear to oblanceolate or obovate, 2-10mm long, 0.5-3.5mm wide; sepals 1-2mm long; petals 2.5-3.5mm long, white; fruits linear to elliptic or oblong-lanceolate, glabrous or not. Rocky areas in the high mountains. NW, SW, SE.

Draba oligosperma Hook., Fl. Bor. Am. 1:51. 1830. Caespitose, scapose perennial to 10cm high, with simple and branched hairs; leaves linear to linear-spatulate, 2-8mm long, 0.5-1.5mm wide; sepals 1-2.5mm long; petals 3-5mm long, yellow, fading to white; fruits ovate to obovate, usually pubescent. Rocky plains, ridges, and slopes. Statewide. D. andina (Nutt.) A. Nels., D. saximontana A. Nels.

Draba paysonii Macbr., Contr. Gray Herb. n. s. 56:52. 1918. Scapose, matted perennial to 6cm high, with simple and forked hairs; leaves linear to spatulate, 2-14mm long, 0.7-2mm wide; sepals 1-2mm long; petals 2-5mm long, yellow; fruits ovate to ovate-lanceolate, usually pubescent. Alpine and subalpine. Park and Fremont Cos. D. vestita Payson.

Draba pectinipila Rollins, Rhodora 55:231. 1953. Caespitose perennial to 10cm high, with branched and simple hairs; leaves mostly basal, the blades linear to linear-oblanceolate, 3-12mm long, 0.7-1.5mm wide; sepals 2-3.5mm long; petals 4-5mm long, white or yellow; fruits oblong-oblanceolate or lanceolate, pubescent. Rocky areas. Park Co.

Draba praealta Greene, Pittonia 3:306. 1898. Annual to perennial to 3dm high, with simple and forked hairs; basal leaves rosulate, oblanceolate, 5-30mm long, 1-9mm wide, entire or remotely toothed; sepals 1-2mm long; petals 2-4mm long, white or rarely yellowish; fruits lanceolate, pubescent. Woods and slopes. NW, SW. *D. lapilutea* A. Nels., *D. yellowstonensis* A. Nels.

Draba reptans (Lam.) Fern., Rhodora 36:368. 1934. Annual to 20cm high, with simple and branched hairs; leaves mostly basal and on lower part of stem, spatulate-obovate, 3-20mm long, 2-12mm wide, entire; sepals about 2mm long; petals 3-5mm long, white; fruits linear, glabrous or not. Plains and hills. NW, NE, SE. *D. caroliniana* Walt.

Draba stenoloba Ledeb., Fl. Ross. 1:154. 1841. Annual to perennial to 4dm high, with simple and branched hairs; basal leaves rosulate, obovate to oblanceolate, 0.5-5cm long, 2-15mm wide, denticulate; cauline leaves becoming lanceolate or ovate; sepals 1-2mm long; petals 2-4.5mm long, yellow to white; fruits linear to oblong, usually glabrous. Meadows, hills, and slopes. Statewide. *D. nitida* Greene, *D. deflexa* Greene (as *D. reflexa* in Coult. & Nels., 1909).

Draba streptocarpa Gray in Parry, Am. Journ. Sci. II, 33:242. 1862. Tufted perennial to 3dm high, with simple and branched hairs; leaf blades ovate or elliptic-lanceolate to oblanceolate, 8-35mm long, 2-10mm wide, mostly entire; sepals 2.5-4mm long; petals mostly 6-8mm long, yellow; fruits lanceolate, glabrous or pubescent along margins. Woods, meadows, and slopes. Albany and Laramie Cos.

Draba ventosa Gray in Parry, Am. Nat. 8:212. 1874. Caespitose, scapose perennial to 8cm high, with simple and branched hairs; leaves ovate-elliptic to oblanceolate, 4-15mm long, 1-4mm wide; sepals 2.5-3.5mm long; petals 4-6mm long, yellow; fruits oval to ovate, pubescent. Ridges and slopes of high mountains. NW, SW.

Erucastrum Presl

Erucastrum gallicum (Willd.) Schulz, Bot. Jahrb. 54 (Beibl. 119): 56. 1916. Sparsely pubescent annual or biennial to 8dm high, with simple hairs; leaves pinnatifid, the blades oblanceolate to obovate, 2-20cm long, 1-8cm wide; flowers in racemes; sepals 2-4mm long; petals pale yellow, 4-7mm long; fruits linear, subterete; seeds uniseriate. Disturbed areas. Weston Co.

Erysimum L. Wallflower

Annual to perennial herbs with 2 or 3 branched hairs and entire to sinuate-dentate or lyrate-pinnatifid leaves; flowers in racemes; petals yellow to reddish or purplish; fruits linear, usually terete or quadrangular; seeds uniseriate.

1. Petals over 11mm long; style usually over 1.5mm long *E. asperum*
1. Petals mostly 3.5-11mm long; style rarely over 1.5mm long
   2. Pedicel as thick as the fruit or nearly so; fruits mostly spreading, 5-10cm long; plants annual *E. repandum*
   2. Pedicel either about half as thick as fruit or the fruits ascending to erect and usually less than 5cm long and the plants biennial or perennial

3. Plants annual; petals 3.5-5mm long; fruits 1.5-3cm long E. cheiranthoides

3. Plants biennial or perennial; petals 5-11mm long; fruits mostly 2.5-5cm long E. inconspicuum

Erysimum asperum (Nutt.) DC., Regni Veg. 2:505. 1821. Biennial or perennial to 1m high; basal leaves somewhat rosulate, the blades mostly oblanceolate, 1-12cm long, 2-15mm wide, entire to dentate or lyrate-pinnatifid; cauline leaves similar; sepals 6-14mm long; petals 15-25mm long, yellow to reddish or purplish; styles mostly 1.5-4mm long. Plains, hills, and slopes. Statewide. Cheiranthus aridus A. Nels., E. aridum (A. Nels.) A. Nels., E. asperrimum (Greene) Rydb., E. argillosum (Greene) Rydb., E. capitatum (Dougl.) Greene, E. nivale (Greene) Rydb. The latter 3 are usually considered distinct species but they all intergrade. Until they are studied in detail, it seems best to consider them all synonymous with E. asperum.

Erysimum cheiranthoides L., Sp. Pl. 661. 1753. Annual to 1m high; leaves cauline, the blades linear, lanceolate, or oblanceolate, mostly 2-11cm long, 2-18mm wide, entire to denticulate; sepals 2-3mm long; petals 3.5-5mm long, yellow; styles about 1mm or less long. Open areas. Statewide.

Erysimum inconspicuum (Wats.) MacM., Metasp. Minn. Valley 268. 1892. Biennial or perennial to 6dm high; leaves basal and cauline, the blades linear to lanceolate or oblanceolate, 1-8cm long, 1-7mm wide, entire to denticulate; sepals 4-7mm long; petals 5-11mm long, yellow; styles about 1mm long. Dry, open areas. Statewide.

Erysimum repandum L., Demonstr. Pl. 17. 1753. Annual to 5dm high; basal leaves somewhat rosulate but early deciduous, the blades mostly lanceolate to oblanceolate or oblong, 0.3-10cm long, 1-10mm wide, usually sinuate-dentate; cauline leaves similar but reduced; sepals 3-6mm long; petals 4-10mm long, yellow; styles about 1mm long or apparently longer. Plains, hills, and slopes. SW, SE, NE.

Euclidium R. Br.

Euclidium syriacum (L.) R. Br. in Ait., Hort. Kew. II, 4:74. 1812. Branched annual to 4dm high, with branched hairs; leaves mostly cauline, the blades oblong-oblanceolate or elliptic, 1-5cm long, 3-15mm wide, remotely dentate; flowers in spike-like racemes with the flowers remote; sepals about 1mm long; petals about 1mm long, white; fruit obovoid, 2-3mm long, with a curved beak nearly as long, pubescent, 2 seeded. Disturbed areas. Natrona and Johnson Cos.

<u>Halimolobos</u> Tausch

<u>Halimolobos</u> <u>virgata</u> (Nutt.) Schulz, Pflanzenr. IV, 105, Heft 86: 290. 1924. Annual to perennial to 5dm high, with simple and branched hairs; basal leaves rosulate, the blades lanceolate to oblanceolate or obovate, 1-6cm long, 3-20mm wide, dentate; cauline leaves reduced, becoming sessile and clasping; flowers in racemes; sepals 1.5-3mm long; petals 2-4mm long, white; fruits linear, terete-quadrangular, glabrous; seeds irregularly biseriate. Plains and hills. Albany Co. <u>Arabis</u> <u>brebneriana</u> A. Nels., <u>Stenophragma</u> <u>virgata</u> (Nutt.) Greene.

<u>Hesperis</u> L. Sweet Rocket

<u>Hesperis</u> <u>matronalis</u> L., Sp. Pl. 663. 1753. Perennial to 13dm high, with simple and some branched hairs; leaf blades lanceolate to elliptic, 4-20cm long, 1-5cm wide, serrate-dentate; flowers in compound racemes; sepals 5-9mm long; petals 16-25mm long, white or rose to purple; fruits linear, usually somewhat torulose; seeds uniseriate. Disturbed areas. NE, SE.

Hutchinsia R. Br.

Hutchinsia procumbens (L.) Desv., Journ. Bot. 3:168. 1814. Glabrous or glabrate annual to 15cm high; leaf blades obovate or oblanceolate, 2-20mm long, 1-7mm wide, entire to pinnatifid, becoming linear above; flowers in racemes; sepals 1.5mm or less long; petals about 1mm long, white; fruits elliptic to obovate, flattened, 2-4mm long; seeds mostly biseriate. Moist, often alkaline areas. SW, SE.

Isatis L. Woad

Isatis tinctoria L., Sp. Pl. 670. 1753. Biennial or perennial to 1m high, with simple hairs; basal leaves rosulate, deciduous, the blades oblanceolate or obovate to elliptic, 4-15cm long, 2-4cm wide, entire or somewhat crenulate; cauline leaves lanceolate to elliptic or oblong, auriculate; flowers in compound racemes; sepals 1.5-2.5mm long; petals 2.5-4mm long, yellow; fruits oblong to club-shaped, flattened, the partition lacking; seeds 1 per fruit. Disturbed areas. SW, SE.

<u>Lepidium</u> L. Pepper Grass

Annual to perennial herbs with simple or no hairs and simple to compound leaves; flowers in simple or compound racemes; sepals usually early deciduous; petals white to yellow, sometimes lacking; stamens 2, 4, or 6; fruits mostly ovate to obovate, flattened contrary to partition; seeds 1 per locule.

1. Stem leaves auriculate to cordate-clasping or perfoliate
  2. Upper leaves cordate-clasping to perfoliate; petals yellow; fruits about 4mm long — <u>L. perfoliatum</u>
  2. Upper leaves merely auriculate; petals white; fruits 5-6mm long — <u>L. campestre</u>
1. Stem leaves not auriculate nor cordate-clasping or perfoliate
  3. Plants perennial with mostly pinnate or pinnatifid leaves; style 0.3-1mm long, exceeding the sinus of fruit — <u>L. montanum</u>
  3. Plants annual, or perennial with the leaves various; style usually less than 0.3mm long or included in sinus of fruit
    4. Fruits about 2-2.5mm long; plants perennial with well developed rootstocks and large entire or dentate basal leaves to 30cm long and 8cm wide — <u>L. latifolium</u>
    4. Fruits usually at least 2.5mm long; plants annual, or biennial or perennial without rootstocks, the leaves mostly less than 10cm long and 3cm wide

5. Fruits oblong-obovate to obovate, usually averaging widest slightly above middle, the fruits rounded at tip; petals usually lacking or vestigial   L. densiflorum
5. Fruits elliptic to oval, averaging widest at or below middle; petals conspicuous, or if lacking, the fruits usually with somewhat pointed lobes at tip
    6. Fruits slightly if at all longer than wide, usually rounded at tip; petals usually conspicuous, 1-3mm long   L. virginicum
    6. Fruits definitely longer than wide, usually with somewhat pointed lobes at tip; petals lacking or vestigial   L. ramosissimum

Lepidium campestre (L.) R. Br. in Ait., Hort. Kew. II, 4:88. 1812. Pubescent annual to 5dm high; basal leaves rosulate, early deciduous, the blades oblanceolate, 3-10cm long, 5-12mm wide, entire to lyrate; cauline leaves oblong-lanceolate and auriculate; sepals 1-2mm long; petals 2-2.5mm long, white; stamens 6; fruits oblong-ovate. Disturbed areas. Teton Co.

Lepidium densiflorum Schrad., Ind. Sem. Hort. Gött. 4. 1832. Pubescent annual to 5dm high; leaf blades linear to oblanceolate, 1-8cm long, 1-20mm wide, entire to pinnatifid; sepals 1mm or less long; petals white, as long as sepals to lacking; stamens mostly 2, rarely 4; fruits oblong-obovate to obovate. Disturbed areas. Statewide. L. pubecarpum A. Nels., L. ramosum A. Nels., L. apetalum Willd. of authors.

Lepidium latifolium L., Sp. Pl. 644. 1753. Glabrous or glabrate perennial to 1.5m high; leaf blades ovate or lanceolate to elliptic, 2-30cm long, 0.5-8cm wide, entire to dentate; racemes often compound; sepals 1-2mm long, white margined; petals 1.5-2mm long, white; stamens 6; fruits ovate-rotund. Disturbed areas. Natrona and Uinta Cos.

Lepidium montanum Nutt. in T. & G., Fl. N. Am. 1:116. 1838. Pubescent perennial to 4dm high; basal leaf blades pinnately compound or pinnatifid, ovate to obovate, 1-5cm long, 5-25mm wide; cauline leaves reduced, sometimes becoming entire; sepals 1-2mm long; petals 2-4mm long, white or drying yellowish; stamens 6; fruits ovate to ovate-elliptic. Deserts and plains. SW, SE.

Lepidium perfoliatum L., Sp. Pl. 643. 1753. Annual to 6dm high, usually pubescent below; basal leaf blades much dissected into linear segments, oblong or elliptic, 2-12cm long, 1-4cm wide; upper cauline leaves cordate-clasping or perfoliate and mostly entire; sepals about 1mm long; petals about 1.5mm long, yellowish; stamens usually 6; fruits rhombic-ovate to orbicular. Disturbed areas. Statewide.

Lepidium ramosissimum A. Nels., Bull. Torrey Club 26:124. 1899. Pubescent annual or biennial to 5dm high; basal leaf blades pinnatifid but usually deciduous before flowering; lower cauline leaves entire to lobed, the blades elliptic to oblanceolate, 0.5-5cm long, 2-15mm wide; sepals about 1mm long; petals vestigial, white; stamens 2; fruits mostly elliptic. Plains, woods, and disturbed areas. NW, SW, SE.

Lepidium virginicum L., Sp. Pl. 645. 1753. Pubescent annual or biennial to 6dm high; basal leaf blades oblanceolate, 2-10cm long, 0.5-3cm wide, toothed to pinnate; cauline leaves reduced and often becoming entire; sepals about 1mm long; petals 1-3mm long, white, rarely vestigial; stamens 2, 4, or 6; fruits elliptic-rotund to orbicular. Disturbed areas. Lincoln and Albany Cos. L. medium Greene.

Lesquerella Wats.

Biennial or perennial herbs with branched hairs and simple, mostly basal, remotely toothed or entire, non-auriculate leaves; flowers in simple or compound racemes; petals yellow, sometimes red or purple tinged; fruits ovate to orbicular or oblong, flattened or inflated; seeds 1-10 per chamber.

Reference: Rollins, R. C. & E. A. Shaw. 1973. The genus Lesquerella (Cruciferae) in North America. Harvard Univ. Press. 288pp.

1. Fruits flattened at a right angle to the partition (sometimes not strongly so)
  2. Fruits strongly keeled on the flattened sides L. carinata
  2. Fruits not keeled on the flattened sides (the suture often slightly raised)
    3. Pedicels in fruit somewhat sigmoid; fruits strongly flattened; styles mostly 2-4mm long; petals 8-12mm long; Teton, Sublette, and Lincoln Cos. L. paysonii
    3. Pedicels in fruit recurved; fruits only somewhat flattened; styles 1.5-2mm long; petals 5-8mm long; Fremont Co. L. fremontii
1. Fruits inflated or sometimes flattened parallel to partition but only along the margins
  4. Pedicels in fruit uniformly recurved, not sigmoid; fruits mostly globose or nearly so
    5. Basal leaf blades, or some of them, ovate or obovate to orbicular, rarely elliptic to rhombic; styles 1.5-2mm long
      6. Fruits slightly flattened at a right angle to the partition; valves densely pubescent on outside, sparsely so inside; Fremont Co. L. fremontii
      6. Fruits inflated; valves somewhat sparsely pubescent on outside, glabrous inside; Sweetwater Co. L. macrocarpa
    5. Basal leaf blades linear to oblanceolate; styles over 2mm long

7. Stems often over 20cm long; fruiting racemes not secund; petals yellow — *L. ludoviciana*

7. Stems rarely over 20cm long; fruiting racemes usually secund; petals sometimes purplish or reddish — *L. arenosa*

4. Pedicels in fruit sigmoid or uniformly curved upward, or rarely straight; fruits often not globose

8. Fruits globose or nearly so, not flattened on margins or near tip — *L. multiceps*

8. Fruits not globose, either distinctly longer than wide or the margins or tip somewhat flattened

9. Petals 7-9mm long; plants not mat-forming; mature fruits mostly 5.5-8mm long, the styles 3-6mm long — *L. montana*

9. Petals 4.5-7 or rarely 8mm long; plants often mat-forming; mature fruits mostly 4-5.5mm long, the styles 2-4mm long

10. Basal leaf blades oblanceolate or narrower, rarely wider; widespread — *L. alpina*

10. Basal leaf blades, or some of them, rhombic, ovate, or subhastate; Uinta Co. — *L. prostrata*

<u>Lesquerella</u> <u>alpina</u> (Nutt.) Wats., Proc. Am. Acad. 23:251. 1888. Stems mostly ascending to erect or the plants mat forming, to 20cm high; leaves mostly basal and rosulate, the blades linear to oblanceolate or rarely obovate, 0.3-4cm long, 0.3-6mm wide; sepals 3-5mm long; petals 4.5-7mm long; fruits ovate, inflated but usually flattened on margins. Hills and slopes. Statewide. <u>L</u>. <u>condensata</u> A. Nels., <u>L</u>. <u>curvipes</u> A. Nels.

<u>Lesquerella</u> <u>arenosa</u> (Richards.) Rydb., Bull. Torrey Club 29:236. 1902. Stems mostly decumbent, to 25cm long; basal leaf blades linear to oblanceolate or rarely ovate, 0.5-4cm long, 1-7mm wide; cauline leaves becoming linear; sepals 4-6mm long; petals 6-9mm long, often tinged with red or purple; fruits globose or slightly elongated. Plains and hills. Central counties from Fremont Co. to E border.

<u>Lesquerella</u> <u>carinata</u> Rollins, Contr. Gray Herb. n. s. 171:42. 1950. Stems prostrate to ascending, to 15cm long; basal leaves rosulate, the blades rhombic or ovate to obovate, 5-25mm long, 2-25mm wide; cauline leaves few and reduced; sepals 4-6mm long; petals 6-9mm long; fruits oblong-elliptic, flattened contrary to partition and keeled on the margins and on the sutures. Hills and slopes. Teton Co.

Lesquerella fremontii Rollins & Shaw, Lesq. N. Am. 228. 1973. Stems prostrate or decumbent, to 15cm long; basal leaf blades elliptic to rhombic, 0.5-4cm long, 2-6mm wide; cauline leaves obovate to elliptic; sepals 3-5mm long; petals 5-8mm long; fruits subglobose to ellipsoid, slightly flattened contrary to partition. Hills and slopes. Endemic in Fremont Co.

Lesquerella ludoviciana (Nutt.) Wats., Proc. Am. Acad. 23:252. 1888. Stems decumbent to erect, to 4dm long; basal leaf blades linear or oblong-oblanceolate to oblanceolate, 1-8cm long, 1-10mm wide; cauline leaves reduced and becoming linear; sepals 4-6mm long; petals 6-9mm long; fruits globose or oblong-rotund. Plains, hills, and slopes. Statewide. L. argentea (Pursh) MacM., not of (Schauer) Wats.

Lesquerella macrocarpa A. Nels., Bot. Gaz. 34:366. 1902. Stems mostly decumbent, to 20cm long; basal leaf blades ovate or obovate to orbicular or some oblanceolate, 0.5-3cm long, 3-20mm wide; cauline leaves numerous, not reduced; sepals 3-5mm long; petals 4-7mm long; fruits globose. Plains and hills. Endemic in Sweetwater Co. This species approaches the genus Physaria.

Lesquerella montana (Gray) Wats., Proc. Am. Acad. 23:251. 1888. Stems decumbent to erect, to 3.5dm long; basal leaves rosulate, the blades ovate to oblanceolate or obovate, 0.5-6cm long, 2-30mm wide; cauline leaves often only slightly reduced; sepals 4-6mm long; petals 7-9mm long; fruits ovate to oblong, inflated. Plains and hills. SE. *L. rosulata* A. Nels.

Lesquerella multiceps Maguire, Am. Midl. Nat. 27:465. 1942. Stems prostrate to ascending, to 20cm long; basal leaf blades elliptic to oblanceolate or obovate, 0.5-3cm long, 3-15mm wide; cauline leaves reduced, oblanceolate; sepals 4-5mm long; petals 7-10mm long, sometimes drying pink; fruits globose. Rocky hills and slopes. Lincoln Co.

Lesquerella paysonii Rollins, Contr. Gray Herb. n. s. 171:44. 1950. Stems decumbent, to 12cm long; basal leaves rosulate, the blades ovate or rhombic to elliptic or oblanceolate, 3-25mm long, 2-17mm wide; cauline leaves reduced; sepals 5-7mm long; petals 8-12mm long, sometimes purplish; fruits elliptic, somewhat flattened contrary to partition. Rocky slopes and ridges. NW, SW.

Lesquerella prostrata A. Nels., Bull. Torrey Club 26:124. 1899. Stems prostrate to ascending, to 15cm long; basal leaf blades ovate or rhombic to subhastate or some oblanceolate, 4-15mm long, 2-14mm wide; cauline leaves becoming linear and reduced; sepals 4-5mm long; petals 5-8mm long; fruits ovate, inflated. Plains, hills, and slopes. Uinta Co.

Lobularia Desv. Sweet Alyssum

Lobularia maritima (L.) Desv., Journ. Bot. 3:162. 1814. Annual to 3dm high, with appressed, 2 branched hairs; leaves linear to linear-oblanceolate, 1-4cm long, 1-4mm wide, entire; flowers in racemes; sepals 1.5-2mm long; petals 2.5-4mm long, white or bluish tinged; fruits elliptic, inflated. Disturbed areas. Albany Co.

Malcolmia R. Br.

Malcolmia africana (L.) R. Br. in Ait., Hort. Kew. II, 4:121. 1812. Annual to 4dm high, with branched hairs; leaf blades oblanceolate to elliptic, 1-8cm long, 5-30mm wide, remotely dentate; flowers in racemes; sepals 3.5-5mm long; petals 6-11mm long, pinkish; fruits linear, terete; seeds uniseriate. Plains, hills, and disturbed areas. NW, SW.

<u>Nasturtium</u> R. Br. Water Cress

<u>Nasturtium</u> <u>officinale</u> R. Br. in Ait., Hort. Kew. II, 4:110. 1812. Plants glabrous; stems usually floating or creeping and rooting at the nodes, to 6dm long; leaf blades obovate to ovate or oblong, 2-15cm long, 1-7cm wide, mostly pinnately compound but often a few simple; sepals 2-3mm long; petals 3-5mm long, white or purple tinged; fruits linear or oblong, terete, usually curved; seeds mostly biseriate. Shallow water and shores. NW, NE, SE. <u>Rorippa</u> <u>nasturtium-aquaticum</u> (L.) Schinz & Thell.

<u>Parrya</u> R. Br.

<u>Parrya</u> <u>nudicaulis</u> (L.) Boiss., Fl. Orient. 1:159. 1867. Usually rhizomatous, glabrous or glandular-pubescent perennial to 2dm high; leaves mostly basal, the blades mostly oblanceolate to obovate, entire to coarsely dentate, 0.5-10cm long, 5-25mm wide; flowers in racemes; sepals 6-8mm long; petals 12-25mm long, white or purple; fruits oblong, flattened parallel to partition, 4mm or more wide when mature; seeds uniseriate or biseriate, winged. Alpine and subalpine, often on limestone. NW, SW.

__Physaria__ (Nutt.) Gray Bladderpod

Perennial herbs with branched hairs and mostly basal, rosulate leaves; flowers in racemes; petals usually yellow, rarely purplish; fruits didymous, mostly orbicular to reniform in outline, inflated; seeds 1-4 per locule, biseriate.

References: Mulligan, G. A. 1967. Canad. Journ. Bot. 45:1887-1898.
1968. Canad. Journ. Bot. 46:735-740.

1. Plants very compact, the stems 3cm long or less, the leaf blades all less than 1cm long; SW Wyoming __P. condensata__
1. Plants not as above
  2. Partition of fruit usually obovate; ovules and funiculi 4 in each locule, 1 or 2 ovules sometimes aborting __P. didymocarpa__
  2. Partition of fruit narrowly oblong to linear; ovules and funiculi 1 or 2 in each locule
    3. Sinuses of mature fruit usually about equal above and below __P. australis__
    3. Sinuses of fruit unequal, the upper one very deep, the lower one absent or nearly so
      4. Fruits obcordate, acute at base __P. brassicoides__
      4. Fruits somewhat rectangular in outline, obtuse or truncate at base __P. vitulifera__

Physaria australis (Payson) Rollins, Rhodora 41:408. 1939. Stems decumbent, to 18cm long; basal leaf blades mostly obovate to orbicular, 0.7-5cm long, 5-35mm wide, entire or with a few teeth; sepals 5-7mm long; petals about 1cm long; fruits with a deep, narrow notch at base and tip. Plains and hills. Statewide. These plants have been called *P. acutifolia* Rydb. by Mulligan but there is still some question as to proper interpretation.

Physaria brassicoides Rydb., Bull. Torrey Club 29:237. 1902. Stems decumbent to ascending, to 15cm long; basal leaf blades mostly orbicular to obovate, 1-5cm long, 5-35mm wide, usually entire or nearly so; sepals 6-8mm long; petals about 1cm long; fruits obcordate, acute or with an obscure notch at base, deeply notched at tip. Plains and hills. Crook Co.

Physaria condensata Rollins, Rhodora 41:407. 1939. Stems decumbent to ascending, to only 3cm long; basal leaf blades obovate, 3-10mm long, 2-8mm wide, entire; sepals 4-6mm long; petals 4-7mm long; fruits deeply notched at base and tip. Plains and hills. Uinta Co.

Physaria didymocarpa (Hook.) Gray, Gen. Pl. U. S. 1:162. 1848. Stems decumbent to ascending, to 2dm long; basal leaf blades obovate or oblanceolate to rhombic or ovate, 1-5cm long, 3-35mm wide, usually few-toothed; sepals 6-8mm long; petals 8-12mm long; fruits deeply notched at base and tip. Banks and slopes. SW, NW, NE.

Physaria _vitulifera_ Rydb., Bull. Torrey Club 28:278. 1901. Stems decumbent to ascending, to 2dm long; basal leaf blades obovate to oblanceolate, 1-5cm long, 0.5-3cm wide, usually toothed; sepals 5-8mm long; petals about 8mm long; fruits obtuse or truncate or very slightly notched at base, deeply and broadly notched at tip. Hills and slopes. Carbon Co.

Raphanus L. Radish

Annual or biennial herbs with lyrate-pinnatifid or pinnate, non-auriculate leaves, at least below, and simple hairs; flowers in simple or compound racemes; petals purple, yellow, or white; fruits linear to lance-ovoid, terete, often torulose, prominently beaked; seeds uniseriate.

Fruits usually 4-12 seeded, 1.5-4mm wide _R_. _raphanistrum_

Fruits usually 1-3 seeded, 5-10mm wide _R_. _sativus_

Raphanus _raphanistrum_ L., Sp. Pl. 669. 1753. Stems to 8dm high; basal leaf blades usually lyrate-pinnatifid or compound, 4-15cm long, 1.5-6cm wide; cauline leaves reduced and becoming merely toothed; sepals 5-9mm long; petals 1-2cm long, usually yellow or white; fruits prominently grooved lengthwise, eventually breaking transversely between the 1 seeded segments. Disturbed areas. Sheridan Co.

Raphanus sativus L., Sp. Pl. 669. 1753. Stems to 1m high; basal leaf blades lyrate-pinnatifid or pinnate with unequal leaflets, 4-20cm long, 2-15cm wide; cauline leaves reduced and becoming merely toothed; sepals 7-10mm long; petals 12-20mm long, usually purple or white; fruits only slightly grooved, usually not breaking transversely. Disturbed areas. Sheridan Co.

Rorippa Scop. Yellow Cress

Annual to perennial herbs with simple hairs or glabrous; leaves simple or compound; flowers in racemes; petals yellow, often drying white; fruits oval to linear, mostly terete; seeds usually numerous, biseriate or sometimes irregularly uniseriate.

Reference: Stuckey, R. L. 1972. Sida 4:279-430.

1. Plants perennial with rhizomes; petals mostly 3.5-5mm long ... *R. sinuata*
1. Plants annual or biennial or short-lived perennials without rhizomes; petals mostly 0.5-3.5mm long
    2. Fruits oval to ovate, mostly 1-1.4 times as long as wide; petals 0.6-1.2mm long ... *R. sphaerocarpa*
    2. Fruits oblong to linear or lanceolate, 1.5-7 times as long as wide, or if less, the petals mostly over 1.2mm long

3. Pedicels mostly 3-13mm long, usually as long as or longer than the fruits; stems mostly erect, (1.5) 3-10dm long R. palustris

3. Pedicels mostly 1-5mm long, usually shorter than the fruits; stems often spreading to decumbent, rarely over 5dm long

4. Valves of fruit minutely papillate; petals 0.6-0.8mm long R. tenerrima

4. Valves of fruit smooth; petals 0.5-3.5mm long

5. Replum margin minutely hirsute; pedicels and stem often with short, retrorse hairs; petals spreading; NW Wyoming R. curvisiliqua

5. Replum margin glabrous; pedicels and stem without retrorse hairs; petals erect; widespread R. curvipes

Rorippa curvipes Greene, Pittonia 3:97. 1896. Glabrous annual or short-lived perennial with prostrate to erect stems to 5dm long; leaf blades obovate to oblong, 1-10cm long, 3-15mm wide, entire to pinnatifid; sepals 0.5-1.5mm long; petals 0.5-2.5mm long; fruits oblong or lanceolate. Moist areas. Statewide. R. obtusa (Nutt.) Britt. of authors.

Rorippa curvisiliqua (Hook.) Bessey ex Britt., Mem. Torrey Club 5:169. 1894. Annual or biennial with prostrate to erect stems to 5dm long; leaf blades mostly oblong-oblanceolate, 1-10cm long, 5-30mm wide, toothed to pinnately compound; sepals 1-2.5mm long, often early deciduous; petals 0.6-3mm long; fruits oblong or lanceolate to linear. Moist areas. NW. *R. lyrata* (Nutt.) Greene.

Rorippa palustris (L.) Besser, Enum. Pl. Volh. 27. 1822. Annual or biennial with mostly erect stems to 1m high; leaf blades ovate-oblong to obovate, 2-17cm long, 0.5-6cm wide, toothed to pinnatifid; sepals 1-2mm long, usually early deciduous; petals 1-2mm long; fruits oblong to orbicular. Moist areas. Statewide. *R. islandica* (Oed.) Borbas of authors, *R. hispida* (Desv.) Britt.

Rorippa sinuata (Nutt.) Hitchc., Spr. Fl. Manhat. 18. 1894. Rhizomatous perennial to 5dm high; leaf blades oblong-oblanceolate, 2-8cm long, 4-18mm wide, deeply pinnately lobed; sepals 2-4mm long; petals 3.5-5mm long; fruits oblong or lanceolate. Woods, ditches, and meadows where moist. Statewide.

<u>Rorippa</u> <u>sphaerocarpa</u> (Gray) Britt., Mem. Torrey Club 5:170. 1894. Annual with erect to decumbent stems to 4dm long; leaf blades oblong-oblanceolate, 2-6cm long, 7-13mm wide, shallowly lobed to pinnately compound; sepals 0.5-1.5mm long, early deciduous; petals 0.6-1.2mm long; fruits oval to ovate. Moist areas. SW, SE. <u>R</u>. <u>obtusa</u> (Nutt.) Britt. of authors in part.

<u>Rorippa</u> <u>tenerrima</u> Greene, Erythea 3:46. 1895. Glabrous annual with decumbent to prostrate stems to 2dm long; leaf blades oblanceolate or oblong, 2-5cm long, 8-15mm wide, lyrate-pinnatifid; sepals 0.5-1.5mm long; petals 0.6-0.8mm long; fruits oblong or lanceolate. Moist areas. NE, SE.

<u>Sisymbrium</u> L.

Annual to perennial, hirsute to glabrous herbs with simple or compound leaves; flowers in racemes; petals yellow, white, or purple; fruit linear, terete; seeds uniseriate.

Rhizomes present; upper stem leaves mostly entire — <u>S</u>. <u>linifolium</u>

Rhizomes lacking; upper stem leaves not entire

  Pedicels nearly as thick as the fruits, the fruits mostly 5-10cm long — <u>S</u>. <u>altissimum</u>

  Pedicels much thinner than the fruits, the fruits mostly 2-3.5cm long — <u>S</u>. <u>loeselii</u>

Sisymbrium altissimum L., Sp. Pl. 659. 1753. Annual or biennial to 15dm high; leaf blades mostly lanceolate to oblanceolate, 3-20cm long, 1-6cm wide, mostly pinnately compound or pinnatifid; sepals 3-5mm long; petals yellow to white, 6-8mm long. Disturbed areas. Statewide.

Sisymbrium linifolium (Nutt.) Nutt. ex T. & G., Fl. N. Am. 1:91. 1838. Rhizomatous perennial to 7dm high; leaves mostly linear or oblanceolate, 1-7cm long, 1-20mm wide, entire to pinnatifid; sepals 3.5-5mm long; petals yellow, 5-10mm long. Plains, hills, and slopes. NW, SW, SE. *Schoenocrambe linifolia* (Nutt.) Greene.

Sisymbrium loeselii L., Cent. Pl. 1:18. 1755. Annual or biennial to 12dm high; leaf blades deltoid to lanceolate or elliptic, 3-15cm long, 1-8cm wide, irregularly pinnate or pinnatifid; sepals 3-4mm long; petals yellow, 5-8mm long. Disturbed areas. Fremont Co.

Smelowskia Meyer

Smelowskia calycina (Steph. ex Willd.) Meyer in Ledeb., Fl. Alt. 3:170. 1831. Somewhat matted perennial from a branching caudex, to 2dm high, with simple and branched hairs; leaves mostly basal, the blades ovate to obovate, 0.5-5cm long, 3-16mm wide, pinnatifid to pinnately compound; flowers in racemes; sepals 2.5-3.5mm long, early deciduous; petals 4-8mm long, white to purple tinged; fruits linear-oblong to elliptic, terete or flattened, 5-11mm long; seeds uniseriate. Alpine or subalpine. NW, SW. *S. americana* Rydb.

Stanleya Nutt. Prince's Plume

Biennial or perennial, glabrous to pilose herbs with simple or compound leaves; flowers in racemes; petals yellow; fruits linear, long-stipitate, terete or flattened; seeds uniseriate.

Stem leaves mostly sessile and auriculate *S. viridiflora*
Stem leaves mostly petioled, not auriculate
  Lower leaves and stem long hairy; claws of petals glabrous; NW Wyoming *S. tomentosa*
  Lower leaves and stem glabrous or nearly so, or if (as rarely) hairy, the claws of petals pubescent; widespread *S. pinnata*

Stanleya pinnata (Pursh) Britt., Trans. N. Y. Acad. Sci. 8:62. 1889. Perennial to 15dm high; lower leaves often pinnately compound, deciduous, the blades ovate to elliptic, 1-15cm long, 0.1-6cm wide; upper leaves becoming merely toothed or entire; sepals 8-16mm long; petals 8-16mm long; fruits spreading to ascending, the stipe 7-20mm long. Plains, hills, and slopes. Statewide. *S. integrifolia* James, *S. bipinnata* Greene.

Stanleya tomentosa Parry, Am. Nat. 8:212. 1874. Biennial or perennial to 18dm high; basal leaf blades mostly lanceolate to oblanceolate, 5-16cm long, 2-6cm wide, usually with several isolated lobes or leaflets at base and becoming entire toward tip; cauline leaves becoming entire; sepals 10-16mm long; petals 10-20mm long; fruits spreading to erect, the stipe 1-3cm long. Sagebrush and plains. NW.

Stanleya viridiflora Nutt. in T. & G., Fl. N. Am. 1:98. 1838. Glabrous perennial to 12dm high; basal leaves usually rosulate, simple, the blades oblanceolate to obovate, 2-25cm long, 1-6cm wide, entire to toothed or irregularly lobed; cauline leaves becoming sessile and auriculate, oblong to lanceolate; sepals 12-20mm long; petals 15-20mm long; fruits mostly spreading, the stipe 1-2.5cm long. Plains, hills, and slopes. NW, SW, SE.

Streptanthella Rydb.

Reference: Payson, E. B. 1922. Ann. Mo. Bot. Gard. 9:233-324.

Streptanthella longirostris (Wats.) Rydb., Fl. Rocky Mts. 364, 1062. 1917. Glabrous annual to 5dm high; leaf blades linear to oblanceolate, 1-6cm long, 1-12mm wide, sinuate-dentate to entire; flowers in racemes; sepals 3-6mm long; petals 5-7mm long, white or yellow; fruits linear, reflexed, flattened parallel to partition, beaked; seeds uniseriate. Sagebrush and desert. NW, SW, SE. Streptanthus longirostris (Wats.) Wats.

Streptanthus Nutt. Twistflower

Streptanthus cordatus Nutt. in T. & G., Fl. N. Am. 1:77. 1838. Glabrous perennial to 9dm high; basal leaf blades spatulate, 1-8cm long, 5-30mm wide, usually dentate near tip; cauline leaves becoming cordate and clasping, to 4cm wide; flowers in racemes; sepals 6-11mm long; petals 9-16mm long, yellow or cream to purple; fruits linear, flattened parallel to partition, erect to ascending; seeds uniseriate. Dry areas. SW.

Subularia L. Awlwort

Reference: Mulligan, G. A. & J. A. Calder. 1964. Rhodora 66:127-135.

Subularia aquatica L., Sp. Pl. 642. 1753. Glabrous, subscapose, aquatic annual to 15cm high; leaves linear and subterete, 1-6cm long, 0.2-2mm wide; flowers in racemes, the perianth less than 1.5mm long; fruits oval to elliptic, about 3mm long, terete or slightly flattened contrary to partition; seeds several, irregularly biseriate. Usually submerged in ponds. NW.

Thelypodiopsis Rydb.

Thelypodiopsis elegans (Jones) Rydb., Bull. Torrey Club 34:432. 1907. Glabrous to villous or hirsute biennial or short-lived perennial to 1m high; leaf blades ovate to oblong, 1-9cm long, 3-18mm wide, entire to lyrate or dentate, mostly auriculate; sepals 3.5-6mm long; petals purple or white, 7-14mm long; fruits linear, terete, spreading to ascending; seeds uniseriate. Plains and hills. SW, SE. *Thelypodium elegans* Jones, *Sisymbrium elegans* (Jones) Payson, *Streptanthus wyomingensis* A. Nels., *Thelypodiopsis wyomingensis* (A. Nels.) Rydb.

Thelypodium Endl.

Biennial to perennial, glabrous to hirsute herbs with entire or sinuate-dentate leaves, the basal petioled, the cauline often sessile; flowers in corymbiform racemes; petals white to rose or purple; fruits linear, usually terete, often torulose; seeds uniseriate.

Reference: Al-Shehbaz, I. A. 1973. Contr. Gray Herb. 204:3-148.

Stem leaves not auriculate T. *integrifolium*

Stem leaves mostly auriculate

Mature fruits 0.5-1.2mm wide; fresh seeds somewhat flattened, 0.7-1.3(1.5)mm long; petals 0.5-3(4)mm wide T. *sagittatum*

Mature fruits 1.3-2.3mm wide; fresh seeds not flattened, (1.3)1.5-2mm long; petals 2-6mm wide T. *paniculatum*

Thelypodium integrifolium (Nutt.) Endl. in Walpers, Repert. 1:172. 1842. Glabrous biennial to 15dm high; basal leaves early deciduous, the blades ovate-lanceolate to oblanceolate, 4-30cm long, 1-8cm wide; cauline leaves lanceolate to oblanceolate or linear, mostly not auriculate; sepals 3-5mm long; petals 5-9mm long; fruits mostly ascending, usually torulose, the stipe to 2mm long. Sagebrush, slopes, and streambanks. NW, SW, SE. T. *lilacinum* Greene.

Thelypodium paniculatum A. Nels., Bull. Torrey Club 26:126. 1899. Glabrous or rarely hirsute biennial or short-lived perennial to 7dm high; basal leaf blades oblanceolate or oblong, 1-15cm long, 0.5-3cm wide; cauline leaves linear-lanceolate to lanceolate or oblong, auriculate; sepals 3-6mm long; petals 6-12mm long; fruits erect to ascending, often somewhat torulose, the stipe 1mm or less long. Plains, hills, and meadows. NW, SW, SE.

Thelypodium sagittatum (Nutt.) Endl. in Walpers, Repert. 1:172. 1842. Glabrous to hirsute biennial or short-lived perennial to 7dm high; basal leaf blades oblanceolate to elliptic, 1-12cm long, 5-30mm wide; cauline leaves mostly lanceolate, auriculate; sepals 3-7mm long; petals 6-16mm long; fruits erect to ascending, somewhat torulose, the stipe 0.5mm or less long. Plains, hills, and meadows. Carbon and Albany Cos. *T. torulosum* Heller.

Thlaspi L. Penny Cress

Annual or perennial, glabrous herbs with simple, entire or toothed to lobed leaves, the upper leaves auriculate; flowers in racemes; petals white; fruits orbicular to oblong-obcordate or elliptic to oblanceolate, flattened contrary to the partition; seeds 2 to several per locule, uniseriate.

Reference: Holmgren, P. K. 1971. Mem. N. Y. Bot. Gard. 21(2): 1-106.

1. Plants annual; fruits oval to oblong-obcordate, notched 1mm or more deep at tip *T. arvense*
1. Plants perennial; fruits elliptic to oblanceolate, slightly if at all notched at tip

2. Petals 2.5-4mm long; styles about 0.5mm long or less — T. parviflorum

2. Petals 4-7mm long; styles 1-3mm long — T. montanum

Thlaspi arvense L., Sp. Pl. 646. 1753. Annual to 5dm high; leaf blades oblanceolate to lanceolate or oblong, 1-9cm long, 3-28mm wide, sinuate to lobed or sometimes entire; sepals 1.5-2.5mm long; petals 2-4mm long; fruits oval to oblong-obcordate, the sinus at tip 1-2.5mm deep; styles nearly obsolete. Disturbed areas. Statewide.

Thlaspi montanum L., Sp. Pl. 647. 1753. Perennial to 3dm high; basal leaves rosulate, the blades oblanceolate or obovate to ovate, 3-25mm long, 1-14mm wide, entire or toothed; cauline leaves ovate or cordate to oblong-elliptic; sepals 2-3mm long; petals 4-7mm long; fruits elliptic, cuneate, or oblanceolate; styles 1-3mm long. Hills and slopes. NW, SW, SE. T. fendleri Gray, T. glaucum (A. Nels.) A. Nels., T. coloradense Rydb.

Thlaspi parviflorum A. Nels., Bull. Torrey Club 27:265. 1900. Similar to above species but the sepals 1-2mm long, the petals 2.5-4mm long, and the styles about 0.5mm long or less. Meadows and streambanks. NE, NW.

Tendril-bearing and climbing annual herbs with simple, alternate, palmately lobed leaves; flowers regular, unisexual, the plants monoecious, the staminate flowers in axillary panicles, the pistillate solitary or paired in the same axils; sepals usually 6, bristle-like, at sinuses of corolla; petals usually 6, united; stamens 3, somewhat connate; ovary 1, inferior; carpels 2 or 3; style 1; locules 2; ovules 2 per locule; placentation axile or parietal; fruit a weakly spiny pepo.

Echinocystis T. & G. Wild Cucumber

Reference: Stocking, K. M. 1955. Madroño 13:84-100.

Echinocystis lobata (Michx.) T. & G., Fl. N. Am. 1:542. 1840. Annual climbing vine with branched tendrils; leaf blades cordate and palmately 5 lobed, 2-15cm long; calyx lobes bristle-like, in sinuses of the greenish-white corolla; corolla lobes 3-6mm long when expanded. Bottomlands, thickets, and disturbed areas. NW, NE, SE.

A
B
C
D
E
F
G
H
I
J
JLD

Grass-like or rush-like herbs with simple, alternate or basal, sheathing, usually 3 ranked leaves, rarely reduced to the sheath; ligule often lacking; flowers bisexual or unisexual, arranged in spikelets or spikes, each flower subtended by a chaffy bract; perianth bristly or scaly or absent; stamens 1-3, rarely 6; ovary 1, superior; carpels 2-3; style 1, usually branched above; locule 1; ovule 1; placentation basal; fruit an achene, enclosed in a sac (perigynium) in <u>Carex</u> and <u>Kobresia</u>.

1. Ovary and fruit enclosed in a sac (perigynium); flowers unisexual
   2. Perigynium closed except the very tip; plants alpine or not — <u>Carex</u>
   2. Perigynium slit down one side nearly to the base; plants mostly alpine — <u>Kobresia</u>
1. Ovary and fruit not enclosed in a sac; flowers bisexual (stamens often deciduous)
   3. Spikelets flattened, the scales in 2 ranks; perianth lacking — <u>Cyperus</u>

Figure 8. Cyperaceae. A. Spike of <u>Carex nardina</u> (X 3). B. Perigynium of <u>Carex</u> (X 3). C. Achene of <u>Carex</u> (X 3). D. One staminate and three pistillate spikes of <u>Carex raynoldsii</u> (X 1.3). E. Spikelet of <u>Cyperus rivularis</u> (X 3). F. Perigynium and flower of <u>Kobresia</u> (X 3). G. Spikelet (left X 3) and achene and perianth (right X 6) of <u>Eleocharis palustris</u>. H. Four spikelets of <u>Eriophorum</u> (X 0.7). I. Spike of <u>Hemicarpha</u> (X 4). J. Achene and perianth (left X 8) and spikelet (right X 3) of <u>Scirpus</u>.

3. Spikelets not flattened, the scales spirally arranged; perianth usually present, of 1 to several scales or bristles
   4. Perianth bristles more than 10, long-exserted and cotton-like at maturity — Eriophorum
   4. Perianth bristles 0-6, not exserted nor cotton-like
      5. Base of style often enlarged, forming a cap at the tip of ovary; spikelet solitary at tip of a bladeless stem, the leaves reduced to the sheaths — Eleocharis
      5. Base of style not enlarged; spikelets sometimes appearing lateral along stem, usually at least 3, rarely solitary; stems leafy or not
         6. Plants perennial, often over 1dm high; flowers without an inner scale, perianth bristles usually present — Scirpus
         6. Plants annual, usually less than 1dm high; flowers with a small scale on the inner side at the base (rarely lacking), perianth bristles lacking — Hemicarpha

Carex L. — Sedge

Plants perennial, monoecious or occasionally dioecious; flowers lacking a perianth, borne in spikes, each flower subtended by a scale (chaffy bract); spikes 1 or more per culm, each often subtended by a bract; pistillate flowers each enclosed by a sac called the perigynium, the style or stigmas protruding from an opening in the top; stigmas 2 and the achene lenticular, or 3 and the achene trigonous.

References: Murray, D. F. 1969. Brittonia 21:55-76.
1970. Canad. Journ. Bot. 48:313-324.
Hermann, F. J. 1970. USDA For. Serv. Agr. Handb. 374.

1. Spikes solitary at tip of each stem GROUP I

1. Spikes more than 1 per stem, sometimes closely aggregated to appear like 1

  2. Plants with all staminate flowers; spikes usually crowded in an ovoid head (see *C. simulata* if head is elongate) *C. douglasii*

  2. Plants with at least some pistillate flowers; spikes various

    3. Stigmas mostly 2 (rarely a few flowers with 3 stigmas); achenes lenticular

      4. Lateral spikes sessile or nearly so, usually not much longer than wide; terminal spike usually with both staminate and pistillate flowers, or the plants rarely dioecious GROUP II

      4. Lateral spikes peduncled, or if sessile, then elongate; terminal spike usually staminate, rarely both staminate and pistillate GROUP III

    3. Stigmas mostly 3; achenes trigonous or rarely nearly terete

5. Perigynia pubescent, puberulent, or prominently ciliate-scabrous at least on margins (do not mistake for papillate) GROUP IV

5. Perigynia glabrous GROUP V

GROUP I

1. Stigmas 2; achenes lenticular, or the flowers all staminate and the leaves 0.4-1mm wide
  2. Spike all staminate or all pistillate *C. gynocrates*
  2. Spike with staminate and pistillate flowers, the pistillate usually below the staminate
    3. Perigynia tapered at the base, often stipitate or substipitate, prominently nerved
      4. Stems single or few together along a slender rhizome or stolon; plants usually of wet, mossy areas *C. gynocrates*
      4. Stems densely caespitose, lacking rhizomes or stolons; plants mostly of dry alpine areas *C. nardina*
    3. Perigynia rounded at base, sessile, lacking nerves *C. capitata*
1. Stigmas 3; achenes trigonous or rarely nearly terete, or the flowers all staminate and the leaves 1-4mm wide
  5. Flowers all staminate
    6. Scales with a prominent green midrib *C. parryana*
    6. Scales lacking a green midrib *C. scirpoidea*

5. Flowers all pistillate or both staminate and pistillate present
  7. Perigynia usually strongly inflated; rachilla alongside achene at least half as long as achene; plants with creeping rhizomes
    8. Scales bract-like, partly enveloping perigynia; leaves usually much exceeding inflorescence — C. backii
    8. Scales not bract-like; leaves mostly not exceeding inflorescence
      9. Perigynia mostly 4-7mm long, much larger than the achene — C. breweri
      9. Perigynia 2.5-4mm long, only slightly larger than the achene — C. subnigricans
  7. Perigynia not inflated, often plump but then 3 angled; rachilla obsolete, or if well developed, then rhizomes lacking
    10. Pistillate scales deciduous; at least the lower perigynia often reflexed at maturity
      11. Leaves channeled, 0.2-1.3mm wide; plants densely caespitose, lacking creeping rhizomes — C. pyrenaica
      11. Leaves flat, mostly 1.5-3mm wide; plants with creeping rhizomes — C. nigricans
    10. Pistillate scales persistent; perigynia not reflexed at maturity

12. Perigynia rounded and beakless at apex, the apex appearing bluntly 2 toothed from the side
  13. Perigynia mostly 5-7mm long, 2-3.5mm wide; staminate portion of spike most conspicuous *C. geyeri*
  13. Perigynia 2.5-5mm long, 1-1.5mm wide; pistillate portion of spike most conspicuous *C. leptalea*
12. Perigynia pointed at apex, usually the result of a beak
  14. Spike usually all pistillate
    15. Perigynia pubescent *C. scirpoidea*
    15. Perigynia glabrous or nearly so *C. parryana*
  14. Spike with staminate flowers above pistillate flowers, the pistillate sometimes only 1
    16. Perigynia 1-3 per spike, mostly 5-7mm long; lower leaves with poorly developed blades *C. geyeri*
    16. Perigynia usually 3 or more per spike, rarely fewer, mostly 5mm long or less; lower leaves with poorly developed blades or not
      17. Stems solitary or few together along a creeping rhizome
        18. Leaves 0.3-1mm wide; perigynia mostly 4-7mm long, usually 10 or more; alpine or subalpine *C. breweri*
        18. Leaves 0.8-3mm wide; perigynia 3-4mm long, often fewer than 10; alpine or not

19. Perigynium beak 0.5-1mm long; leaves 0.8-1.5mm wide *C. obtusata*
19. Perigynium beak about 0.2mm long; leaves mostly 1.5-3mm wide *C. rupestris*

17. Stems densely caespitose, creeping rhizomes lacking

20. Perigynia finely striate; staminate portion of spike very short and inconspicuous; plants usually alpine or subalpine; perigynia beaks not hyaline *C. nardina*
20. Perigynia usually 2 keeled, otherwise nerveless; staminate portion of spike usually long and conspicuous; plants, if alpine, with perigynia beaks hyaline

21. Leaf blades flattened-canaliculate, 1-2mm wide toward base; culms stout, often roughened below the spikes; lowest scale usually awned *C. oreocharis*
21. Leaf blades acicular, 0.2-0.8mm wide at base; culms filiform, smooth below the spikes; lowest scale rarely awned

22. Perigynia rounded on the angles, the beak 0.2-0.4mm long; basal sheaths usually strongly filamentose; mostly at middle elevations *C. filifolia*

22. Perigynia more sharply triangular, the beak 0.5-1mm long; basal sheaths usually not filamentose; mostly at higher elevations *C. elynoides*

GROUP II

1. Culms single or few together from long creeping rhizomes
  2. Perigynia strongly wing-margined, the beak deeply bidentate *C. foenea*
  2. Perigynia not wing-margined, the beak usually obliquely cut dorsally, becoming bidentulate (rarely slightly winged but then the perigynia stipitate)
    3. Spikes densely aggregated into a globose or ovoid head, appearing like a single spike; plants alpine or subalpine *C. foetida*
    3. Spikes, at least the lower ones, easily recognizable, or the plants lower than subalpine
      4. Upper sheaths usually hyaline ventrally; perigynia not thin-margined above
        5. Plants dioecious or nearly so; perigynium beak nearly as long as body, the body nerved ventrally; heads often over 15mm wide *C. douglasii*
        5. Plants usually not dioecious; perigynium beak usually shorter, or if not, the body not nerved ventrally and the heads 12mm or less wide

6. Leaves narrowly involute, at least above; culms obtusely angled, usually smooth *C. stenophylla*

6. Leaves flat or channeled; culms usually sharply triangular, often rough above

7. Spikes with mostly 1-4 perigynia, usually distant *C. disperma*

7. Spikes with mostly 6 or more perigynia, usually crowded

8. Perigynia 1.7-2.7mm long, the beak 0.2-0.5mm long; rootstocks and lowest sheaths light brown *C. simulata*

8. Perigynia 3-4mm long, the beak 0.6-1.8mm long; rootstocks and lower sheaths dark brown to black *C. praegracilis*

4. Upper sheaths green striate ventrally except near the mouth; perigynia with the body thin-margined above *C. sartwellii*

1. Culms caespitose or the rhizomes short with very short internodes and not long creeping

9. Spikes with staminate flowers above the pistillate

10. Perigynia very gradually tapering into a beak

11. Perigynia 4-5.2mm long, the beak about as long as the body; heads usually not globose, compound *C. stipata*

11. Perigynia 3-4.5mm long, the beak mostly shorter than the body; heads usually globose or ovoid and single

12. Leaves all clustered near base; sheaths not cross-rugulose ventrally *C. jonesii*

12. Leaves not all clustered near base; sheaths cross-rugulose ventrally *C. neurophora*

10. Perigynia somewhat abruptly contracted into a beak

13. Spikes few, usually 10 or less, often greenish; sheaths not red dotted ventrally

14. Leaf blades 0.5-3.5mm wide; sheaths tight, inconspicuously or not at all mottled with green and white, not septate dorsally

15. Beak of perigynium obliquely cleft dorsally, minutely bidentulate; leaf blades 0.5-2mm wide *C. vallicola*

15. Beak of perigynium conspicuously bidentate; leaf blades mostly 1.5-3.5mm wide

16. Perigynia ovate, glossy brown with a wide green margin; heads mostly ovate *C. hoodii*

16. Perigynia elliptic, greenish straw colored to brown-centered with a narrow green margin; heads mostly elongate *C. occidentalis*

14. Leaf blades 3.5-6mm wide; sheaths loose, mottled with green and white, usually septate dorsally *C. gravida*

13. Spikes often more than 10, usually brownish or yellowish; sheaths often red dotted or banded near the mouth ventrally

17. Leaf sheaths copper colored at mouth; head interrupted, 3-8cm long; some leaves 3-6mm wide *C. cusickii*

17. Leaf sheaths not copper colored at mouth; head little interrupted, mostly 2-3.5cm long; leaves mostly 1-3mm wide *C. diandra*

9. Spikes with pistillate flowers above the staminate

18. Perigynia lacking winged margins, at most thin-edged, 1.5-3.5(4)mm long

19. Spikes in a dense head *C. illota*

19. Spikes somewhat scattered

20. Perigynia widely spreading to descending, the beaks 0.5-2mm long; spikes mostly about as long as wide

21. Perigynium beak 1/4-1/3 the length of the body, shallowly bidentate *C. interior*

21. Perigynium beak more than 1/2 to about the length of the body, deeply bidentate *C. muricata*

20. Perigynia mostly erect to ascending, the beaks 0.2-0.7(1)mm long; spikes mostly longer than wide

22. Scales strongly brown or chestnut tinged; perigynia with mostly 6 or more nerves on each side, the beak 0.2-0.5mm long *C. praeceptorum*

22. Scales hyaline or greenish, often light brownish tinged at maturity; perigynia often with fewer nerves, the beak sometimes over 0.5mm long

23. Perigynia 2.5-4mm long, the beak 0.5-1mm long, the body gradually tapering to the beak *C. laeviculmis*

23. Perigynia 1.7-2.7(3)mm long, the beak 0.2-0.7mm long, the body somewhat abruptly narrowed to beak

24. Beaks of perigynia 0.5-0.7mm long; spikes mostly 5-10 flowered; leaves green, 1-2.5mm wide *C. brunnescens*

24. Beaks of perigynia less than 0.5mm long; spikes 9-20 flowered; leaves glaucous, some often over 2.5mm wide *C. canescens*

18. Perigynia with winged margins, 2.5-8mm long

25. Bracts either leaflike or conspicuously exceeding the head; body of achene usually over 1 1/2 times as long as wide *C. athrostachya*

25. Bracts not leaflike nor conspicuously exceeding the head, or if so, the body of achene about as long as wide

26. Beak of perigynium flat and somewhat winged at tip, usually serrulate to tip

27. Perigynia 6-8mm long

28. Scales conspicuously narrower and shorter than perigynia *C. egglestonii*

28. Scales only slightly if at all narrower and shorter than perigynia — *C. xerantica*

27. Perigynia 2.5-6mm long

29. Perigynia 2.2-3.5mm wide, the body about as long as wide — *C. brevior*

29. Perigynia 1-2mm wide, or if wider, the body about twice as long as wide

30. Perigynia 2.5-3.7mm long, 1-1.5mm wide; achenes 0.6-0.8mm wide — *C. bebbii*

30. Perigynia 2.8-6mm long, 1.5-2.8mm wide, or if narrower, the achenes 0.9mm or more wide

31. Spikes densely clustered in a somewhat ovoid head, dark brown at maturity — *C. arapahoensis*

31. Spikes loosely clustered, not in a dense head, greenish or straw colored at maturity

32. Culms essentially smooth on the angles; perigynia 4-7mm long, 1.9-2.8mm wide — *C. xerantica*

32. Culms conspicuously scabrous on the angles; perigynia 2.8-4.5mm long, 1.4-2mm wide — *C. tenera*

26. Beak of perigynium slender and terete, sometimes scarcely winged at tip, the upper 0.2-2mm often little if at all serrulate

33. Perigynia mostly 6-8mm long, the scales light reddish-brown; spikes loosely aggregated in an erect, elongate inflorescence *C. petasata*

33. Perigynia 2.5-6mm long, or if longer, the scales very dark or the spikes in a flexuous or moniliform inflorescence or a dense ovate head

34. Scales about the same length as perigynia, concealing them above or nearly so

35. Perigynia 2.5-4mm long, 0.8-1.3mm wide

36. Perigynia 2.5-3.5mm long; some leaves often over 2mm wide *C. limnophila*

36. Perigynia 3.2-4mm long; leaves 0.7-2mm wide *C. leporinella*

35. Perigynia mostly 4-6.5mm long, 1.3-2.5mm wide

37. Plants mostly 1-3dm high, often near or above timberline; leaves 0.5-2mm wide *C. phaeocephala*

37. Plants 2-8dm high, mostly below timberline; leaves often over 2mm wide

38. Leaves mostly clustered near base; spikes in a moniliform or flexuous inflorescence, usually not overlapping below or barely so; perigynia mostly over 4.5mm long *C. praticola*

38. Leaves mostly covering lower third of culms; spikes, except sometimes lowermost, approximate in a tight inflorescence; perigynia mostly less than 4.5mm long
C. pachystachya

34. Scales shorter and usually narrower than perigynia, the perigynia conspicuous in the spikes

39. Perigynia much flattened, thin and scale-like except where distended by the achene, rarely slightly planoconvex

40. Perigynia 2.5-3.5mm long; heads mostly 5-10mm wide *C. limnophila*

40. Perigynia 3.2-7mm long; heads often over 10mm wide

41. Perigynia mostly 6-7mm long, narrowly lanceolate; anthers about 2.5mm long
C. ebenea

41. Perigynia mostly 3.5-6mm long, ovate or lanceolate-ovate; anthers 1.3-2mm long

42. Perigynia mostly 3.2-5mm long; plants often below timberline *C. microptera*

42. Perigynia mostly 5-6.2mm long; plants near or above timberline *C. haydeniana*

39. Perigynia planoconvex, not thin and scale-like

43. Perigynia 2.5-3.2mm long *C. illota*

43. Perigynia 3.5-5mm long

44. Perigynia oblong-lanceolate to narrowly ovate-lanceolate, 1-1.5mm wide, straw colored *C. stenoptila*

44. Perigynia ovate, 1.5-2.5mm wide, copper colored at maturity *C. pachystachya*

GROUP III

1. Perigynia 3-5mm long; achenes continuous with the style, the style often flexuous at maturity *C. saxatilis*
1. Perigynia mostly 1.5-3.5mm long; achenes jointed with the style, the style mostly straight
   2. Lowest bract usually sheathing; perigynia usually whitish-pulverulent or golden-yellow at maturity, often inflated; plants 4dm or less high *C. aurea*
   2. Lowest bract usually sheathless, occasionally short-sheathing; perigynia not pulverulent nor golden-yellow at maturity, inflated or not; plant height various
      3. Lowest bract definitely shorter than the inflorescence; pistillate scales usually with obsolete or slender midveins
         4. Pistillate spikes 2-6mm wide; perigynia flattened to planoconvex *C. bigelowii*
         4. Pistillate spikes often over 6mm wide; perigynia somewhat inflated to triangular *C. scopulorum*

3. Lowest bract almost equaling or exceeding the inflorescence; pistillate scales with conspicuous midveins or with a broader, light colored center

5. Perigynia conspicuously veined or ribbed ventrally

6. Perigynia early deciduous, slenderly nerved, the beak apiculate and entire; lowest bract usually exceeding the inflorescence *C. lenticularis*

6. Perigynia persistent, strongly ribbed, the beak broad and at least shallowly bidentate; lowest bract about equaling the inflorescence *C. nebraskensis*

5. Perigynia nerveless ventrally, or with obscure impressed nerves *C. aquatilis*

GROUP IV

1. Staminate and pistillate spikes on different culms *C. scirpoidea*

1. Staminate and pistillate spikes on the same culm

2. Bracts reduced to bladeless sheaths or sometimes with very short hyaline blades

3. Staminate spikes 3-6mm long, sessile or very short peduncled; pistillate spikes 5-12 flowered, 4-8mm long at maturity; bracts 7mm long or less, green or brown *C. concinna*

3. Staminate spikes 10-25mm long, peduncled; pistillate spikes usually 10-25 flowered, mostly 8-22mm long at maturity; bracts 10-20mm long, reddish with white-hyaline margins ... C. richardsonii

2. Bracts sheathing or sheathless, the blades well developed

4. Perigynia closely enveloping the achene, strongly tapering at base; bracts sheathless or nearly so; pistillate spikes mostly less than 15mm long

5. Fertile culms of two types, some 1-5cm high and partly hidden among the tufted leaf bases and bearing mostly pistillate spikes, others elongate, 5-30cm high and bearing staminate and pistillate spikes or some only pistillate ... C. rossii

5. Fertile culms all alike, elongate, 5-40cm high, bearing staminate and pistillate spikes; basal spikes absent

6. Perigynia inflated, 2.5-4.5mm long, pubescent all over, the beak 0.5-1.5mm long ... C. pensylvanica

6. Perigynia planoconvex, 1.8-3mm long, pubescent only near tip, the beak 0.1-0.6mm long ... C. parryana

4. Perigynia not as above, the top part empty, or if as above, the lowest bract strongly sheathing; pistillate spikes often over 15mm long

7. Bracts, or at least some of them, long sheathing, the sheaths usually 1/3 their total length or more

8. Spikes, at least the lower ones, often drooping on capillary peduncles, the terminal with pistillate flowers above staminate; perigynia about 1mm wide, the upper half long tapering *C. misandra*

8. Spikes erect, the terminal staminate (rarely with a few pistillate flowers); perigynia 1.2-2mm wide, contracted into the beak *C. luzulina*

7. Bracts sheathless or very short sheathing *C. lanuginosa*

GROUP V

1. Pistillate scales, except sometimes the uppermost, leaflike or bractlike, concealing and partly enveloping the perigynia, often over twice as long as the perigynia; leaves usually much exceeding the inflorescence; achenes strongly constricted at base, rounded at tip *C. backii*

1. Pistillate scales not leaflike or bractlike, mostly not longer than perigynia; leaves usually not exceeding the inflorescence; achenes mostly not as above

2. Roots clothed with yellow felt; terminal spike staminate or rarely with pistillate flowers above the staminate; lateral spikes usually drooping on long slender peduncles

3. Terminal spike 13-27mm long; pistillate scales obtuse to minutely awn-tipped, at least as broad as perigynia and barely longer, persistent *C. limosa*

3. Terminal spike 4-12mm long; pistillate scales long-attenuate, narrower and longer than the perigynia, deciduous *C. paupercula*

2. Roots usually not clothed with a yellow felt; terminal spike sometimes pistillate or with staminate flowers above the pistillate; lateral spikes often not as above

4. Beak of achene bent or recurved; scales mucronate to long-awned, white-hyaline with a narrow greenish center; perigynia tapering at the base, closely enveloping the achenes

5. Beak of perigynium 0.5mm long or less *C. blanda*

5. Beak of perigynium 2-3mm long *C. hystricina*

4. Beak of achene usually straight or nearly so; scales and perigynia various

6. Beak of perigynia about as long as or longer than the body, the perigynia mostly 5-7mm long, usually strongly 2 nerved; rootstocks and base of culms heavily fibrillose *C. sprengelii*

6. Beak of perigynia usually much shorter than the body, if as long, the perigynia 7mm long or more or less than 5mm long or strongly many nerved; rootstocks and base of culms usually not fibrillose

7. Plants with body of perigynia 3.5mm or more long, the beak at least 1mm long; lowest bract equaling or exceeding the inflorescence; style continuous with the achene, indurated, not withering

8. Beak of perigynium with teeth about as long as body of beak; leaf sheaths hairy; ligule longer than wide — *C. atherodes*

8. Beak of perigynium with shorter teeth; leaf sheaths usually glabrous; ligule often as wide as or wider than long

9. Staminate spike solitary; beak of perigynia 2mm or more long; pistillate scales with a body about 2mm long tipped by an awn 2-6mm long — *C. hystricina*

9. Staminate spikes usually 2 or more per culm; beak of perigynia mostly 1-2mm long; pistillate scales not as above

10. Perigynia usually spreading at maturity, the body somewhat abruptly contracted to the beak; some leaves often over 6mm wide; rhizomes usually long-creeping — *C. rostrata*

10. Perigynia ascending at maturity, the body gradually tapering to the beak; leaves mostly less than 6mm wide, rarely to 8mm; rhizomes usually short — *C. vesicaria*

7. Plants with body of perigynia 2-4.5mm long, the beak usually about 1mm long or less; lowest bract shorter than or exceeding the inflorescence; style jointed with the achene, not indurated, finally withering and deciduous

11. Lower bracts, or some of them, long-sheathing
  12. Pistillate spikes on slender, usually drooping peduncles, 3-20 flowered — *C. capillaris*
  12. Pistillate spikes usually erect, the peduncles relatively stout, 10-45 flowered
    13. Plants caespitose; beak of perigynium usually about 1/3 the length of the body
      14. Lowest bract exceeding the inflorescence — *C. oederi*
      14. Lowest bract shorter than inflorescence
        15. Leaves 3-9mm wide; mostly below timberline — *C. luzulina*
        15. Leaves 1-3.5mm wide; mostly above timberline — *C. misandra*
    13. Plants with long creeping rootstocks; beak of perigynium much less than 1/3 the length of the body — *C. crawei*
11. Lower bracts sheathless or only slightly sheathing
  16. Terminal spike staminate
    17. Perigynia 3.5-4.5mm long
      18. Perigynia plump, slightly if at all flattened — *C. raynoldsii*
      18. Perigynia strongly flattened — *C. spectabilis*
    17. Perigynia 1.8-3.5mm long

19. Beak of perigynia 0.7-1.2mm long; lowest bract much exceeding inflorescence *C. oederi*

19. Beak of perigynia 0.1-0.6mm long; lowest bract equaling or shorter than inflorescence

20. Lower spikes drooping on long slender peduncles *C. paupercula*

20. Lower spikes mostly erect on short stout peduncles or sessile

21. Perigynia inflated *C. torreyi*

21. Perigynia flattened or planoconvex

22. Perigynia elliptic to suborbicular, very short stipitate, planoconvex *C. parryana*

22. Perigynia ovate to orbicular, sessile, strongly flattened *C. spectabilis*

16. Terminal spike pistillate or both pistillate and staminate

23. Plants with long creeping rhizomes, the culms of the year arising singly or few together and not surrounded by dried leaves of the previous year; pistillate scales longer than the perigynia and awned *C. buxbaumii*

23. Plants with short rhizomes or none, culms of year clustered and surrounded by dried leaves (at least sheaths) of previous year; pistillate scales not as above

24. Perigynia 1.8-3mm long; spikes approximate, not in a head
  25. Terminal spikes mostly 6-14mm long; pistillate scales black-purple or brownish-black *C. norvegica*
  25. Terminal spikes, or some of them, 14-30mm long; pistillate scales brown or stramineous, not blackish *C. parryana*
24. Perigynia 3-5.5mm long, or if shorter, the spikes in a dense or loose head
  26. Spikes approximate, not in a dense head, peduncle of lowest spike often nearly as long as spike or longer; plants montane to alpine *C. atrata*
  26. Spikes usually in a dense head, peduncle of lowest spike much shorter than spike; plants mostly alpine or subalpine
    27. Perigynia not strongly flattened, with a beak 0.6-0.9mm long *C. nelsonii*
    27. Perigynia strongly flattened, the beak less than 0.7mm long
      28. Spikes all sessile or nearly so, very crowded, the lateral ones spreading or widely ascending; achenes often stipitate *C. nova*
      28. Spikes not so crowded, the lowest one often peduncled, mostly ascending; achenes sessile or nearly so *C. albonigra*

Carex albonigra Mack. in Rydb., Fl. Rocky Mts. 137, 1060. 1917. Culms to 3dm high, loosely tufted on short rhizomes; leaves 2-7mm wide; spikes 2-4, the lateral pistillate, the terminal gynaecandrous; scales about equaling the perigynia; perigynia flattened, 2.7-3.5mm long, about 2mm wide, the beak scarcely 0.5mm long; achenes trigonous, 1.2-2mm long. High mountain slopes. Statewide.

Carex aquatilis Wahl., Sv. Vet.-Akad. Handl. 24:165. 1803. Culms to 1m high, on creeping rhizomes; leaves 2-8mm wide; spikes mostly 3-7, the terminal mostly staminate, the others pistillate or androgynous; scales shorter or longer than perigynia; perigynia flattened, 2-3.3mm long, 1.2-1.8mm wide, the beak 0.1-0.3mm long; achenes lenticular, 1.2-1.7mm long. Wet areas. Statewide. *C. variabilis* and *C. acutina* Bailey of authors.

Carex arapahoensis Clokey, Rhodora 21:83. 1919. Culms to 4dm high, from very short rootstocks; leaves 1.5-3mm wide; spikes 3-6, gynaecandrous; scales nearly as long as perigynia; perigynia flattened, 4-5.5mm long, 1.5-2.5mm wide, the beak about 1mm long; achenes lenticular, 1.5-2mm long. High mountain slopes and meadows. Albany Co.

Carex atherodes Spreng., Syst. Veg. 3:828. 1826. Culms to 15dm high, from creeping rhizomes; leaves 3-12mm wide; spikes 4-10, the lower pistillate, the upper staminate, the middle ones sometimes androgynous; scales shorter or longer than perigynia, usually awned; perigynia somewhat inflated, 7-10mm long, about 2mm wide, the beak 1.2-3mm long; achenes trigonous, 2-2.5mm long. Wet areas, often in water. Albany and Uinta Cos. *C. aristata* R. Br.

Carex athrostachya Olney in Gray, Proc. Am. Acad. 7:393. 1868. Culms to 6dm high, without rhizomes; leaves 1-5mm wide; spikes 3-20, gynaecandrous; scales mostly shorter than perigynia; perigynia flattened, 3-5mm long, 1-1.8mm wide, the beak 1mm or less long; achenes lenticular, 1-1.6mm long. Moist areas. NW, SW, SE. *C. tenuirostris* Olney.

Carex atrata L., Sp. Pl. 976. 1753. Culms to 1m high, without creeping rhizomes; leaves 2-7mm wide; spikes mostly 2-5, the terminal gynaecandrous, the lateral pistillate or gynaecandrous; scales longer or shorter than perigynia; perigynia flattened, 3-4.5mm long, rarely shorter, 1.5-4mm wide, the beak 0.2-0.5mm long; achenes trigonous, 1.3-2mm long. Moist areas in the mountains. Statewide. *C. chalciolepis* Holm, *C. epapillosa* Mack., *C. heteroneura* Boott.

Carex aurea Nutt., Gen. Pl. 2:205. 1818. Culms to 4dm high, from creeping rhizomes; leaves mostly 1-4mm wide; spikes 2-6, the terminal staminate or sometimes gynaecandrous, the lateral pistillate; scales as long as or shorter than perigynia; perigynia inflated or only slightly flattened, 1.7-3mm long, about 1.5mm wide, the beak obsolete or nearly so; achenes mostly lenticular, 1.2-2mm long. Moist areas. Statewide. *C. hassei* Bailey.

Carex backii Boott in Hook., Fl. Bor. Am. 2:210. 1839. Culms to 35cm high, without creeping rhizomes; leaves 2-6mm wide; spikes 1-3, androgynous, few flowered; scales longer than perigynia, leaf-like; perigynia turgid, 4-6mm long, 2-2.5mm wide, the beak mostly 1-2mm long; achenes rounded-trigonous, about 3mm long; stigmas 3. Woods and thickets. NW, NE, SE. *C. saximontana* Mack., *C. durifolia* Bailey.

Carex bebbii Olney ex Fern., Proc. Am. Acad. 37:478. 1902. Culms to 9dm high, without creeping rhizomes; leaves 2-4.5mm wide; spikes 3-12, gynaecandrous; scales as long as or shorter than perigynia; perigynia planoconvex, 2.5-3.7mm long, 1-1.5mm wide, the beak 0.7-1mm long; achenes lenticular, 1-1.5mm long. Moist areas. Crook Co.

Carex bigelowii Torr. ex Schwein., Ann. Lyc. N. Y. 1:67. 1824. Culms to 4dm high, from creeping rhizomes; leaves 2-8mm wide; spikes 2-7, the terminal usually staminate, the lateral pistillate; scales longer or shorter than perigynia; perigynia somewhat flattened, 2.5-3.5mm long, 1-2mm wide, the beak 0.1-0.3mm long; achenes mostly lenticular, 1.5-2mm long; stigmas 2 or occasionally 3. Moist areas. Park Co.

Carex blanda Dewey, Am. Journ. Sci. 10:45. 1826. Culms to 6dm high, without creeping rhizomes; leaves 3-15mm wide; spikes 3-6, the terminal usually staminate, the lateral pistillate; scales shorter than perigynia, usually awned; perigynia somewhat trigonous, 3-4mm long, about 1.7mm wide, the beak 0.5mm or less long; achenes trigonous, about 2.5mm long. Woods and thickets. Crook Co.

Carex brevior (Dewey) Mack. ex Lunell, Am. Midl. Nat. 4:235. 1915. Culms to 1m high, without creeping rhizomes; leaves 1-4mm wide; spikes 3-10, gynaecandrous; scales as long as or shorter than perigynia; perigynia flattened, 3.5-5.5mm long, 2.2-3.5mm wide, the beak 0.8-1.5mm long; achenes lenticular, 1.6-2.2mm long. Moist to dry areas. NW, NE, SE. *C. festucacea* Willd. of authors.

Carex breweri Boott, Ill. Carex 4:142. 1867. Culms to 3dm high, from creeping rhizomes; leaves 0.3-1mm wide; spike solitary, androgynous; scales as long as or shorter than perigynia; perigynia mostly inflated, 4-7mm long, 1.8-4mm wide, the beak to 0.5mm long; achenes mostly trigonous, 1.2-2mm long. Near or above timberline. NW, SW. *C. engelmannii* Bailey.

Carex brunnescens (Pers.) Poir. in Lam., Encyc. Meth. Suppl. 3:286. 1813. Culms to 5dm high, without creeping rhizomes; leaves 1-2.5mm wide; spikes 4-10, gynaecandrous; scales mostly shorter than perigynia; perigynia planoconvex, 1.7-2.7mm long, 1-1.5mm wide, the beak 0.5-0.7mm long; achenes lenticular, 1.2-1.5mm long. Wet areas in the mountains. Albany Co.

Carex buxbaumii Wahl., Sv. Vet.-Akad. Handl. 24:163. 1803. Culms to 1m high, from creeping rhizomes; leaves 1-4mm wide; spikes 2-5, the terminal gynaecandrous, the lateral pistillate; scales longer than perigynia, awn-tipped; perigynia triangular-biconvex, papillate, 2.5-4.3mm long, 1.5-2mm wide, the beak 0.5mm or less long; achenes trigonous, 1.4-2mm long. Wet areas. Yellowstone and Teton Parks.

Carex canescens L., Sp. Pl. 974. 1753. Culms to 8dm high, from very short rhizomes; leaves 1.5-4mm wide; spikes 4-8, gynaecandrous; scales mostly shorter than perigynia; perigynia planoconvex, 1.8-3mm long, 1-1.7mm wide, the beak less than 0.5mm long; achenes lenticular, 1.2-1.5mm long. Wet areas. Statewide.

Carex capillaris L., Sp. Pl. 977. 1753. Culms to 6dm high, without creeping rhizomes; leaves 0.5-4mm wide; spikes 2-5, the terminal usually staminate, the lateral pistillate; scales shorter than perigynia; perigynia slightly inflated, 2-4mm long, 0.6-1.2mm wide, the beak 1mm or less long; achenes trigonous, 1.2-1.5mm long. Wet areas in the mountains. NW, SE.

Carex capitata L., Syst. Nat. 10th ed. 1261. 1759. Culms to 35cm high, from very short rhizomes; leaves 1mm or less wide; spike solitary, androgynous; scales as long as or shorter than perigynia; perigynia flattened to planoconvex, 1.7-3.5mm long, 1.2-2mm wide, the beak 0.3-0.7mm long; achenes lenticular, about 1.5mm long. Moist or dry areas in the high mountains. NW, SW, NE. *C. arctogena* Smith.

Carex concinna R. Br. in Richards. in Frankl., Narr. 1st Journ. 751. 1823. Culms to 2dm high, from creeping rhizomes; leaves 1-3mm wide; spikes 2-4, the terminal usually staminate, the lateral pistillate; scales shorter than perigynia; perigynia somewhat trigonous, 2-3.5mm long, 1-1.3mm wide, the beak less than 0.5mm long; achenes trigonous, 1.4-2mm long. Woods. Yellowstone Park and Sublette Co.

Carex crawei Dewey, Am. Journ. Sci. II, 2:246. 1846. Culms to 3dm high, from creeping rhizomes; leaves 1.5-3mm wide; spikes 3-6, the terminal staminate, the lateral pistillate or sometimes androgynous; scales shorter than perigynia; perigynia subinflated, 2.3-3.8mm long, 1.2-2mm wide, the beak 0.4mm or less long; achenes trigonous, 1.3-2mm long. Wet places in plains and foothills, especially on limestone. Goshen Co.

Carex cusickii Mack. ex Piper & Beattie, Fl. N. W. Coast 72. 1915. Culms to 12dm high, from very short rhizomes; leaves 2.5-6mm wide; spikes numerous and few-flowered, androgynous; scales as long as or longer than perigynia; perigynia biconvex or planoconvex, 2.5-4mm long, about 1.5mm wide, the beak less than 1.5mm long; achenes lenticular, 1-1.4mm long. Wet areas. Yellowstone and Teton Parks.

Carex diandra Schrank, Cent. Bot. Anmerk. 57. 1781. Culms to 1m high, without creeping rhizomes; leaves 1-3mm wide; spikes numerous, few-flowered, androgynous; scales mostly shorter than perigynia; perigynia somewhat biconvex or planoconvex, 2-3mm long, 1-1.5mm wide, the beak about 1mm long; achenes lenticular, 1-1.5mm long. Wet areas. NW, SE. C. teretiuscula Gooden.

Carex disperma Dewey, Am. Journ. Sci. 8:266. 1824. Culms to 6dm high, from creeping rhizomes; leaves 0.7-2mm wide; spikes 2-6, androgynous or pistillate, 1-8 flowered; scales as long as or shorter than perigynia; perigynia planoconvex to inflated, 2-3mm long, about 1.5mm wide, the beak minute; achenes lenticular, about 1.7mm long. Woods and moist areas. Statewide. C. tenella Schkuhr.

Carex douglasii Boott in Hook., Fl. Bor. Am. 2:213. 1839. Culms to 3dm high, from creeping rhizomes; leaves 0.5-2.5mm wide; plants usually dioecious, the spikes several to many; scales longer than perigynia; perigynia planoconvex, 3.5-4.5mm long, about 1.7mm wide, the beak 1-2mm long; achenes lenticular, 1.6-2mm long. Moist or dry, often alkaline areas. Statewide.

<u>Carex</u> <u>ebenea</u> Rydb., Bull. Torrey Club 28:266. 1901. Culms to 6dm high, from short rootstocks; leaves 2-4mm wide; spikes 5-10, gynaecandrous; scales shorter than perigynia; perigynia flattened, 5-7mm long, 1-1.8mm wide, the beak about 2mm long; achenes lenticular, about 2-3mm long. Moist areas in the mountains. SE, SW. <u>C</u>. <u>festiva</u> Dewey of authors.

<u>Carex</u> <u>egglestonii</u> Mack., Bull. Torrey Club 42:614. 1915. Culms to 8dm high, from short rootstocks; leaves 2-6mm wide; spikes 3-6, gynaecandrous; scales shorter than perigynia; perigynia flattened, 6-8mm long, 2-3.5mm wide, the beak about 1.5mm long; achenes lenticular, about 2mm long. Dry, open areas in the mountains. Albany Co.

<u>Carex</u> <u>elynoides</u> Holm, Am. Journ. Sci. IV, 9:356. 1900. Culms to 20cm high, without creeping rhizomes; leaves 0.2-0.5mm wide; spike solitary, androgynous; scales longer than perigynia; perigynia trigonous, 2.5-4.5mm long, 1-2mm wide, the beak 0.5-1mm long; achenes trigonous, 1.5-3mm long. High mountain slopes. Statewide.

Carex filifolia Nutt., Gen. Pl. 2:204. 1818. Culms to 3dm high, without creeping rhizomes; leaves 0.2-0.8mm wide; spike solitary, androgynous; scales about as long as perigynia; perigynia somewhat turgid, 3-4.5mm long, 1.8-2.3mm wide, the beak 0.2-0.5mm long; achenes trigonous, 2.2-3.5mm long. Plains, hills, and mountain slopes. Statewide.

Carex foenea Willd., Enum. Pl. 957. 1809. Culms to 9dm high, from creeping rhizomes; leaves 1-3mm wide; spikes 4-12, most gynaecandrous or androgynous but the lowest often staminate and the upper sometimes pistillate; scales shorter than perigynia; perigynia flattened, 3-6mm long, 1.5-2mm wide, the beak 1-3mm long; achenes lenticular, 1.7-2.2mm long. Moist or dry areas. NE, SE, SW. *C. siccata* Dewey.

Carex foetida Allioni, Fl. Pedem. 2:265. 1785. Culms to 3dm high, from creeping rhizomes; leaves 1.5-4mm wide; spikes numerous but usually crowded and indistinguishable, androgynous; scales about as long as perigynia; perigynia flattened, 3.3-4.8mm long, 1-1.8mm wide, the beak 0.8-1.7mm long; achenes lenticular, 1.2-1.5mm long. Moist areas in the mountains. Albany and Carbon Cos. *C. vernacula* Bailey.

Carex geyeri Boott, Trans. Linn. Soc. 20:118. 1846. Elk Sedge. Culms to 5dm high, from short or elongate rhizomes; leaves 1.5-4mm wide; spike solitary, androgynous, with 1-3 perigynia at base; scales mostly longer than perigynia; perigynia triangular-inflated, 5-7mm long, 2-3.5mm wide, the beak nearly obsolete; achenes trigonous, 4-5mm long. Woods, slopes, and meadows in the mountains and foothills. NW, SW, SE.

Carex gravida Bailey, Mem. Torrey Club 1:5. 1889. Culms to 6dm high, from short creeping rhizomes; leaves 3.5-6mm wide; spikes 6-12, androgynous, crowded; scales shorter than or about as long as perigynia; perigynia planoconvex, 3.5-5mm long, 1.5-3mm wide, the beak 1-2mm long; achenes lenticular, 1.5-3mm long. Plains and hills. Crook and Johnson Cos.

Carex gynocrates Wormsk. ex Drejer, Nat. Tidssk. 3:434. 1841. Culms to 3dm high, from creeping rhizomes; leaves 0.4-1mm wide; spike solitary, androgynous or rarely pistillate or staminate; scales shorter than or about equal to perigynia; perigynia somewhat inflated, 2.5-4mm long, 1.5-2mm wide, the beak about 0.5mm long; achenes lenticular, about 1.5mm long. Wet areas. NW, SE. C. redowskyana Meyer.

Carex haydeniana Olney in Wats., Bot. King Exp. 366. 1871. Culms to 4dm high, without creeping rhizomes; leaves 1.5-4mm wide; spikes 4-7, gynaecandrous, crowded in a head; scales mostly shorter than perigynia; perigynia flattened, 4-6.2mm long, 1.7-3mm wide, the beak 1-3mm long; achenes lenticular, 1.4-1.7mm long. Moist or dry areas in the mountains. Statewide. *C. nubicola* Mack.

Carex hoodii Boott in Hook., Fl. Bor. Am. 2:211. 1839. Culms to 8dm high, without creeping rhizomes; leaves 1.5-3.5mm wide; spikes 4-8, androgynous, crowded in a head; scales as long as or shorter than perigynia; perigynia planoconvex, 3.5-5mm long, 1.5-2.5mm wide, the beak 0.7-2mm long; achenes lenticular, 1.7-2.1mm long. Woods, slopes, and meadows in the mountains. Statewide.

Carex hystricina Muhl. ex Willd., Sp. Pl. 4:282. 1805. Culms to 1m high, from a short rhizome; leaves 2-10mm wide; spikes 2-5, the terminal staminate, the lateral pistillate; scales awned, the body shorter than perigynia; perigynia somewhat inflated, 5-7mm long, 1.5-2mm wide, the beak 2-3mm long; achenes trigonous, 1.2-1.7mm long. Wet areas. NE, SE.

<u>Carex</u> <u>illota</u> Bailey, Mem. Torrey Club 1:15. 1889. Culms to 35cm high, with short creeping rhizomes or without rhizomes; leaves 1-3mm wide; spikes 3-6, gynaecandrous, crowded in a head; scales shorter than perigynia; perigynia planoconvex, 2.5-3.2mm long, 1-1.5mm wide, the beak about 0.8-1.2mm long; achenes lenticular, 1.2-1.5mm long. Wet areas in the mountains. NW, SW, SE.

<u>Carex</u> <u>interior</u> Bailey, Bull. Torrey Club 20:426. 1893. Culms to 5dm high, without creeping rhizomes; leaves 1-3mm wide; spikes 2-6, gynaecandrous or some pistillate, few-flowered; scales shorter than perigynia; perigynia planoconvex, 2.2-3.3mm long, 1.2-2mm wide, the beak 0.5-1mm long; achenes lenticular, 1.2-1.6mm long. Wet areas. NW, NE, SE.

<u>Carex</u> <u>jonesii</u> Bailey, Mem. Torrey Club 1:16. 1889. Culms to 6dm high, with short creeping rhizomes; leaves 1-3mm wide; spikes 4-8, androgynous, crowded in a head; scales shorter than perigynia; perigynia planoconvex, 3-4.5mm long, 1.2-1.5mm wide, the beak 1-1.5mm long; achenes lenticular, 1.1-1.3mm long. Moist areas in the mountains. Albany and Carbon Cos.

<u>Carex</u> <u>laeviculmis</u> Meinsh., Bot. Centralb. 55:195. 1893. Culms to 7dm high, from short creeping rootstocks; leaves 1-2mm wide; spikes 3-8, the terminal gynaecandrous, the lateral pistillate or gynaecandrous; scales shorter than perigynia; perigynia planoconvex, 2.5-4mm long, 1-1.5mm wide, the beak 0.5-1mm long; achenes lenticular, about 1.5mm long. Moist areas. Reported from Teton Co.

<u>Carex</u> <u>lanuginosa</u> Michx., Fl. Bor. Am. 2:175. 1803. Culms to 1m high, from creeping rhizomes; leaves 1.5-5mm wide; spikes 3-5, the terminal 1 or 2 staminate, the lower ones pistillate; scales shorter or longer than perigynia; perigynia somewhat inflated, 2.5-5mm long, 1.7-2mm wide, the beak 0.8-1.5mm long; achenes trigonous, 1.5-2.2mm long. Wet areas. Statewide.

<u>Carex</u> <u>lenticularis</u> Michx., Fl. Bor. Am. 2:172. 1803. Culms to 8dm high, from short creeping rhizomes or without rhizomes; leaves 1-4mm wide; spikes 3-6, the terminal staminate or gynaecandrous, the lateral mostly pistillate; scales as long as or shorter than perigynia; perigynia flattened-biconvex, 1.5-3mm long, about 1.2mm wide, the beak 0.1-0.4mm long; achenes lenticular, 1-1.5mm long. Wet areas. NW, SW, SE. <u>C</u>. <u>kelloggii</u> Boott.

Carex leporinella Mack., Bull. Torrey Club 43:605. 1917. Culms to 3dm high, without creeping rhizomes; leaves 0.7-2mm wide; spikes 3-8, gynaecandrous; scales usually as long as perigynia; perigynia planoconvex, 3.2-4mm long, 0.8-1.2mm wide, the beak about 1mm long; achenes lenticular, 1.4-2mm long. Slopes and meadows. Yellowstone Park and Fremont Co.

Carex leptalea Wahl., Sv. Vet.-Akad. Handl. 24:139. 1803. Culms to 7dm high, from short creeping rhizomes; leaves 0.5-1.5mm wide; spike solitary, androgynous, few-flowered; scales mostly shorter than perigynia; perigynia flattened-orbicular, 2.5-5mm long, 1-1.5mm wide, the beak obsolete; achenes trigonous, 1.3-1.8mm long. Wet areas. Teton Co.

Carex limnophila Hermann, Leafl. West. Bot. 8:28. 1956. Culms to 45cm high, with short creeping rhizomes; leaves 1-3.5mm wide; spikes 4-8, gynaecandrous, usually crowded in a head; scales as long as or shorter than perigynia; perigynia flattened to planoconvex, 2.5-3.5mm long, 1-1.3mm wide, the beak about 1.5mm long; achenes lenticular, 1-1.4mm long. Moist areas in the mountains. NW, SW, SE.

Carex _limosa_ L., Sp. Pl. 977. 1753. Culms to 6dm high, from creeping rhizomes; roots with a yellow-brown tomentum or looser pubescence; leaves 1-3mm wide; spikes 2-4, the terminal staminate, the lateral mostly pistillate; scales mostly about as long as perigynia or slightly longer; perigynia flattened-biconvex, 2.3-4.2mm long, about 2mm wide, the beak 0.2mm or less long; achenes trigonous, 1.5-2.3mm long. Wet areas. Big Horn Co.

Carex _luzulina_ Olney in Gray, Proc. Am. Acad. 7:395. 1868. Culms to 9dm high, without creeping rhizomes; leaves 3-9mm wide; spikes 3-7, the terminal usually staminate, the lateral usually pistillate; scales mostly shorter than perigynia; perigynia flattened-trigonous, 3-5mm long, 1-2mm wide, the beak 0.5-1mm long; achenes trigonous, 1.4-2mm long. Wet areas in the mountains. Yellowstone Park and Teton Co. _C_. _ablata_ Bailey.

Carex _microptera_ Mack., Muhlenbergia 5:56. 1909. Culms to 1m high, without creeping rhizomes; leaves mostly 2-5mm wide; spikes 5-10, gynaecandrous; scales shorter than perigynia; perigynia flattened, 3.2-5mm long, 1-2mm wide, the beak 0.7-2mm long; achenes lenticular, 1-1.5mm long. Moist to dry areas in the mountains. Statewide. _C_. _festivella_ Mack.

<u>Carex</u> <u>misandra</u> R. Br., Chlor. Melv. cclxxxiii. 1823. Culms to 3dm high, without creeping rhizomes; leaves 1-3.5mm wide; spikes 2-4, the terminal gynaecandrous or rarely pistillate, the lateral pistillate; scales shorter than perigynia; perigynia flattened, 3.3-5mm long, about 1mm wide, the beak 1-2.5mm long; achenes trigonous, 1.4-2mm long. Mostly alpine. Park and Fremont Cos.

<u>Carex</u> <u>muricata</u> L., Sp. Pl. 974. 1753. Culms to 6dm high, without creeping rhizomes; leaves 0.7-2mm wide; spikes 2-6, gynaecandrous or some lateral ones pistillate; scales shorter than perigynia; perigynia planoconvex, 2.5-3.5mm long, 1-1.4mm wide, the beak 1-2mm long; achenes lenticular, 1-2mm long. Wet areas. Yellowstone Park and Teton Co. <u>C</u>. <u>angustior</u> Mack., <u>C</u>. <u>sterilis</u> Willd.

<u>Carex</u> <u>nardina</u> Fries, Nov. Fl. Suec. Mant. 2:55. 1839. Culms to 15cm high, without creeping rhizomes; leaves 0.2-0.5mm wide; spike solitary, androgynous; scales about as long as perigynia; perigynia planoconvex, 3-4.5mm long, 1.2-2mm wide, the beak 0.4-0.8mm long; achenes lenticular or trigonous, 2-2.5mm long. Rocks, slopes, and meadows in the high mountains. NW, SW, SE. <u>C</u>. <u>hepburnii</u> Boott.

Carex nebraskensis Dewey, Am. Journ. Sci. II, 18:102. 1854. Culms to 12dm high, from creeping rhizomes; leaves 3-12mm wide; spikes 3-6, the terminal 1 or 2 staminate, the rest mostly pistillate; scales shorter or longer than perigynia; perigynia slightly inflated, 2.8-4mm long, about 2mm wide, the beak 0.2-1mm long; achenes lenticular, 1.2-2mm long. Wet areas. Statewide.

Carex nelsonii Mack. in Rydb., Fl. Rocky Mts. 137, 1060. 1917. Culms to 3dm high, from short creeping rhizomes; leaves 1-4mm wide; spikes 2-4, the terminal gynaecandrous, the lateral pistillate; scales shorter than perigynia; perigynia somewhat inflated, about 4mm long and 1.5mm wide, the beak 0.6-0.9mm long; achenes trigonous, about 1.4mm long. Meadows and slopes in the high mountains. NW, SE.

Carex neurophora Mack. in Abrams, Ill. Fl. Pac. St. 1:298. 1923. Culms to 8dm high, without long creeping rhizomes; leaves 1-3.5mm wide; spikes 5-10, androgynous; scales as long as or shorter than perigynia; perigynia planoconvex, 2.9-4mm long, 0.8-1.5mm wide, the beak to 2mm long; achenes lenticular, 1.1-1.5mm long. Moist areas. NW, SE.

Carex nigricans Meyer, Mem. Acad. St. Petersb. Sav. Etr. 1:211. 1830. Culms to 5dm high, from creeping rhizomes; leaves mostly 1.5-3mm wide; spike solitary, androgynous or sometimes staminate or pistillate; scales as long as or shorter than perigynia; perigynia biconvex, 3-4.5mm long, 1-1.5mm wide, the beak 0.5-1mm long; achenes trigonous, 1.5-2mm long. Wet areas in the high mountains. NW, SE.

Carex norvegica Retz., Fl. Scand. Prodr. 179. 1779. Culms to 8dm high, from short rhizomes; leaves 1-3mm wide; spikes 2-5, the terminal gynaecandrous, the lateral pistillate; scales as long as or shorter than perigynia; perigynia slightly inflated, 2-3mm long, about 1.3mm wide, the beak 0.3-0.7mm long; achenes trigonous, 1.2-1.8mm long. Moist areas in the mountains. Statewide. *C. media* R. Br., *C. alpina* Sw.

Carex nova Bailey, Journ. Bot. 26:322. 1888. Culms to 6dm high, from short rhizomes or without rhizomes; leaves 1.5-5mm wide; spikes 2-5, the terminal gynaecandrous, the lateral pistillate; scales shorter or longer than perigynia; perigynia flattened, 2.8-4.5mm long, 2-3.5mm wide, the beak 0.3-0.7mm long; achenes trigonous, 1.5-1.9mm long. Moist areas in the high mountains. Statewide. *C. pelocarpa* Hermann.

Carex _obtusata_ Lilj., Sv. Vet.-Akad. Handl. 14:69. 1793. Culms to 2dm high, from creeping rhizomes; leaves 0.8-1.5mm wide; spike solitary, androgynous; scales shorter or longer than perigynia; perigynia slightly inflated, 3-4mm long, 1.3-2mm wide, the beak 0.5-1mm long; achenes trigonous, 1.6-2mm long. Open, somewhat dry areas. NW, NE.

Carex _occidentalis_ Bailey, Mem. Torrey Club 1:14. 1889. Culms to 7dm high, from short rhizomes; leaves mostly 1.5-2.5mm wide; spikes 4-10, androgynous; scales about as long as perigynia; perigynia planoconvex, 2.5-4.5mm long, 1.5-1.8mm wide, the beak 0.5-1mm long; achenes lenticular, about 2mm long. Mostly dry areas. Albany Co.

Carex _oederi_ Retz., Fl. Scand. Prodr. 179. 1779. Culms to 4dm high, without creeping rhizomes; leaves 1-3mm wide; spikes 3-7, the terminal staminate, the lateral pistillate or rarely some androgynous; scales shorter than perigynia; perigynia slightly inflated, 2-3.3mm long, about 1.2mm wide, the beak 0.7-1.2mm long; achenes trigonous, 1.1-1.4mm long. Wet areas. NW, SW. _C._ _viridula_ Michx.

Carex oreocharis Holm, Am. Journ. Sci. IV, 9:358. 1900.
Culms to 35cm high, without creeping rhizomes; leaves 1-2mm wide; spike solitary, androgynous; scales shorter or longer than perigynia; perigynia planoconvex, 3.5-4.5mm long, about 2mm wide, the beak 0.5-1mm long; achenes trigonous, about 3mm long. Dry hills and slopes. Albany Co.

Carex pachystachya Cham. ex Steud., Syn. Cyp. 197. 1855.
Culms to 1m high, without creeping rhizomes; leaves mostly 2-5mm wide; spikes 3-12, gynaecandrous; scales mostly shorter than or equal to perigynia; perigynia planoconvex, 3.5-5mm long, 1.5-2.5mm wide, the beak 1-1.5mm long; achenes lenticular, 1.4-1.9mm long. Meadows, woods, and slopes. Statewide. *C. preslii* Steud., *C. platylepis* Mack., *C. multinoda* Bailey.

Carex parryana Dewey, Am. Journ. Sci. 27:239. 1835. Culms to 6dm high, from short creeping rhizomes; leaves 1-4mm wide; spikes 1-5, all pistillate or the terminal staminate or gynaecandrous; scales shorter or longer than perigynia; perigynia planoconvex, 1.8-3mm long, 1-2mm wide, the beak 0.1-0.6mm long; achenes trigonous, 1.2-1.9mm long. Moist areas. NW, SW, SE. *C. hallii* Olney.

Carex paupercula Michx., Fl. Bor. Am. 2:172. 1803. Culms to 8dm high, from rhizomes or without rhizomes; leaves 1-4mm wide; spikes 2-5, the terminal staminate or gynaecandrous, the lateral usually pistillate; scales longer than perigynia; perigynia somewhat flattened, 2.2-3.2mm long, 1.7-2.3mm wide, the beak obsolete or nearly so; achenes trigonous, 1.4-2mm long. Wet areas. Albany Co.

Carex pensylvanica Lam., Encyc. Meth. 3:388. 1792. Culms to 5dm high, from creeping rhizomes; leaves 1-3mm wide; spikes 2-4, the terminal staminate, the lateral pistillate; scales longer or shorter than perigynia; perigynia inflated, 2.5-4.5mm long, 1.5-2.2mm wide, the beak 0.5-1.5mm long; achenes trigonous, 1.8-3mm long. Plains, hills, and open woods. NW, NE, SE. C. heliophila Mack.

Carex petasata Dewey, Am. Journ. Sci. 29:246. 1836. Culms to 9dm high, without creeping rhizomes; leaves 1-4mm wide; spikes 3-6, gynaecandrous; scales about as long as perigynia; perigynia planoconvex, 5.8-8mm long, 1.5-2.3mm wide, the beak 0.5-2mm long; achenes lenticular, 2-2.7mm long. Moist or dry areas. Statewide. C. liddonii Boott.

Carex phaeocephala Piper, Contr. U. S. Nat. Herb. 11:172. 1906. Culms to 3dm high, without creeping rhizomes; leaves 0.5-2mm wide; spikes 2-7, gynaecandrous; scales about as long as perigynia; perigynia planoconvex, 4-6mm long, 1.3-2.5mm wide, the beak 0.5-1mm long; achenes lenticular, 1.5-2mm long. Rocks, meadows, and open woods in the high mountains. Statewide. *C. eastwoodiana* Stacey.

Carex praeceptorum Mack., N. Am. Fl. 18:95. 1931. Culms to 3dm high, from very short rhizomes; leaves 1-2.5mm wide; spikes 4-6, gynaecandrous or some pistillate; scales shorter than perigynia; perigynia planoconvex, 1.5-2.5mm long, 1-1.5mm wide, the beak 0.2-0.5mm long; achenes lenticular, 1.1-1.4mm long. Wet areas in the high mountains. Statewide.

Carex praegracilis Boott, Bot. Gaz. 9:87. 1884. Culms to 7dm high, from creeping rhizomes; leaves 1-3mm wide; spikes 5-25, androgynous or some pistillate; scales shorter to longer than perigynia; perigynia planoconvex, 3-4mm long, about 1.5mm wide, the beak 0.6-1.8mm long; achenes lenticular, 1.2-1.9mm long. Open, moist areas. Statewide. *C. marcida* Boott.

<u>Carex</u> <u>praticola</u> Rydb., Mem. N. Y. Bot. Gard. 1:84. 1900.
Culms to 8dm high, without creeping rhizomes; leaves 1-4mm wide; spikes 2-7, gynaecandrous; scales as long as perigynia or nearly so; perigynia planoconvex, 4.3-6.5mm long, 1.3-2.5mm wide, the beak 1-2mm long; achenes lenticular, 1-7-2.2mm long. Moist areas. NW, NE. <u>C</u>. <u>piperi</u> Mack.

<u>Carex</u> <u>pyrenaica</u> Wahl., Sv. Vet.-Akad. Handl. 24:139. 1803.
Culms to 3dm high, without creeping rhizomes; leaves 0.2-1.3mm wide; spike solitary, androgynous; scales shorter than or as long as perigynia; perigynia biconvex, 2.5-4.5mm long, 0.7-1.5mm wide, the beak about 0.5mm long or less; achenes trigonous, 1.1-1.5mm long. Rocks, slopes, and meadows in the high mountains. Statewide.

<u>Carex</u> <u>raynoldsii</u> Dewey, Am. Journ. Sci. II, 32:39. 1861.
Culms to 8dm high, from short rhizomes; leaves 2-8mm wide; spikes 3-6, the terminal staminate, the lateral pistillate; scales shorter than perigynia; perigynia somewhat inflated, 3.5-4.5mm long, 1.7-2mm wide, the beak 0.5-1mm long; achenes trigonous, 1.8-2.5mm long. Woods, slopes, and meadows. Statewide.

Carex richardsonii R. Br. in Richards. in Frankl., Narr. 1st Journ. 751. 1823. Culms to 35cm high, with creeping rhizomes; leaves 1-3mm wide; spikes usually 3, the terminal staminate, the lateral pistillate; scales longer than perigynia; perigynia planoconvex, 2.5-3mm long, 1.2-1.5mm wide, the beak about 0.5mm long or less; achenes trigonous, 1.5-2mm long. Mostly dry areas. Crook Co.

Carex rossii Boott in Hook., Fl. Bor. Am. 2:222. 1839. Culms to 3dm high, without creeping rhizomes; leaves 1-4mm wide; spikes 2-6, the terminal staminate or sometimes pistillate, the lateral pistillate; some spikes usually near base of plant; scales shorter or longer than perigynia; perigynia inflated, 2.5-4.5mm long, 1.2-2.5mm wide, the beak 0.3-1.8mm long; achenes usually trigonous, 1.5-2mm long. Open woods, meadows, and slopes. Statewide. *C. brevipes* Boott, *C. deflexa* Hornem. of authors.

Carex rostrata Stokes ex With., Brit. Pl. 2(2):1059. 1787. Culms to 12dm high, from creeping rhizomes; leaves 2-12mm wide; spikes 4-10, the upper ones staminate, the lower pistillate, the middle sometimes androgynous; scales shorter or longer than perigynia; perigynia inflated, 3.5-8mm long, 2-3.5mm wide, the beak 1-2mm long; achenes trigonous, 1.3-2mm long. Wet areas. Statewide. *C. utriculata* Boott.

Carex rupestris Allioni, Fl. Pedem. 2:264. 1785. Culms to 15cm high, from creeping rhizomes; leaves 1-3mm wide; spike solitary, androgynous; scales shorter or longer than perigynia; perigynia somewhat flattened to triangular, 3-4mm long, about 1.7mm wide, the beak about 0.2mm long; achenes trigonous, about 2.5mm long. Mostly alpine or subalpine, rarely lower. Fremont and Albany Cos. *C. drummondiana* Dewey.

Carex sartwellii Dewey, Am. Journ. Sci. 43:90. 1842. Culms to 8dm high, from creeping rhizomes; leaves 2-5mm wide; spikes often 20 or more, androgynous; scales shorter than or equal to perigynia; perigynia planoconvex, 2.3-4.2mm long, 1.5-1.9mm wide, the beak 0.5-1.5mm long; achenes lenticular, 1.2-2mm long. Moist areas. Sheridan Co.

Carex saxatilis L., Sp. Pl. 976. 1753. Culms to 8dm high, from creeping rhizomes; leaves 1.5-5mm wide; spikes 2-4, the upper staminate, the lower pistillate; scales mostly shorter than perigynia; perigynia biconvex, 3-5mm long, 1.7-2.3mm wide, the beak about 0.5mm long; achenes lenticular or rarely trigonous, 1.7-2mm long. Moist areas in the mountains. Sublette and Albany Cos. *C. physocarpa* Presl.

<u>Carex</u> <u>scirpoidea</u> Michx., Fl. Bor. Am. 2:171. 1803. Culms to 4dm high, from short creeping rhizomes; leaves 1-3mm wide; spike solitary, or rarely with 1 or 2 shorter ones at base of main one, pistillate or staminate; scales shorter or longer than perigynia; perigynia planoconvex, 2-4.5mm long, 1-1.3mm wide, the beak 0.2-0.6mm long; achenes trigonous, 1.4-2mm long. Moist to dry areas. NW, SW, NE. <u>C</u>. <u>pseudoscirpoidea</u> Rydb.

<u>Carex</u> <u>scopulorum</u> Holm, Am. Journ. Sci. IV, 14:422. 1902. Culms to 4dm high, from creeping rhizomes; leaves 2-7mm wide; spikes 3-7, the terminal staminate or sometimes also with pistillate flowers, the others pistillate or the upper ones androgynous; scales shorter or longer than perigynia; perigynia triangular to slightly inflated, 1.8-4mm long, 1.5-2mm wide, the beak 0.1-0.4mm long; achenes lenticular or some trigonous, 1.3-1.6mm long. Moist areas in the mountains. Statewide. <u>C</u>. <u>rigida</u> Gooden. of authors, <u>C</u>. <u>chimaphila</u> Holm.

<u>Carex</u> <u>simulata</u> Mack., Bull. Torrey Club 34:604. 1908. Culms to 9dm high, from creeping rhizomes; leaves 1-4mm wide; spikes 5-25, androgynous or sometimes all pistillate or all staminate; scales longer than perigynia; perigynia planoconvex, 1.7-2.7mm long, about 1.5mm wide, the beak 0.2-0.5mm long; achenes lenticular, 1-1.5mm long. Wet areas in the mountains. Statewide. <u>C</u>. <u>gayana</u> Desv.

Carex spectabilis Dewey, Am. Journ. Sci. 29:248. 1836. Culms to 9dm high, from short rhizomes; leaves 2-7mm wide; spikes 3-8, the terminal staminate, the lateral pistillate or the upper ones staminate or androgynous; scales about as long as perigynia; perigynia flattened, 2-5mm long, 1.5-3mm wide, the beak 0.2-0.5mm long; achenes trigonous, 1.2-2.5mm long. Moist areas in the mountains. Statewide. *C. tolmiei* Boott, *C. paysonis* Clokey.

Carex sprengelii Dewey ex Spreng., Syst. Veg. 3:827. 1826. Culms to 9dm high, from fibrillose rhizomes; leaves 1.5-4mm wide; spikes 3-7, the 1 or 2 terminal staminate or rarely androgynous, the rest pistillate or the upper rarely androgynous; scales about as long as perigynia or shorter; perigynia somewhat inflated, 4.5-7.5mm long, 1.2-2mm wide, the beak 1.7-4mm long; achenes trigonous, 1.7-2.5mm long. Moist areas. NE, SE.

Carex stenophylla Wahl., Sv. Vet.-Akad. Handl. 24:142. 1803. Culms to 2dm high, from creeping rhizomes; leaves 0.3-1.5mm wide; spikes several, androgynous, crowded; scales about as long as perigynia; perigynia planoconvex, 2.5-3.3mm long, 1.5-2mm wide, the beak 0.5-1mm long; achenes lenticular, 1.5-2mm long. Open, often dry areas. Statewide. *C. eleocharis* Bailey.

Carex stenoptila Hermann, Leafl. West. Bot. 4:194. 1945. Culms to 7dm high, from very short rhizomes; leaves 1-3.5mm wide; spikes 7-10, gynaecandrous, usually crowded in a head; scales mostly shorter than perigynia; perigynia planoconvex, 4-5mm long, 1-1.5mm wide, the beak about 1.5mm long; achenes lenticular, about 1.5mm long. Woods, rocks, and slopes in the mountains. Carbon Co. and Yellowstone Park.

Carex stipata Muhl. ex Willd., Sp. Pl. 4:233. 1805. Culms to 1m high, without creeping rhizomes; leaves 4-11mm wide; spikes numerous, androgynous, crowded; scales shorter than or equal to perigynia; perigynia planoconvex, 4-5.2mm long, 1.3-1.8mm wide, the beak about 2-2.5mm long; achenes lenticular, 1.3-2mm long. Wet areas. Sheridan Co.

Carex subnigricans Stacey, Leafl. West. Bot. 2:167. 1939. Culms to 2dm high, from creeping rhizomes; leaves 0.2-1.2mm wide; spike solitary, androgynous; scales equaling or shorter than perigynia; perigynia somewhat inflated, 2.5-4mm long, 1-1.5mm wide, the beak about 0.2mm long; achenes trigonous, about 1.2mm long. Moist meadows and slopes in the high mountains. Sublette Co.

Carex tenera Dewey, Am. Journ. Sci. 8:97. 1824. Culms to 8dm high, without creeping rhizomes; leaves 0.7-4mm wide; spikes mostly 4-8, gynaecandrous; scales mostly shorter than perigynia; perigynia planoconvex to flattened, 2.8-4.5mm long, 1.4-2mm wide, the beak about 1mm long; achenes lenticular, 1.2-2.1mm long. Woods and meadows. Crook Co. *C. festucacea* Willd. of authors.

Carex torreyi Tuckerm., Enum. Caric. 21. 1843. Culms to 5dm high, from very short rhizomes; leaves 1.5-4mm wide; spikes 2-4, the terminal staminate, the lateral pistillate; scales shorter than or as long as perigynia; perigynia inflated, 2.5-3.5mm long, 1.8-2.2mm wide, the beak 0.2-0.4mm long; achenes trigonous, 2-2.5mm long. Moist or dry areas of plains or open woods. Crook and Sheridan Cos. *C. abbreviata* Boott.

Carex vallicola Dewey, Am. Journ. Sci. II, 32:40. 1861. Culms to 6dm high, from very short rhizomes; leaves 0.5-2mm wide; spikes 4-9, androgynous, crowded; scales shorter than or equal to perigynia; perigynia planoconvex, 3.5-4mm long, 1.7-2.3mm wide, the beak 0.6-1mm long; achenes lenticular, 1.6-2.5mm long. Moist or dry slopes. Statewide.

Carex vesicaria L., Sp. Pl. 979. 1753. Culms to 1m high, from short creeping rhizomes; leaves 1-8mm wide; spikes 3-7, the upper ones staminate, the lower pistillate or some androgynous; scales shorter than or equal to perigynia; perigynia inflated, 3.5-9mm long, 2-3mm wide, the beak 1-2mm long; achenes trigonous, 1.7-2.5mm long. Wet areas. NW, SW. *C. monile* Tuckerm.

Carex xerantica Bailey, Bot. Gaz. 17:151. 1892. Culms to 7dm high, without creeping rhizomes; leaves 1-4mm wide; spikes 3-6, gynaecandrous; scales shorter than or equal to perigynia; perigynia flattened to planoconvex, 4-7mm long, 1.9-2.8mm wide, the beak about 1mm long; achenes lenticular, 2-3mm long. Plains, slopes, and parks. Albany Co.

Cyperus L. Galingale

Spikelets several to many, arranged in capitate to spicate clusters, usually borne on rays from the axils of sheathless involucral bracts; scales in 2 vertical rows, the flowers bisexual, single in the axil of each scale; stigmas 2 or 3; stamens 1-3.

Plants perennial; culms with bulbs at base; stamens 3 *C. schweinitzii*

Plants annual; culms lacking bulbs at base; stamens 1 or 2 or rarely 3

Stigmas 3; achenes trigonous; scales with long slender tips *C. aristatus*

Stigmas 2; achenes lenticular; scales mostly broad and blunt at tip *C. rivularis*

Cyperus aristatus Rottb., Descr. Pl. Icon. 22. 1772. Annual to 15cm high; leaves mostly 0.5-2.5mm wide; spikelets 4-10mm long; scales 1-2mm long, the awn-tip 0.3-1mm long; stamen 1; stigmas 3; achenes 0.6-1mm long. Wet areas. Albany and Platte Cos. *C. inflexus* Muhl.

Cyperus rivularis Kunth, Enum. Pl. 2:6. 1837. Annual to 2dm high; leaves 0.5-2mm wide; spikelets 3-15mm long; scales 2-2.5mm long, blunt at tip; stamens 2 or rarely 3; stigmas 2; achenes 1-1.3mm long. Wet areas. Platte and Goshen Cos.

Cyperus schweinitzii Torr., Ann. Lyc. N. Y. 3:276. 1836. Rhizomatous perennial to 7dm high; leaves mostly 1-4mm wide; spikelets 5-25mm long; scales mostly 3-4mm long, acute to acuminate; stamens 3; stigmas 3; achenes 2-2.5mm long. Sandy areas. Laramie Co.

<u>Eleocharis</u> R. Br. Spike Rush

Leaves basal or nearly so and reduced to sheaths or scales; spikelet solitary and terminal, not subtended by bracts although the lowest scales of inflorescence sometimes lack flowers in their axils; perianth of 6 or fewer bristles or lacking; style usually thickened near base, the thickening persistent on the achene as a tubercle.

1. Stigmas 3; achenes trigonous or suborbicular in cross section
 2. Achenes longitudinally ribbed, abruptly contracted into tubercle; culms usually capillary, often in dense mats — <u>E. acicularis</u>
 2. Achenes not longitudinally ribbed, gradually tapering into tubercle; culms not capillary, usually not matted
  3. Spikelets 2-12 flowered, 8mm long or less; culms less than 1mm in diameter, cylindrical or 3 angled — <u>E. pauciflora</u>
  3. Spikelets 10 or more flowered, mostly 8mm long or more; culms 1-2mm in diameter, flattened — <u>E. rostellata</u>
1. Stigmas 2 or very rarely 3; achenes lenticular
 4. Plants annuals; tubercle about 3/4 or more as wide as achene — <u>E. ovata</u>
 4. Plants rhizomatous perennials; tubercle about 1/2 or less as wide as achene
  5. Culms mostly less than 15cm high; sheaths often with prolonged scarious tips; spikelets 2-4mm long — <u>E. flavescens</u>

5. Culms mostly over 15cm high; sheaths without prolonged scarious tips; spikelets 5mm or more long *E. palustris*

Eleocharis acicularis (L.) R. & S., Syst. Veg. 2:154. 1817. Slender, rhizomatous, perennial often forming dense tufts, the culms to 15cm high; spikelets 1.5-7mm long, 3-15 flowered; scales 1.3-2.2mm long; stigmas 3; achenes 0.7-1.1mm long including tubercle. Wet areas. Statewide.

Eleocharis flavescens (Poir.) Urban, Symb. Antill. 4:116. 1903. Rhizomatous perennial with tufted culms to 15cm high; spikelets 2-4mm long, less than 20 flowered; scales mostly 1.7-2.5mm long; stigmas 2; achenes 0.9-1.1mm long including tubercle. Wet areas near hot springs and geysers. Yellowstone Park. *E. thermalis* Rydb.

Eleocharis ovata (Roth) R. & S., Syst. Veg. 2:152. 1817. Tufted annual to 5dm high; spikelets mostly 4-13mm long, often over 40 flowered; scales 1.5-2.5mm long; stigmas 2 or sometimes 3; achenes lenticular, 1-1.5mm long including tubercle. Wet areas. Teton and Crook Cos. *E. engelmannii* Steud., *E. obtusa* (Willd.) Schultes.

Eleocharis palustris (L.) R. & S., Syst. Veg. 2:151. 1817. Rhizomatous perennial with culms to 1m high; spikelets 5-25mm long, many flowered; scales mostly 2-4.5mm long; stigmas 2; achenes mostly 1.5-2.5mm long including tubercle. Wet areas. Statewide. *E. macrostachya* Britt.

Eleocharis pauciflora (Lightf.) Link, Hort. Berol. Desc. 1:284. 1827. Rhizomatous perennial to 3dm high; spikelets 4-8mm long, mostly 3-9 flowered; scales 2.5-5.5mm long; stigmas 3; achenes 1.9-2.6mm long including tubercle. Wet areas. NW, SW, SE. *Scirpus pauciflorus* Lightf.

Eleocharis rostellata (Torr.) Torr., Fl. N. Y. 2:347. 1843. Perennial, usually rhizomatous, to 1m high; spikelets mostly 8-13mm long, mostly 10-20 flowered; scales 2-4mm long; stigmas 3; achenes 1.9-2.8mm long including tubercle. Alkaline or calcareous areas especially around hot springs and geysers. Yellowstone Park.

Eriophorum L. Cotton Grass

Plants perennial; leaves grass-like with closed sheaths, some often bladeless; spikelets 1 to several in a terminal inflorescence which is often subtended by 1 to several leafy or scale-like involucral bracts; perianth of numerous white or buffy bristles forming a cottony tuft in fruit; stamens 3; stigmas 3.

1. Spikelet solitary at tip of stem
   2. Plants with rhizomes or stolons, the culms not tufted
      3. Anthers 1mm long or less; bristles of perianth white — E. scheuchzeri
      3. Anthers 1mm long or more; bristles usually cinnamon colored for 2/3 or more of their length — E. chamissonis
   2. Plants lacking rhizomes or stolons, the culms densely tufted — E. callitrix
1. Spikelets several, some or all of them pedicelled
   4. Leaf blades flat below the middle, triangular near tip, some over 2mm wide; leafy bracts usually 2 or more
      5. Midrib of scales attenuate at tip, not reaching end of scale; anthers mostly over 2mm long — E. polystachion
      5. Midrib of scales expanded at tip, reaching end of scale or nearly so; anthers mostly less than 1.5mm long — E. viridicarinatum
   4. Leaf blades triangular-channeled throughout, 1-2mm wide; leafy bract usually 1 — E. gracile

Eriophorum callitrix Cham. ex Meyer, Mem. Acad. St. Petersb. Sav. Etr. 1:203. 1830. Plants densely tufted, to 2.5dm high; leaf blades mostly 1-2mm wide; spikelet solitary and terminal; anthers 0.7-1mm long; achenes 1.5-2.5mm long. Alpine. Park and Fremont Cos.

<u>Eriophorum</u> <u>chamissonis</u> Meyer in Ledeb., Fl. Alt. 1:70. 1829. Plants rhizomatous, to 7dm high; leaf blades to 2mm wide; spikelet solitary and terminal; anthers mostly 1-2.5mm long when dry; achenes 1.8-2.2mm long. Wet areas in the mountains. Yellowstone Park.

<u>Eriophorum</u> <u>gracile</u> Koch in Roth, Cat. Bot. 2:259. 1800. Plants rhizomatous, to 6dm high; leaf blades to 2mm wide; spikelets 2-5; anthers mostly 1-3mm long when dry; achenes 2-3.5mm long. Wet areas in the mountains. Teton Co.

<u>Eriophorum</u> <u>polystachion</u> L., Sp. Pl. 52. 1753. Plants rhizomatous, to 6dm high; leaf blades 2-7mm wide; spikelets 2-8; anthers mostly 2.5-4mm long when dry; achenes mostly 2-4mm long. Swamps and bogs in the mountains. NW, NE, SE. <u>E</u>. <u>ocreatum</u> A. Nels., <u>E</u>. <u>angustifolium</u> Honck.

<u>Eriophorum</u> <u>scheuchzeri</u> Hoppe, Bot. Taschenb. 1800:104. 1800. Plants rhizomatous, to 4dm high; leaf blades to 2mm wide; spikelet solitary and terminal; anthers mostly 0.5-1mm long when dry; achenes about 2mm long. Wet areas in the high mountains. Sublette Co.

Eriophorum viridicarinatum (Engelm.) Fern., Rhodora 7:89. 1905. Plants rhizomatous, to 7dm high; leaf blades mostly 2-6mm wide; spikelets 2-20; anthers mostly 1-1.5mm long when dry; achenes mostly 2.5-4mm long. Swamps and bogs in the mountains. Teton and Yellowstone Parks.

Hemicarpha Nees & Arn. Dwarf Bulrush

Reference: Svenson, H. K. 1957. N. Am. Fl. 18:508-509.

Hemicarpha drummondii Nees in Mart., Fl. Bras. 2(1):62. 1842. Plants annual, to 15cm high; leaves 0.5mm wide or less; spikes mostly 1-3 per culm, sessile, appearing lateral, 2-6mm long; each flower subtended by an outer scale and a smaller inner (toward spike axis) scale which is sometimes obsolete; stamen usually solitary; achenes 0.4-0.7mm long. Wet, usually sandy areas. Albany Co. H. aristulata (Cov.) Smyth.

Kobresia Willd.

Tufted perennials resembling Carex; flowers unisexual, solitary in axils of scarious bracts called scales, in spikelets which are aggregated into a spike or spikes, each spikelet subtended by a scale and having 1 pistillate flower below and 1 or more staminate flowers above or the spikelet with flowers of only 1 sex; pistillate scales wrapped around achene to form a perigynium, the perigynium open down 1 side; stamens 3; stigmas 3.

Spikes 2-3mm wide; scales of spikelets 2-4mm long; perigynia 3-3.5mm long — *K. bellardii*

Spikes mostly 4-5mm wide; scales of spikelets 4-5mm long; perigynia about 5.5mm long — *K. macrocarpa*

Kobresia bellardii (All.) Degl. ex Loisel., Fl. Gall. 626. 1807. Culms to 3dm high; leaves mostly 0.3-0.6mm wide; spike solitary and terminal, 0.8-3cm long, 2-3mm thick; perigynia 3-3.5mm long; achenes about 2mm long. Mostly alpine. Park and Sublette Cos. *K. myosuroides* (Vill.) Fiori & Paol., *Elyna bellardii* (All.) Hartm.

Kobresia macrocarpa Clokey ex Mack., N. Am. Fl. 18:5. 1931. Culms to 2dm high; leaves 0.2-0.6mm wide; spike solitary and terminal, 0.8-2cm long, mostly 4-5mm thick; perigynia about 5.5mm long; achenes about 4mm long. Mostly alpine. Park Co.

Scirpus L. Bulrush

Perennials with grass-like or sometimes obsolete leaves; spikelets 1 to many, the inflorescence subtended by 1 to several leafy or scale-like bracts; flowers bisexual, each in an axil of a scale; perianth of 2-6 bristles; stamens 3 or sometimes fewer; stigmas 2 or 3; achenes commonly beaked.

References: Schuyler, A. E. 1967. Proc. Acad. Phila. 119:295-323.
1969. Not. Naturae 423:1-12.
1974. Rhodora 76:51-52.

Schuyler (1969) lists a possible collection of S. saximontanus Fern. from Wyoming. It differs from our other species in being an annual.

1. Inflorescence subtended by a single, well developed, green involucral bract, often appearing like a prolongation of the stem so the inflorescence appears lateral rather than terminal (other bracts often present but these scale-like and not green); leaves various but rarely over 5mm wide

2. Spikelets solitary; plants usually submerged or floating S. subterminalis

2. Spikelets usually 2 or more; plants usually emergent or terrestrial

3. Culms terete, usually about 1m high or more; spikelets in a branching inflorescence, the inflorescence sometimes compact

4. Middle and lower scales mostly 3.5-4mm long; red-brown striolae usually prominent on the gray-white background of the scales; inflorescence rarely over 6cm long and 4cm wide S. acutus

4. Middle and lower scales mostly 2.5-3mm long; red-brown striolae usually not prominent on the dark reddish-brown background of many scales; inflorescence sometimes over 6cm long or 4cm wide *S. validus*

3. Culms triangular or subterete, usually less than 1m high; spikelets sessile or nearly so, the inflorescence not branched

5. Culms subterete; scales without an awn *S. nevadensis*

5. Culms triangular; scales with a short awn from a notched tip

6. Bract solitary, mostly 1-3cm long; achenes 1.8-2.5mm long, 1.4-1.7mm wide *S. americanus*

6. Bracts 2 or 3, the largest mostly 3-10cm long, the smaller resembling large scales but not subtending individual flowers; achenes 2.2-3.3mm long, 1.6-2.3mm wide *S. pungens*

1. Inflorescence usually subtended by 2 or more green involucral bracts, the inflorescence definitely terminal; leaves flat, some often over 5mm wide

7. Spikelets usually 10mm or more long, sessile or nearly so in a crowded cluster, rarely a few terminal ones on a stalk *S. maritimus*

7. Spikelets usually less than 8mm long, in an open or branched inflorescence

8. Stigmas usually 2; achenes lenticular; sheaths of basal leaves usually reddish tinged *S. microcarpus*

8. Stigmas usually 3; achenes trigonous; sheaths of basal leaves usually not reddish tinged *S. pallidus*

Scirpus acutus Muhl. ex Bigel., Fl. Bost. 15. 1814. Plants rhizomatous, mostly 1-3m high, in large colonies; culms terete; leaves near base, the blades short or lacking; spikelets numerous, mostly 8-15mm long; scales mostly 3-4mm long; achenes 2.2-2.5mm long. Shores and marshes. Statewide. *S. lacustris* L. of authors.

Scirpus americanus Pers., Syn. 1:68. 1805. Plants rhizomatous, to 15dm high; culms triangular; leaves near base, the blades mostly reduced but to 1cm wide; spikelets mostly 2-15, 6-15mm long; scales 3-5mm long; achenes 1.8-2.5mm long. Wet areas. Yellowstone Park and Teton Co. *S. chilensis* Nees & Mey. of authors, *S. olneyi* Gray ex Engelm. & Gray.

Scirpus maritimus L., Sp. Pl. 51. 1753. Plants rhizomatous, to 15dm high; culms triangular; leaves well developed, the blades to 1cm wide; spikelets 3 to many, 1-2cm long; scales 6-8mm long; achenes 2.5-4mm long. Marshes and shores especially where alkaline. Statewide. *S. paludosus* A. Nels., *S. campestris* Britt.

Scirpus microcarpus Presl, Rel. Haenk. 1:195. 1828. Plants rhizomatous, to 15dm high; culms obscurely triangular; leaves well developed, the blades 3-20mm wide; spikelets many, 3-8mm long; scales about 1.5mm long; achenes mostly 0.9-1.2mm long. Wet areas. NW, NE, SE. *S. rubrotinctus* Fern.

Scirpus nevadensis Wats., Bot. King Exp. 360. 1871. Plants rhizomatous, to 6dm high; culms subterete; leaves near base, the blades mostly 1.5-2mm wide; spikelets 1-10, 7-20mm long; scales 3-4mm long; achenes 2-3mm long. Wet, usually alkaline areas. Albany and Fremont Cos.

Scirpus pallidus (Britt.) Fern., Rhodora 8:163. 1906. Plants rhizomatous, to 15dm high; culms triangular; leaves well developed, 6-16mm wide; spikelets many, mostly 3-4mm long; scales 1.5-2mm long; achenes mostly 0.9-1.2mm long. Wet areas. NW, NE, SE. S. atrovirens Muhl. of authors.

Scirpus pungens Vahl, Enum. Pl. 2:255. 1805. Plants rhizomatous, mostly 1.5-10dm high; culms triangular; leaves near base, the blades mostly 1-4mm wide; spikelets mostly 1-6, 7-20mm long; scales about 4mm long; achenes 2.2-3.3mm long. Marshes and ditches. Statewide. S. americanus Pers. of authors.

Scirpus subterminalis Torr., Fl. N. & Mid. U. S. 47. 1823. Plants rhizomatous, to 8dm long; culms subterete; leaves near base, 0.2-1.5mm wide; spikelet solitary, 7-12mm long; scales 4-6mm long; achenes 2.5-3.8mm long. Floating or submerged in shallow water or rarely on shores. Teton Co.

Scirpus validus Vahl, Enum. Pl. 2:268. 1805. Plants rhizomatous, mostly 1-3m high; culms terete; leaves near base, the blades short or lacking; spikelets many, mostly less than 1cm long; scales mostly 2-3mm long; achenes mostly 2-2.3mm long. Shores and marshes. Statewide. *S. lacustris* L. of authors.

DROSERACEAE Sundew Family

Herbs with simple, mostly basal leaves bearing long, gland-tipped hairs which catch insects; flowers bisexual, regular, in an apparent raceme; sepals 4 or 5, usually united; petals mostly 5, separate; stamens 4-8; ovary 1, superior; carpels 3-5; styles 3-5 or apparently 6-10; locule 1; ovules many; placentation parietal; fruit a capsule.

Drosera L. Sundew

Drosera anglica Huds., Fl. Angl. 2:135. 1778. Perennial herb to 2dm high; leaves mostly basal and rosulate, the blades oblong-oblanceolate to spatulate, 5-25mm long, 2-5mm wide, with long, reddish, gland-tipped hairs; flowers 1-7, in a somewhat 1 sided raceme; sepals 3-6mm long; petals white, 3-7mm long; capsule 3-5 valved; seeds numerous. Swamps and bogs. Yellowstone and Teton Parks.

Trees or shrubs with simple, entire, alternate or opposite leaves which are usually covered with silvery or rusty scale-like hairs; flowers bisexual, or unisexual and the plants dioecious, regular, clustered in the leaf axils or solitary; sepals 4, united; petals lacking; stamens 4 or 8; ovary 1, superior or sometimes appearing inferior; carpel 1; style 1; locule 1; ovule 1; placentation basal; fruit an achene, often with a fleshy covering, or drupe-like.

Leaves and branches alternate — *Elaeagnus*

Leaves and branches opposite or subopposite — *Shepherdia*

Elaeagnus L.

Tree or shrub with alternate, usually silvery leaves; flowers axillary, mostly bisexual; stamens 4; fruit drupe-like.

Leaves usually all less than 4 times as long as wide; leafy twigs brown except when young — *E. commutata*

Leaves, or some of them, over 4 times as long as wide; leafy twigs silvery — *E. angustifolia*

Elaeagnus angustifolia L., Sp. Pl. 121. 1753. Russian Olive. Tree or shrub to 10m high; branches sometimes spiny; leaf blades mostly oblong-lanceolate to linear-lanceolate, 2-10cm long, silvery-scurfy beneath, green or somewhat scurfy above; flowers 1-3 per axil, 7-12mm long. Introduced and becoming established along streams and ditches at lower elevations.

Elaeagnus commutata Bernh. ex Rydb., Fl. Rocky Mts. 582. 1917. Silverberry. Shrub to 4m high; leaf blades ovate, lanceolate, or elliptic, 1.5-8cm long, silvery-scurfy at least beneath; flowers 1-3 per axil, 6-12mm long. Streambanks and slopes. NW, SW. *E. argentea* Pursh.

Shepherdia Nutt. Buffaloberry

Dioecious trees or shrubs with opposite leaves; flowers axillary; stamens mostly 8; fruit fleshy.

Plants with spiny spur branches; leaves usually somewhat silvery on both surfaces *S. argentea*

Plants without spiny spur branches; leaves green above, silvery or brownish dotted beneath *S. canadensis*

Shepherdia argentea (Pursh) Nutt., Gen. Pl. 2:240. 1818. Tree or shrub to 6m high, usually with spine-tipped branches; leaf blades oblong or oblanceolate, or some elliptic or lanceolate, 1-6cm long, silvery-scurfy; flowers 3-7mm long. Streambanks. Statewide. *Elaeagnus utilis* A. Nels.

Shepherdia canadensis (L.) Nutt., Gen. Pl. 2:241. 1818. Shrub to 4m high; leaf blades ovate to ovate-lanceolate, 1-7cm long, usually dark green above, brown dotted on a whitish surface beneath; flowers 3-7mm long. Woods and slopes. Statewide. *Elaeagnus canadensis* (L.) A. Nels.

ELATINACEAE Waterwort Family

Aquatic or terrestrial herbs with simple, opposite leaves; flowers bisexual, regular or irregular, solitary and axillary; sepals 2 or 3, separate or not; petals 2 or 3, separate; stamens 3 or rarely 6; ovary 1, superior; carpels 2 or 3; styles usually 2 or 3; locules 2 or 3; ovules numerous; placentation axile; fruit a capsule.

Elatine L. Waterwort

Elatine triandra Schkuhr, Bot. Handb. 1:345. 1791. Prostrate to erect annual herb with stems to 10cm long; leaves linear to spatulate, 1-10mm long; sepals and petals 2 or 3 each, about 1mm long, the petals white or pinkish; seeds reticulate in longitudinal rows. Submerged or on mud flats and shores or rarely in drier areas. NW, SE. *E. rubella* Rydb.

ERICACEAE Heath Family

Shrubs or herbs with simple, alternate, opposite, whorled or basal leaves, the plants rarely saprophytic and lacking green leaves; flowers bisexual, mostly regular, solitary in axils or terminal, or in axillary or terminal clusters, corymbs, or racemes; sepals usually 4 or 5(6), united or not, rarely lacking; petals usually 4 or 5(6), united or not, rarely lacking; stamens (4)5-10(12), the anthers often opening by terminal pores; ovary 1, superior or inferior; carpels 4-10; styles usually solitary; locules usually 4-10; ovules several to many; placentation usually axile; fruit a capsule or berry. This family is sometimes divided into two more families, the Pyrolaceae and Monotropaceae.

Reference: Nowicke, J. W. 1966. Ann. Mo. Bot. Gard. 53:213-219.

1. Plants lacking green leaves, saprophytic, brown to red, pinkish, white, or yellowish in color
   2. Petals united nearly to tip; anthers awned on back — _Pterospora_
   2. Petals separate; anthers not awned — _Hypopitys_
1. Plants with green leaves, not saprophytic, variously colored
   3. Leaves linear, mostly 2-13mm long and 1-2mm wide, resembling the needles of a fir tree; pedicels glandular-pubescent — _Phyllodoce_
   3. Leaves not as above; pedicels various

4. Plants herbaceous, leafy only at or near the base; flowers solitary or in a raceme; styles mostly linear, often curved
 5. Flowers solitary and terminal — <u>Moneses</u>
 5. Flowers several in a raceme
  6. Raceme secund; style straight or nearly so, usually over 2mm long — <u>Orthilia</u>
  6. Raceme not secund; style curved, or if straight, then shorter — <u>Pyrola</u>

4. Plants woody, or herbaceous with leafy stems; flowers various
 7. Plants upright, herbaceous or woody only at base; some leaves usually whorled; petals separate; flowers usually in a terminal corymbose raceme — <u>Chimaphila</u>
 7. Plants not with the above combination of characters
  8. Petals separate; leaves usually with yellow resinous dots beneath — <u>Ledum</u>
  8. Petals united; leaves not as above
   9. Ovary inferior; anthers dorsally awned; petals united nearly to tip; fruit a berry — <u>Vaccinium</u>
   9. Ovary superior, or if appearing inferior, the anthers and petals not as above; fruit various
    10. Leaves opposite, pale beneath from minute pubescence, the margins often rolled; corolla broadly bowl-shaped — <u>Kalmia</u>
    10. Leaves alternate, often glabrous, flat; corolla usually not as above

11. Plants upright shrubs in NW Wyo. *Menziesia*
11. Plants prostrate and creeping shrubs, widespread
12. Petals united nearly to tip; filaments hairy near the middle; leaves mostly oblanceolate to obovate *Arctostaphylos*
12. Petals united about halfway to tip; filaments glabrous; leaves mostly ovate to orbicular or rarely elliptic *Gaultheria*

Arctostaphylos Adans. Bearberry; Kinnikinnick

Evergreen, prostrate shrubs with alternate, leathery leaves; flowers in terminal racemes; corolla urn-shaped, pinkish; stamens 10; fruit a red berry.

Reference: Packer, J. G. & K. E. Denford. 1974. Canad. Journ. Bot. 52:743-753.

Branchlets viscid-villous with spreading, stipitate, multicellular hairs often 0.5mm or more long *A. adenotricha*
Branchlets not viscid-villous, the hairs mostly curled and not spreading nor glandular, or rarely viscid with minute stipitate hairs about 0.1mm or less long *A. uva-ursi*

Arctostaphylos adenotricha (Fern. & Macbr.) Löve et al., Arctic & Alp. Res. 3:154. 1971. Plants to 15cm high; leaf blades oblanceolate to obovate, entire, 0.5-3cm long; calyx 1-2.5mm long; corolla 4-6mm long. Woods and slopes. NE, SE.

Arctostaphylos uva-ursi (L.) Spreng., Syst. Veg. 2:287. 1825. Similar to A. adenotricha except the branchlets not viscid-villous but rather closely pubescent, rarely minutely viscid. Woods and slopes. Statewide.

Chimaphila Pursh Pipsissewa

Chimaphila umbellata (L.) Barton, Veg. Mat. U. S. 1:17. 1817. Evergreen semishrub to 3dm high; leaves leathery, mostly whorled, the blades elliptic or oblanceolate or rarely obovate, 1-7cm long, serrate; flowers in a corymbose raceme; calyx 2-4mm long; petals distinct, 4-7mm long, pink or rose; stamens 10; fruit a capsule. Woods. Statewide.

Gaultheria L. Wintergreen

Gaultheria humifusa (Grah.) Rydb., Mem. N. Y. Bot. Gard. 1:300. 1900. Depressed evergreen shrub to 7cm high; leaves leathery, mostly alternate, the blades ovate to orbicular or rarely elliptic, 5-25mm long, entire to serrulate; flowers solitary in leaf axils; calyx 2-4mm long; corolla short campanulate, 3-4mm long, pinkish; stamens 8-10; fruit a capsule surrounded by a fleshy pulp. Moist areas in the mountains. NW, NE, SE.

Hypopitys Hill Pinesap

Hypopitys _monotropa_ Crantz, Inst. Rei Herb. 2:467. 1766. Red, pink, or yellowish saprophyte to 3dm high; leaves not green, somewhat scale-like, 5-20mm long, 3-10mm wide; flowers in racemes, all the parts pubescent; perianth mostly 4-merous; sepals 5-9mm long; petals distinct, 9-16mm long, pinkish or cream colored; stamens twice as many as petals; fruit a capsule. Woods. NW, SE. _H_. _multiflora_ Scop. of authors, _Monotropa_ _hypopitys_ L.

Kalmia L. Laurel

References: Ebinger, J. E. 1974. Rhodora 76:315-398.
Southall, R. M. & J. W. Hardin. 1974. Journ. Elisha Mitchell Soc. 90:1-23.

Kalmia _microphylla_ (Hook.) Heller, Bull. Torrey Club 25:581. 1898. Evergreen shrub to 5dm high; leaves leathery, opposite, the blades broadly elliptic or ovate to linear-elliptic, 0.3-3cm long, entire; flowers in terminal corymbs; calyx 2-4mm long; corolla broadly bowl-shaped, 5-10mm long or 7-16mm wide, rose-pink; stamens 10; fruit a capsule. Bogs and meadows. Statewide. _K_. _polifolia_ Wang. var. _microphylla_ (Hook.) Rehd.

Ledum L. Labrador Tea

Ledum glandulosum Nutt., Trans. Am. Phil. Soc. II, 8:270. 1842. Evergreen shrub to 2m high; leaves coriaceous, mostly alternate, the blades ovate or oblong to elliptic, 1-3.5cm long, entire; flowers in terminal racemes or corymbs; calyx about 1mm long; petals distinct, 4-6mm long, white; stamens mostly 8-12; fruit a capsule. Woods and moist slopes. NW.

Menziesia Smith

Reference: Hickman, J. C. & M. P. Johnson. 1969. Madroño 20:1-11.

Menziesia ferruginea Smith, Pl. Ic. Ined. pl. 56. 1791. Deciduous shrub to 3m high; leaves thin, alternate, the blades ovate-elliptic to elliptic-obovate, 1-6cm long, serrulate; flowers in clusters terminating the old growth, appearing with the leaves; perianth 4-merous; calyx about 1mm long; corolla urn-shaped to campanulate, 6-9mm long, reddish-yellow or pinkish; stamens 8; fruit a capsule. Woods and thickets. Teton Co.

Moneses Salisb.

Moneses uniflora (L.) Gray, Man. 273. 1848. Perennial, rhizomatous herb to 15cm high; leaves all near base, the blades ovate-elliptic to obovate or orbicular, 5-25mm long, serrate-crenate; flowers solitary and terminal; calyx 2-4mm long; petals distinct, 6-13mm long, white; stamens 10; fruit a capsule. Woods. Statewide. *Pyrola uniflora* L.

Orthilia Raf.

Orthilia secunda (L.) House, Am. Midl. Nat. 7:134. 1921. Perennial, rhizomatous herb to 20cm high; leaves mostly on lower third of plant, mostly alternate, the blades mostly ovate or ovate-elliptic, 1-6cm long, crenulate or serrulate; flowers in a secund raceme; calyx about 1-2mm long; petals distinct, mostly 4-5mm long, greenish-white or white; stamens 10; fruit a capsule. Woods. Statewide. *Pyrola secunda* L., *Ramischia secunda* (L.) Garcke.

Phyllodoce Salisb. Mountain Heath

Evergreen shrubs with linear, alternate leaves resembling a fir tree; flowers solitary in the terminal leaf axils, on long pedicels; corolla urn-shaped to campanulate; stamens 10; fruit a capsule. Our two species often hybridize when growing together.

Corolla pink to rose, glabrous on outside *P. empetriformis*
Corolla greenish-white to yellow, glandular-pubescent on outside *P. glanduliflora*

Phyllodoce empetriformis (Sw.) D. Don, Edinb. New Phil. Journ. 17:160. 1834. Plants somewhat matted, to 3dm high; leaves mostly 4-13mm long, 1-2mm wide, minutely glandular-serrulate; calyx about 2mm long; corolla rose or pink, 4-8mm long. Mostly alpine and subalpine. NW, SW.

Phyllodoce glanduliflora (Hook.) Cov., Mazama 1:196. 1897. Plants somewhat matted, to 3dm high; leaves mostly 2-10mm long, 1-2mm wide, minutely glandular-serrulate; calyx 2-4mm long; corolla yellowish to greenish-white, 4-7mm long. Mostly alpine and subalpine. NW.

Pterospora Nutt. Pinedrops

Pterospora andromedea Nutt., Gen. Pl. 1:269. 1818. Reddish-brown to white saprophyte with bracteate stems to 1m high; flowers in an elongate raceme; calyx 4-6mm long, glandular-pubescent; corolla urn-shaped, 5-8mm long, pale yellow to brown; stamens 10; fruit a capsule. Woods. Statewide.

Pyrola L.

Perennial, rhizomatous herbs with the leaves all near the base; flowers in a terminal raceme; petals distinct; stamens 10; fruit a capsule.

References: Křísa, B. 1966. Bot. Jahrb. 85:612-637.
Haber, E. 1972. Rhodora 74:396-397.

1. Style straight or nearly so, 2mm long or less; petals 2-5mm long P. minor
1. Style curved, usually over 2mm long; petals mostly 5mm or more long
  2. Leaf blades with a tapered acute base, mostly oblanceolate or obovate, usually less than 20(25)mm wide; petals greenish-white or cream colored; NW Wyo. P. dentata

2. Leaf blades often rounded at base, usually elliptic or ovate to orbicular, rarely obovate, often over 25mm wide; petals greenish-white or not; widespread

3. Leaves prominently white-mottled or streaked along the veins on upper surface; NW Wyo. *P. picta*

3. Leaves usually not white-mottled or streaked; widespread

4. Petals pink to purplish *P. asarifolia*

4. Petals white or greenish-white to yellowish

5. Leaf blades usually less than 3cm long; sepals rounded to acute *P. chlorantha*

5. Leaf blades, or some of them, usually over 3cm long; sepals acute to acuminate *P. elliptica*

Pyrola asarifolia Michx., Fl. Bor. Am. 1:251. 1803. Plants to 4dm high; leaf blades orbicular to ovate or broadly elliptic, 1.5-7cm long, entire to serrulate; calyx 2.5-4mm long; petals 5-8mm long, pink to purplish. Woods, thickets, and wet meadows. Statewide. *P. uliginosa* T. & G. ex Torr., *P. californica* Křísa.

Pyrola chlorantha Sw., Sv. Vet.-Akad. Handl. 31:190. 1810. Plants to 25cm high; leaf blades elliptic or ovate to orbicular, 0.5-3cm long, crenulate or serrulate; calyx 2-4mm long; petals 5-8mm long, pale yellow or greenish-white. Woods and thickets. Statewide. *P. virens* Schreber (Schweigg. of authors).

<u>Pyrola</u> <u>dentata</u> Smith in Rees, Cycl. 29: Pyrola no. 6. 1814. Plants to 25cm high; leaf blades oblanceolate or obovate to elliptic, 0.5-4cm long, entire to serrulate; calyx 2-4mm long; petals 5-7mm long, cream or greenish-white. Woods and slopes. Teton Co.

<u>Pyrola</u> <u>elliptica</u> Nutt., Gen. Pl. 1:273. 1818. Plants to 25cm high; leaf blades elliptic to obovate, 2-6cm long, crenulate-serrulate; calyx 2-4mm long; petals 5-8mm long, white or greenish-white. Woods and thickets. Sheridan and Crook Cos.

<u>Pyrola</u> <u>minor</u> L., Sp. Pl. 396. 1753. Plants to 25cm high; leaf blades elliptic to orbicular, 0.5-3.5cm long, crenulate-serrulate; calyx 1-3mm long; petals 2-5mm long, pink to rose. Woods, thickets, and moist slopes. NW, SW, SE.

<u>Pyrola</u> <u>picta</u> Smith in Rees, Cycl. 29: Pyrola no. 8. 1814. Plants to 30cm high; leaf blades ovate or elliptic to orbicular, 2-7cm long, entire to serrulate; calyx 2-4mm long; petals 6-9mm long, yellowish or greenish-white, rarely purplish. Woods. Teton Co.

Vaccinium L. Blueberry; Huckleberry

Creeping to erect, deciduous or evergreen shrubs with mostly alternate leaves; flowers solitary or 2-4 and axillary; corolla usually urn-shaped but sometimes globular or campanulate; stamens 10, the anthers bearing 2 awns; ovary inferior; fruit a red, blue, or purple berry.

Reference: Young, S. B. 1970. Rhodora 72:439-459.

1. Flowers 1-4 per axil, arising directly from a bud on a twig from the previous year — _V. occidentale_
1. Flowers usually 1 per leaf axil on twigs of the year
   2. Leaves obovate or oblanceolate or some elliptic; plants mostly less than 3dm high; teeth of leaves usually gradually disappearing toward the base; twigs mostly inconspicuously angled to subterete, puberulent — _V. caespitosum_
   2. Leaves mostly elliptic, oval, ovate, or lanceolate; plant height various; teeth of leaves usually prominent from base to apex; twigs conspicuously angled, puberulent or glabrous
      3. Plants usually less than 3dm high; branches numerous and crowded, most of them green; pedicels usually less than 3mm long

4. Leaves mostly 1-3cm long, mostly 8mm or more wide; young twigs usually finely hairy; mature fruits usually bluish — V. myrtillus

4. Leaves 4-15mm long, mostly less than 8mm wide; young twigs usually glabrous; mature fruit usually reddish, sometimes drying bluish — V. scoparium

3. Plants usually over 3dm high; branches not very crowded, most of them brown; some pedicels usually over 5mm long

5. Corollas averaging as long as wide; leaves averaging about 2/3 as wide as long, rounded to acute at tip — V. globulare

5. Corollas averaging longer than wide; leaves averaging about 1/2 or less as wide as long, acute to acuminate at tip — V. membranaceum

Vaccinium caespitosum Michx., Fl. Bor. Am. 1:234. 1803. Somewhat prostrate shrub to 3dm high; leaf blades mostly obovate or oblanceolate, 0.5-3cm long, serrulate at least at tip; flowers solitary in axils; calyx 0.5-2mm long; corolla 3-6mm long, white to pink; mature berry blue, glaucous. Meadows and slopes in the mountains. NW, SE.

Vaccinium globulare Rydb., Mem. N. Y. Bot. Gard. 1:300. 1900. Shrub to 12dm high; leaf blades mostly ovate to oval, 1-5cm long, serrulate; flowers solitary in axils; calyx 0.5-2mm long; corolla 4-7mm long, pinkish-yellow to white; mature berry blue-purple. Woods and thickets. NW, SW.

Vaccinium membranaceum Dougl. ex Torr., Bot. Wilkes Exp. 377. 1874. Shrub to 2m high; leaf blades mostly elliptic, 1-5cm long, serrulate; flowers solitary in axils; calyx 0.5-2mm long; corolla 4-7mm long, yellowish-pink to white; mature berry purple or red-purple. Woods and slopes. Yellowstone Park and Teton Co. This name was not validly published in Hooker's Fl. Bor. Am.

Vaccinium myrtillus L., Sp. Pl. 349. 1753. Shrub to 3dm high; leaf blades ovate to elliptic, 0.5-3cm long, serrulate; flowers solitary in axils; calyx 0.5-2mm long; corolla mostly 4-6mm long, pinkish; mature berry bluish or rarely dark red. Woods and slopes in the mountains. Yellowstone Park and Carbon Co. *V. oreophilum* Rydb.

Vaccinium occidentale Gray, Bot. Calif. 2nd ed. 1:451. 1880. Shrub to 8dm high; leaf blades mostly oblanceolate or elliptic, 0.5-3cm long, entire or nearly so; flowers 1-4 per axil, from a bud on wood of previous season; calyx lobes about 1mm long; corolla 3-7mm long, pinkish; mature berry blue, glaucous. Bogs, woods, and slopes. NW, SW.

Vaccinium scoparium Leiberg ex Cov., Mazama 1:196. 1897. Somewhat prostrate shrub to 3dm high; leaf blades lanceolate or ovate to elliptic, 4-15mm long, serrulate; flowers solitary in axils; calyx 0.5-2mm long; corolla mostly 3-4mm long, pinkish; mature berry red, sometimes drying bluish. Woods and slopes in the mountains. Statewide.

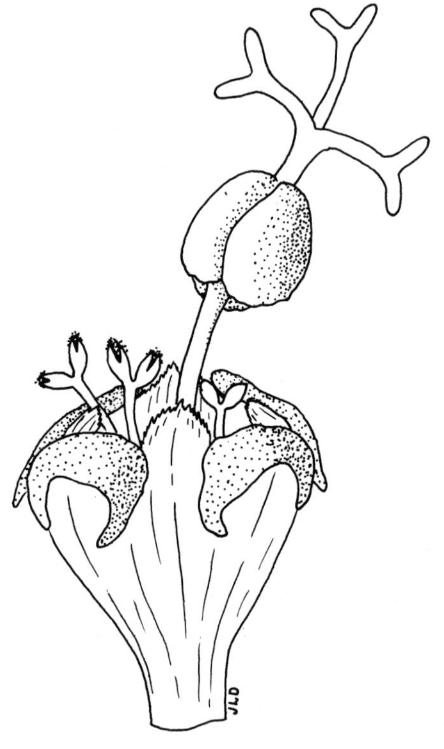

Figure 9. Euphorbiaceae. Portion of inflorescence of *Euphorbia esula* (X 9) with 1 pistillate and 3 staminate flowers protruding from the involucre.

Herbs with alternate, opposite, or rarely whorled, simple leaves; plants monoecious with a single stalked pistil and several to many stamens subtended by each involucre, or occasionally dioecious without involucres; flowers mostly regular; sepals usually 5 or lacking, separate or united at base; petals lacking or 5 and minute; stamens solitary (but several to many per involucre) or 8-12; ovary 1, superior, often stalked; carpels 3; styles usually 3, often lobed; locules 3; ovules usually 1 per locule; placentation axile; fruit a 3 seeded capsule.

Plants stellate-pubescent, dioecious — *Croton*
Plants not stellate-pubescent nor dioecious — *Euphorbia*

Croton L.

Croton texensis (Klotzsch) Muell. in DC., Prodr. 15(2):692. 1866. Annual to 6dm high, stellate-pubescent; leaves alternate, lanceolate to oblong, 1-5cm long, entire; plants dioecious; stamens 8-12; sepals 1-3mm long; petals lacking or minute. Plains, hills, and disturbed areas. NE, SE.

Euphorbia L. Spurge

Annual or perennial herbs with simple, alternate or opposite leaves, the floral bracts usually opposite or whorled; flowers in perianth-like involucres, 1 stalked pistil and several to many stamens in each involucre; involucres solitary or clustered in leaf axils or in terminal cymes or umbels, usually with 4 terminal glands alternating with small tooth-like lobes, the glands often with horns or appendages; perianth lacking.

1. Upper leaves and bracts with prominent white margins or almost completely white, some of these usually at least 2cm long; involucre pubescent — E. marginata
1. Upper leaves and bracts not white-margined, or if so, much less than 2cm long; involucre often glabrous
   2. Plants mostly prostrate and perennial; leaves mostly less than 7mm long and less than 1½ times as wide, deltoid to ovate, opposite and entire, mostly petioled — E. fendleri
   2. Plants not as above
      3. Plants perennial with a thick woody base
         4. Leaves mostly over 5 times longer than wide, linear or oblong to narrowly elliptic
            5. Leaves mostly 3mm or less wide, 2cm or less long — E. cyparissias
            5. Leaves, or some of them, over 3mm wide and over 2cm long — E. esula
         4. Leaves mostly less than 5 times longer than wide, not linear or oblong, rarely elliptic

6. Plants with some leaves over 3cm long E. *agraria*

6. Plants with leaves mostly less than 2cm long E. *robusta*

3. Plants taprooted annuals

7. Leaves scabrous, rough to the touch, the margins toothed, mostly lanceolate, some usually over 1cm wide E. *dentata*

7. Leaves without the above combination of characters

8. Glands of involucre with petaloid appendages, or if appendages lacking, the leaves all strictly opposite with asymmetrical bases and mostly less than 5mm wide

9. Leaves symmetrical at base, entire; stipules minute and gland-like or lacking E. *hexagona*

9. Leaves usually asymmetrical at base, or if symmetrical, the stipules usually well developed although small or the leaves toothed

10. Ovary, capsule, stems, and leaves hairy with mostly soft, loose hairs E. *stictospora*

10. Ovary and capsule glabrous or rarely puberulent with very short, stiff hairs; stems and leaves often glabrous

11. Leaves symmetrical at base or nearly so

12. Leaves entire E. *missurica*

12. Leaves toothed E. *exstipulata*

11. Leaves very asymmetrical at base

13. Seeds with coarse transverse ridges; leaf margins thickened, entire to denticulate *E. glyptosperma*

13. Seeds smooth, punctate, or wrinkled; leaf margins not thickened, often serrulate *E. serpyllifolia*

8. Glands of involucre without petaloid appendages; leaves alternate or opposite with the bases symmetrical or nearly so, often over 5mm wide

14. Floral leaves usually tapering to base; capsule smooth *E. helioscopia*

14. Floral leaves usually rounded or cordate at base; capsule tuberculate *E. spathulata*

Euphorbia agraria Bieb., Fl. Taur.-Cauc. 1:375. 1808. Perennial to 9dm high; leaves mostly alternate, the blades triangular-ovate to oblong or lanceolate, cordate-auriculate at base, 2-5cm long, entire; involucres 2.5-3.5mm long; glands without appendages. Disturbed areas. Converse Co.

Euphorbia cyparissias L., Sp. Pl. 461. 1753. Perennial to 4dm high; leaves alternate, linear, 0.5-2cm long, entire; involucres 1-3mm long; glands with short horns from the 2 lateral ends. Disturbed areas. Albany Co.

Euphorbia _dentata_ Michx., Fl. Bor. Am. 2:211. 1803. Annual to 7dm high; leaves mostly opposite, the blades ovate-lanceolate to lanceolate, 1-8cm long, coarsely toothed; involucres 2-3mm long; glands lacking appendages. Plains, hills, and disturbed areas. Albany and Platte Cos. _E_. _cuphosperma_ (Engelm.) Boiss.

Euphorbia _esula_ L., Sp. Pl. 461. 1753. Perennial to 9dm high; middle leaves oblong to linear or narrowly elliptic, alternate, 2-10cm long, entire; involucres 2-4mm long; glands with short horns from the 2 lateral ends. Disturbed areas. Statewide.

Euphorbia _exstipulata_ Engelm. in Torrey, Bot. Mex. Bound. 189. 1859. Annual to 3dm high; leaves mostly opposite, the blades linear to lanceolate or oblong, 0.5-4cm long, coarsely toothed; involucres 1-2.5mm long; glands with white appendages. Plains and hills. Platte Co. _E_. _aliceae_ A. Nels.

Euphorbia _fendleri_ T. & G., Pac. R. R. Rep. 2(4):175. 1855. Perennial with mostly prostrate stems to 15cm long; leaves opposite, the blades obliquely ovate or deltoid, 2-7mm long, entire; involucres 1.5-2.5mm long; glands with or without appendages. Plains and hills. NE, SE.

Euphorbia _glyptosperma_ Engelm. in Torrey, Bot. Mex. Bound. 187. 1859. Annual with prostrate stems to 4dm long; leaves all opposite, the blades mostly lanceolate, oblanceolate, or oblong, 1-12mm long, entire to denticulate; involucres 1-2mm long; glands usually with appendages. Plains and hills. Statewide.

Euphorbia helioscopia L., Sp. Pl. 459. 1753. Annual to 5dm high; middle leaves oblanceolate to spatulate or obovate, alternate, 1-3cm long, serrate-dentate; involucres 2-3.5mm long; glands without appendages. Disturbed areas. Albany Co.

Euphorbia hexagona Nutt. ex Spreng., Syst. Veg. 3:791. 1826. Annual to 5dm high; leaves opposite, the blades linear, oblong, or oblanceolate, 5-45mm long, entire; involucres 1.5-3mm long; glands with greenish-yellow to white appendages. Plains and hills especially where sandy. Platte Co.

Euphorbia marginata Pursh, Fl. Am. Sept. 607. 1814. Annual to 9dm high; leaves alternate below, opposite or whorled above, the blades ovate, obovate, or oblong-ovate, 1.5-7cm long, entire; floral leaves with white margins; involucres 3-5mm long; glands with white appendages about 2mm long. Plains, hills, and disturbed areas. NW, NE, SE.

Euphorbia missurica Raf., Atl. Journ. 1:146. 1832. Annual with stems to 6dm long; leaves opposite, the blades mostly oblong or linear, 5-25mm long, entire; involucres 1.5-3mm long; glands with white to pink appendages. Plains, hills, and open woods. NE, SE. *E. petaloidea* Engelm.

Euphorbia robusta (Engelm.) Small in Britt. & Brown, Ill. Fl. 2:381. 1897. Perennial to 4dm high; middle leaves mostly ovate, deltoid, or cordate, alternate, 0.5-2cm long, entire; involucres about 3mm long; glands with short horns from each lateral side. Plains, hills, and slopes. Statewide.

Euphorbia serpyllifolia Pers., Syn. 2:14. 1806. Annual with usually prostrate stems to 3dm long; leaves all opposite, oblong-obovate or oblanceolate to oblong-ovate, 3-15mm long, usually serrulate near tip; involucres about 1mm long; glands with whitish appendages. Plains, hills, and bottomlands. NW, NE, SE.

Euphorbia spathulata Lam., Encyc. Meth. 2:428. 1788. Annual to 6dm high; lower leaves alternate, the blades obovate to oblanceolate, 1-4cm long, crenulate; involucres 1-2mm long; glands without appendages. Hills and slopes. NE, SE. E. dictyosperma Fisch. & Meyer, E. arkansana Engelm. & Gray of authors.

Euphorbia stictospora Engelm. in Torrey, Bot. Mex. Bound. 187. 1859. Annual with prostrate to ascending stems to 35cm long; leaves opposite, the blades ovate to obovate or oblong, 2-10mm long, serrate at tip; involucres about 1mm long; glands with white appendages. Plains, hills, and disturbed areas. Laramie and Niobrara Cos.

FAGACEAE Oak Family

Monoecious trees or shrubs with simple, lobed or cleft, alternate, petioled leaves; flowers unisexual, the staminate in catkins, the pistillate solitary or few in a cluster; staminate flowers with a 2-8 lobed perianth and 3-12 stamens; pistillate flowers with an urn-shaped perianth adnate to the ovary, surrounded by an involucre; pistil 1; ovary inferior; carpels 3-6; styles 1-3; locules usually 3; ovules 1 or 2 per locule; placentation axile; fruit a nut enveloped by an involucre and called an acorn.

Quercus L. Oak

Cup of acorn usually fringed around the rim; Black Hills *Q. macrocarpa*

Cup of acorn not fringed around rim; S. Wyo. *Q. gambelii*

Quercus gambelii Nutt., Proc. Acad. Phila. 4:22. 1848. Gambel Oak. Tree or shrub to 10m high; leaf blades mostly obovate, pinnately lobed or parted, 5-12cm long, 3-8cm wide; acorn cup not fringed around rim. Mountain slopes. Carbon Co. *Q. utahensis* (DC.) Rydb.

Quercus macrocarpa Michx., Hist. Chenes Am. No. 2. 1801. Bur Oak. Tree or shrub to 15m high; leaf blades mostly obovate, pinnately lobed or parted, 5-16cm long, 3-9cm wide; acorn cup usually fringed around rim. Hills and canyons. Crook Co.

Herbs with mostly compound, dissected, alternate or basal leaves; flowers bisexual, irregular, in racemes or solitary; sepals 2, separate, often early deciduous; petals 4, separate or united; stamens (anthers) 6, the filaments often united; ovary 1, superior; carpels 2; style 1; locule 1; ovules 2 to many; placentation parietal; fruit a capsule.

Corolla yellow; flowers in racemes; leaves cauline *Corydalis*
Corolla white to pink or lavender; flowers solitary; leaves basal or sometimes lacking *Dicentra*

Corydalis Vent.

Corydalis aurea Willd., Enum. Pl. Hort. Berol. 2:740. 1809. Spreading or ascending annual or biennial with stems to 5dm long; leaves several times pinnately compound, the blades 2-8cm long, 0.5-6cm wide; sepals yellow-white, 1-3mm long; petals yellow, 12-20mm long; capsules resembling a silique or legume. Woods and open areas, frequently in disturbed sites. Statewide. *C. montana* Engelm., *C. wyomingensis* Fedde, *C. curvisiliquaeformis* Fedde, *C. tortisiliqua* Fedde.

Dicentra Bernh.

Reference: Stern, K. R. 1961. Brittonia 13:1-57.

Dicentra uniflora Kell., Proc. Calif. Acad. Sci. 4:141. 1871. Steer's Head. Scapose perennial to 12cm high from fascicled, fleshy roots; leaves petioled, the blades mostly ternate and again lobed or divided, 1-6cm long, 1-5cm wide; flowers solitary on each scape; sepals 4-6mm long; petals white to pink or lavender, the outer ones recurved, the inner straight and connate at tip and 12-16mm long; capsule 9-14mm long. Wet slopes and meadows, flowering in early spring. NW, SW.

GENTIANACEAE Gentian Family

Herbs with simple, mostly opposite or whorled leaves, rarely all basal; flowers usually bisexual, regular, in cymes, thyrses, or solitary; sepals mostly 4 or 5, united or occasionally separate; petals mostly 4 or 5, united, sometimes with plaits between the lobes or bearing fringed appendages or glands on inner surface; stamens mostly 4 or 5, alternate with the corolla lobes; ovary 1, superior; carpels 2; style 1 or rarely lacking; locule 1; ovules usually many; placentation parietal; fruit a capsule.

1. Corolla lobes at least twice as long as the tube
   2. Corolla usually purple or blue, the lobes mostly 3-5cm long; style usually at least 8mm long — Eustoma
   2. Corolla purple or blue or not, the lobes mostly less than 2.5cm long; style much less than 8mm long
      3. Plants annual or biennial; stigmas decurrent on the ovary for about half the length of ovary or more, the style lacking; basal leaves mostly lacking — Lomatogonium
      3. Plants biennial or perennial; stigmas not decurrent on ovary, the style, or at least the terminal stigmas, developed; basal leaves usually present
         4. Corolla lobes 4, greenish-white or yellowish; styles 2mm long or more — Frasera
         4. Corolla lobes 5, blue or purplish; styles about 1mm long or less — Swertia
1. Corolla lobes rarely longer than the tube
   5. Corolla salverform (the tube long and slender and abruptly flaring into a circular, flattened limb), salmon to white in color, the lobes lacking plaits between them and not fringed or erose nor bearing fringed appendages; plants annual — Centaurium
   5. Corolla usually not salverform, usually blue, rarely whitish or yellowish; plants perennial, or if annual, the corolla lobes either with plaits between them, or fringed or erose on the margins, or bearing fringed appendages
      6. Corolla plicate at the sinuses, the folds often extended into teeth or lobes; corolla lobes entire or nearly so, without fimbriae at base; plants perennial except *G. prostrata* and *G. aquatica* — Gentiana

6. Corolla not plicate at the sinuses, lacking teeth or lobes between the corolla lobes; margins of corolla lobes fringed or erose or else the lobes with fimbriae at base inside; plants annual or biennial except *G. barbellata* *Gentianella*

Centaurium Hill

Centaurium exaltatum (Griseb.) Wight ex Piper, Contr. U. S. Nat. Herb. 11:449. 1906. Annual herb to 25cm high; leaves opposite, the blades lanceolate to oblanceolate, 5-25mm long, entire; flowers solitary or several in a terminal cyme; calyx 5-9mm long; corolla salverform with 4 or 5 lobes, 7-20mm long, white to salmon colored. Moist areas, often in mineral or alkaline areas. Fremont and Hot Springs Cos.

Eustoma Salisb. Prairie Gentian

Reference: Shinners, L. H. 1957. Southw. Nat. 2:38-43.

Eustoma grandiflorum (Raf.) Shinners, Southw. Nat. 2:41. 1957. Annual to short lived perennial to 7dm high; leaves opposite, the blades ovate to elliptic-lanceolate, 2-7cm long, entire; flowers in cymose panicles; calyx 1-2cm long; corolla short campanulate to rotate, blue or purple, rarely pink, white, or yellowish, 3-5cm long, lacking glands or appendages or plaits. Moist plains and bottomlands. Natrona and Goshen Cos. *E. russellianum* (Hook.) G. Don ex Sweet.

Frasera Walt. Green Gentian

Frasera _speciosa_ Dougl. ex Griseb. in Hook., Fl. Bor. Am. 2:66. 1837. Biennial or perennial herb to 2m high; leaves whorled, the blades elliptic-oblong to oblanceolate, 5-50cm long, entire; flowers 4-merous (2 carpels), in a thyrse; calyx lobes 8-25mm long; corolla short rotate-campanulate or saucer-shaped, greenish-white or yellowish, the lobes 10-25mm long and bearing paired pits and fimbriate processes. Woods, slopes, and meadows. Statewide. _Swertia_ _radiata_ (Kell.) Kuntze.

Gentiana L. Gentian

Perennial or rarely annual herbs with opposite leaves; flowers solitary or in cymose clusters; corolla usually funnelform or campanulate, mostly 4 or 5 lobed, plicate in the sinuses.

Reference: Löve, A. & D. Löve. 1972. Bot. Notiser 125:255-258.

1. Corolla less than 22mm long; plants annual or biennial
    2. Corolla white or green (rarely purplish tinged); mostly lower and middle elevations ... _G_. _aquatica_
    2. Corolla blue; mostly high elevations ... _G_. _prostrata_
1. Corolla often 22mm or more long; plants perennial

3. Corolla greenish-white to yellowish, streaked with purple or blue; leaves 8mm or less wide, the blades linear or oblong to narrowly lanceolate or oblanceolate, often well over 6 times longer than wide; plants mostly alpine or subalpine *G. algida*

3. Corolla blue or purple, or if, as rarely, yellowish or whitish, the leaves broader or else less than 6 times longer than wide; plants alpine or not

4. Leaves mostly ovate, about twice as long as wide or less; plants glabrous or nearly so, usually with a single flower or occasionally 2; NW Wyo. *G. calycosa*

4. Leaves often not ovate, but if so, the stems finely pubescent in lines below the leaves; leaves usually more than twice as long as wide; flowers often more than 2; statewide *G. affinis*

Gentiana affinis Griseb. in Hook., Fl. Bor. Am. 2:56. 1837. Perennial to 8dm high; leaf blades ovate or elliptic-ovate to oblong, 1-5cm long, usually glandular-ciliolate at least near base; calyx 3-15mm long; corolla 2-5cm long, blue or purple, 5 lobed. Meadows and other moist areas. Statewide. *G. parryi* Engelm., *G. forwoodii* Gray, *G. bigelovii* Gray.

Gentiana algida Pall., Fl. Ross. 1(2):107. 1789. Perennial to 20cm high; leaf blades linear or oblong to narrowly lanceolate or oblanceolate, 1-10cm long, the margins minutely roughened; calyx 15-30mm long; corolla 3-5cm long, greenish-white or yellowish, blotched or streaked with blue or purple, 5 lobed. Mostly alpine and subalpine. NW, SW, SE. *G. romanzovii* Ledeb. ex Bunge, *Gentianodes algida* (Pall.) Löve & Löve.

Gentiana aquatica L., Sp. Pl. 229. 1753. Annual or biennial to 10cm high; leaf blades ovate or lanceolate to orbicular, obovate, or oblanceolate, 1-10mm long, the margins usually whitened and entire; calyx 4-12mm long; corolla 7-15mm long, usually white or green, rarely purplish tinged, 4 or 5 lobed. Plains and meadows. NW, SE.

Gentiana calycosa Griseb. in Hook., Fl. Bor. Am. 2:58. 1837. Perennial to 4dm high; leaf blades mostly ovate, 1-3cm long, entire or minutely roughened; calyx 1-2cm long; corolla 2.5-5cm long, 5 lobed, blue, streaked or mottled with green, rarely yellowish. Meadows and slopes in the mountains. Teton and Sublette Cos.

Gentiana prostrata Haenke in Jacq., Coll. Bot. 2:66. 1789. Annual or biennial to 15cm high; leaf blades ovate to oblanceolate or obovate, 1-10mm long, the margins entire, only occasionally whitened; calyx 4-14mm long; corolla 7-22mm long, blue, 4 or 5 lobed. Meadows and bogs. NW, SW. *Ericoilea prostrata* (Haenke) Borkh., *G. fremontii* Torr., *Chondrophylla fremontii* (Torr.) A. Nels., *C. americana* (Engelm.) A. Nels.

Gentianella Moench Gentian

Annual to rarely perennial herbs with opposite leaves (rarely all basal); flowers solitary or in cymes; corolla tubular, funnelform, or campanulate, mostly 4 or 5 lobed, without plaits between the lobes, either with filiform fimbriae from base of lobes or with margins of lobes fringed or erose.

References: Gillett, J. M. 1957. Ann. Mo. Bot. Gard. 44:195-269.
Iltis, H. H. 1965. Sida 2:129-153.

1. Corolla lobes usually 4 and fringed or erose on the margins; corolla usually over 2cm long, not fringed in the throat from base of lobes
   2. Plants perennial; flowers closely subtended by a pair of bract-like leaves; plants less than 15cm high *G. barbellata*
   2. Plants annual; flowers on naked peduncles, not subtended by bract-like leaves; plants mostly over 15cm high *G. detonsa*
1. Corolla lobes 4 or 5, the margins entire or nearly so; corolla usually less than 2cm long, fringed in the throat from base of lobes
   3. Flowers on long naked pedicels; calyx lobed to the base and somewhat overhanging at base; fringe at base of corolla lobes from 2 distinct scales per lobe *G. tenella*
   3. Flowers usually crowded, often subtended by bracts; calyx not lobed to base; fringe of 1-10 similar filiform setae per lobe *G. amarella*

Gentianella amarella (L.) Börner, Fl. Deut. Volk 543. 1912. Annual or biennial to 5dm high; leaf blades ovate to oblanceolate, 0.5-6cm long, minutely ciliolate; calyx 2-18mm long; corolla 6-20mm long, blue or purple to yellowish and blue tinged, the (4)5 lobes with filiform fimbriae. Meadows and other moist areas. Statewide. *Gentiana amarella* L., *Gentiana heterosepala* Engelm., *Gentiana strictiflora* (Rydb.) A. Nels., *Gentiana plebeja* Cham. ex Bunge.

Gentianella barbellata (Engelm.) Gillett, Ann. Mo. Bot. Gard. 44:230. 1957. Perennial to 15cm high; leaf blades oblanceolate to linear, 1.5-7cm long, entire or minutely roughened; calyx 11-25mm long; corolla 24-45mm long, deep blue, the 4 lobes fimbriate on the margins below and entire to erose above. Woods, meadows, and slopes in the mountains. NW, SE. *Gentiana barbellata* Engelm., *Gentianopsis barbellata* (Engelm.) Iltis.

Gentianella detonsa (Rottb.) G. Don, Gen. Hist. Dichl. Pl. 4:179. 1838. Fringed Gentian. Annual to 5dm high; leaf blades lanceolate or oblong to oblanceolate, 1-5cm long, entire or minutely ciliolate; calyx 15-35mm long; corolla 2.5-6cm long, blue or purple, the 4 lobes erose on the margins. Meadows, bogs, and other moist areas. NW, SW, SE. *Gentiana elegans* A. Nels., *Gentiana thermalis* Kuntze, *Gentiana detonsa* Rottb., *Gentianopsis detonsa* (Rottb.) Ma, *Gentianopsis thermalis* (Kuntze) Iltis.

Gentianella tenella (Rottb.) Börner, Fl. Deut. Volk 542. 1912. Annual to 15cm high; leaf blades oblanceolate to lanceolate, 2-15mm long, entire or minutely roughened; calyx 4-10mm long; corolla 4-13mm long, white to blue-purple, the mostly 4 lobes each with 2 fimbriate scales. Meadows and slopes in the high mountains. Sublette and Albany Cos. *Gentiana tenella* Rottb., *Comastoma tenellum* (Rottb.) Toyokuni.

Lomatogonium A. Br. Felwort

Lomatogonium _rotatum_ (L.) Fries, Summa Veg. Scand. 2:554. 1849. Annual or biennial to 35cm high; leaves opposite, the blades oblanceolate to lanceolate or linear, 0.5-3cm long, entire or minutely roughened; flowers mostly axillary; calyx 6-18mm long, with 4 or 5 separate lobes; corolla 6-15mm long, blue or rarely white, the 4 or 5 lobes with 2 small basal appendages, the tube very short. Moist meadows, bogs, and streambanks. Albany and Carbon Cos. _Pleurogyne_ _fontana_ A. Nels.

Swertia L.

Reference: St. John, H. 1941. Am. Midl. Nat. 26:1-29.

Swertia _perennis_ L., Sp. Pl. 226. 1753. Perennial to 5dm high; leaves mostly opposite but occasionally alternate, the blades obovate to elliptic or lanceolate, 1-12cm long, entire or minutely roughened; flowers solitary in axils or in a thyrse or cyme; calyx 3-10mm long; corolla 5 lobed nearly to base, blue-purple and green or white mottled, 6-14mm long, each lobe with a basal pair of glands which are surrounded by fringed appendages. Moist meadows, bogs, and seeps. NW, SW, SE. _S_. _palustris_ A. Nels., _S_. _congesta_ A. Nels.

GERANIACEAE Geranium Family

Herbs with simple or compound, alternate or opposite leaves; flowers bisexual, mostly regular, in cymes or umbels; sepals usually 5 and separate; petals usually 5, separate; stamens 5 or 10; pistil 1, usually 3-5 lobed; ovary superior; carpels 3-5; styles 3-5 or apparently 1; locules 3-5; ovules 1-2 per locule; placentation axile; fruit capsular, usually splitting from base into 3-5 mericarps.

Leaves pinnately compound; fertile stamens 5 — *Erodium*

Leaves palmately lobed or divided; fertile stamens usually 10, rarely 5 — *Geranium*

Erodium L'Her. Storksbill

Erodium cicutarium (L.) L'Her. ex Ait., Hort. Kew. 2:414. 1789. Annual to 3dm high, the stems often prostrate; leaves pinnately compound, the blades oblong-elliptic to oblanceolate or lanceolate, 1-7cm long; flowers umbellate; sepals 3-6mm long, mucronate or aristate; petals pink or purplish, 2.5-6mm long; filaments 10, only the 5 larger ones with anthers; styles spirally twisted in fruit. Plains, hills, and disturbed areas. NW, NE, SE.

<u>Geranium</u> L. Wild Geranium

Annual to perennial herbs with mostly simple, reniform or cordate-orbicular, palmately lobed or divided leaves; flowers in cymes; stamens 10, usually all with anthers (except 1 species), the filaments often somewhat connate at base.

1. Petals 8mm or less long; plants usually annual or biennial
  2. Fertile stamens 5; sepals not bristle-tipped ..... *G. pusillum*
  2. Fertile stamens 10; sepals bristle-tipped
    3. Beak of stylar column, including stigmas, 4-7mm long; fruiting pedicel usually much longer than calyx ..... *G. bicknellii*
    3. Beak of stylar column, including stigmas, mostly under 3mm long; fruiting pedicel usually slightly if at all longer than calyx ..... *G. carolinianum*
1. Petals over 8mm long; plants perennial
  4. Petals white with pink or purple veins; inflorescence pilose-glandular with usually purple-tipped hairs; plants usually in moist or shaded areas ..... *G. richardsonii*
  4. Petals usually pink or purple; inflorescence not glandular, or glandular with yellow or whitish-tipped hairs; plants often in dry, open areas
    5. Plants not glandular-pubescent ..... *G. caespitosum*
    5. Plants glandular-pubescent at least in inflorescence

6. Leaf blades rarely over 5cm wide, the 3-5 primary divisions usually merely toothed or shallowly lobed; plants often pinkish or reddish tinged, especially below; plains, hills, and slopes in SE Wyoming

G. fremontii

6. Leaf blades, or some of them, often well over 5cm wide, the primary divisions often moderately to deeply lobed or divided; plants only rarely pinkish or reddish tinged; often in somewhat moist areas, statewide

G. viscosissimum

Geranium bicknellii Britt., Bull. Torrey Club 24:92. 1897. Hirsute, somewhat glandular annual or biennial with erect to decumbent stems to 6dm long; leaf blades 2-7cm wide; sepals 4-9mm long; petals pink, 3-8mm long. Woods, meadows, and slopes. Yellowstone Park and Natrona Co. *G. longipes* (Wats.) Goodding.

Geranium caespitosum James emend. Gray, Pl. Fendl. 25. 1849. Perennial to 9dm high, not glandular; leaf blades 1-6cm wide; sepals 8-12mm long; petals rose-purple, 12-18mm long. Hills, slopes, and open woods. Albany and Carbon Cos.

Geranium carolinianum L., Sp. Pl. 682. 1753. Hirsute to pilose annual to 7dm high, glandular or not; leaf blades 1-5cm wide; flowers usually densely crowded; sepals 4-8mm long; petals pink or rose, 4-8mm long. Woods and disturbed areas. SW, NE.

Geranium fremontii Torr. ex Gray, Pl. Fendl. 26. 1849. Perennial to 6dm high, glandular at least above; leaf blades 1-5(8)cm wide; sepals 6-12mm long; petals pink or purplish, 10-18mm long. Plains, hills, and slopes. SE. G. parryi (Engelm.) Heller.

Geranium pusillum L., Syst. Nat. 10th ed. 1144. 1759. Hirsute and somewhat glandular annual or biennial with prostrate to erect stems to 5dm long; leaf blades 1-6cm wide; sepals 2.5-4mm long; petals purple, 2.5-5mm long; fertile stamens 5. Disturbed areas. Albany Co. G. pusillum Burm. f.

Geranium richardsonii Fisch. & Trautv. in Fisch. & Meyer, Ind. Sem. Hort. Petrop. 4:37. 1837. Perennial to 8dm high, glandular-pilose above; leaf blades 2-17cm wide; sepals 6-11mm long; petals white or slightly pinkish, 10-17mm long. Meadows, streambanks, and moist woods. Statewide.

Geranium viscosissimum Fisch. & Mey. ex Mey., Ind. Sem. Hort. Petrop. 11: suppl. 18. 1846. Hirsute or pilose, glandular perennial to 9dm high; leaf blades 2-19cm wide; sepals 6-13mm long; petals pink to purplish, 12-20mm long. Woods, sagebrush, meadows, and slopes. Statewide. G. strigosius St. John, G. nervosum Rydb.

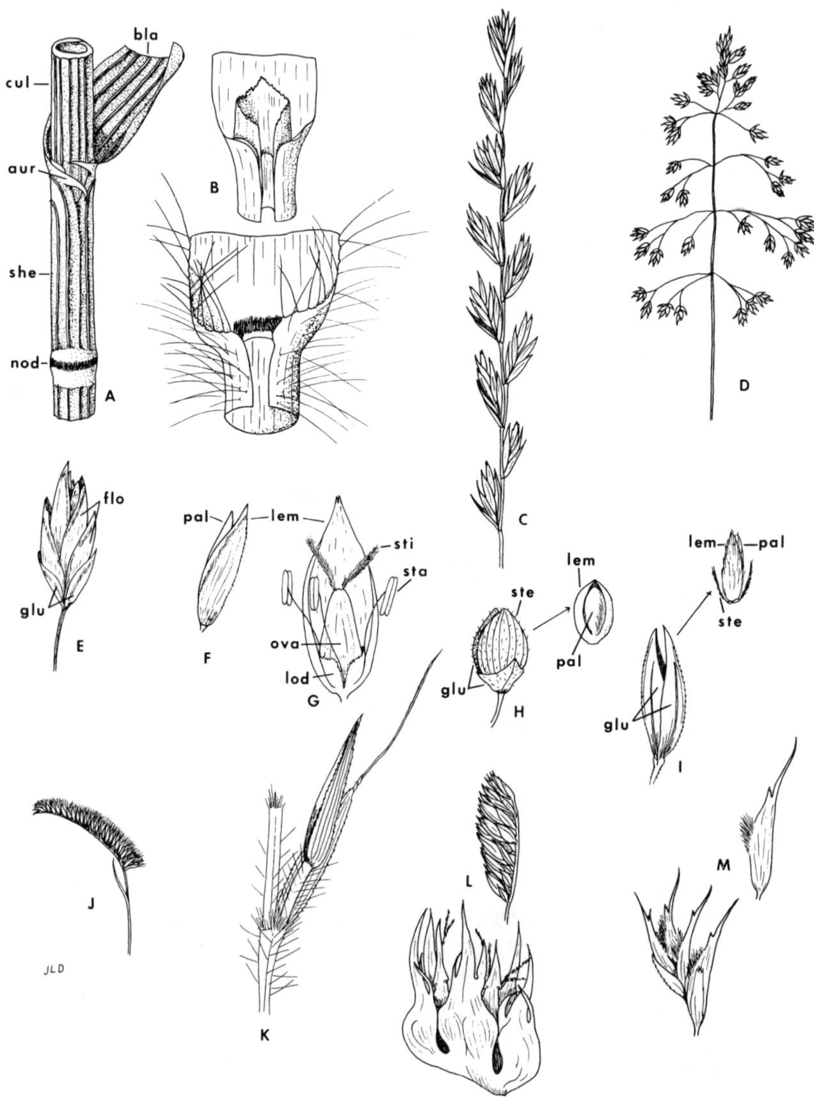

bla
cul
aur
she
nod
A
B
C
D
flo
glu
E
pal
lem
F
sti
sta
ova
lod
G
ste
lem
pal
glu
H
lem
pal
ste
glu
I
J
JLD
K
L
M

Figure 10. Gramineae. A. Portion of grass culm (X 3): nod = node, she = sheath, bla = blade, aur = auricle, cul = culm. B. Ligule types (X 6): Poa palustris with membranous ligule (above), Panicum oligosanthes with hairy ligule (below). C. Agropyron smithii with sessile spikelets (X 0.7). D. Poa nervosa with spikelets on pedicels (X 0.7). E. Spikelet of Poa nervosa (X 3): glu = glumes, flo = florets. F. Floret of Poa nervosa (X 4): lem = lemma, pal = palea. G. Grass flower (X 7): lem = lemma, ova = ovary, sta = stamen, sti = stigma, lod = lodicule. H. Spikelet of Panicum oligosanthes (X 4): glu = glumes, ste = sterile lemma, lem = lemma, pal = palea. I. Spikelet of Phalaris arundinacea (X 3): glu = glumes, ste = sterile lemma, lem = lemma, pal = palea. J. Spike of Bouteloua gracilis with spikelets on one side of rachis only (X 0.7). K. Pair of spikelets of Andropogon gerardii (X 3). L. Staminate (above X 3) and pistillate (below X 4) spikes of Buchloe dactyloides. M. Spikelet (left) and floret (right) of Munroa squarrosa (X 3).

GRAMINEAE Grass Family

Herbs with simple, alternate, sheathing, parallel-veined, 2 ranked leaves; ligule usually present at junction of sheath and blade on inner side, of hairs or membranous; flowers bisexual or unisexual, arranged in spikelets, each spikelet usually consisting of 2 empty lower bracts (glumes) subtending 1 or more florets, each floret composed of 2 bracts (a lemma which usually has a midnerve, and a palea which usually lacks a midnerve) which subtend the flower; perianth greatly reduced; stamens 1-6, usually 3; ovary 1, superior; carpels 3; styles 1-3, usually 2; locule 1; ovule 1; fruit a caryopsis. The traditional genera have been maintained here despite increasing agreement that some should be combined. A number of mostly annual species including wheat, oats, and barley are commonly cultivated or are introduced in other seed, but these rarely, if ever, spread and persist more than a year.

1. Plants dioecious or monoecious, to 20cm high, stoloniferous; pistillate spikelets with the thickened rachis and 2nd glumes forming a rigid, yellow-white, globular structure crowned by green-toothed summits of the glumes; staminate spikelets 2 flowered, sessile, in 2 rows on 1 side of rachis — Buchloe
1. Plants not as above
   2. Plants annual but mat forming by branching, to 10cm high; leaves and spikelets in fascicles, the fascicles separated by mostly naked internodes; lemmas long-pilose toward margin near midlength, awn-tipped — Munroa

2. Plants not as above
   3. Spikelets of crowded scales subtending bulblets rather than flowers or seeds, the bulblets usually purplish; culms usually bulbous at base, densely tufted; leaf tips boat shaped — Poa bulbosa
   3. Spikelets containing flowers or seeds; culms and leaves various
      4. Glumes both lacking; spikelets 1 flowered — Leersia
      4. Glumes both present or only 1 lacking; spikelets 1 or more flowered
         5. Spikelets enclosed by a bur-like involucre bearing coalescent bristles forming spines; sandbur — Cenchrus
         5. Spikelets not enclosed by a bur-like involucre; if bristles are present , they are not coalescent to form spines
            6. Spikelets sessile, forming terminal or lateral spikes, occasionally with the lower spikelets short-pedicelled but then the glumes are usually bristle-like or nearly so
               7. Spikelets with 1 perfect terminal floret and 2 opposite, sterile lemmas below — Phalaris
               7. Spikelets with all perfect florets, or with only 1 sterile lemma below, or with sterile lemmas all above the fertile

8. Spikelets dorsally flattened and falling entire, with 1 perfect terminal floret and 1 sterile lemma (which resembles a glume) or staminate floret below, awnless — <u>Setaria</u>

8. Spikelets usually flattened from the sides, the florets often falling individually with the glumes persistent; spikelets usually with 1 or more perfect florets, with or without sterile or staminate florets, or, the plants dioecious or monoecious; lemmas or glumes awned or not

9. Spikes 1 or more, usually lateral or not directly continuous with the main axis; spikelets often on only 1 side of the rachis — GROUP I

9. Spikes single and terminal; spikelets on opposite sides of rachis

10. Spikelets 3 at a node, the groups with a tuft of long hairs at base and falling entire; central spikelet perfect, 1 flowered (rarely 2 flowered); lateral spikelets staminate, 2 or 3 flowered — <u>Hilaria</u>

10. Spikelets without the above combination of characters — GROUP II

6. Spikelets, or most of them, with very short or long pedicels, the inflorescence a raceme or panicle which is sometimes spike-like; rarely with a single spikelet

11. Spikelets in pairs, 1 sessile and fertile and the other pedicelled and sterile, or staminate, or reduced to only the pedicel; pedicel with long hairs; glumes indurate

12. Primary branches of panicle mostly more than 5 jointed, usually digitate or solitary at each node of the rachis; pedicellate spikelet developed or not <u>Andropogon</u>

12. Primary branches of panicle mostly 1-5 jointed, the panicle relatively open and further branched; pedicellate spikelet reduced to the hairy pedicel <u>Sorghastrum</u>

11. Spikelets not in pairs as above; pedicels hairy or not; glumes usually not indurate

13. Plants stout reeds to 4m high with plume-like panicles; at least some leaves 9mm or more wide; rachilla with long silky hairs as long as the lemmas, the hairs often inconspicuous in young flowers; spikelets several flowered, the florets sometimes poorly differentiated in very young spikelets; glumes shorter than lowest lemma; lowest lemmas mostly 9mm or more long; usually in moist areas or in water <u>Phragmites</u>

13. Plants not as above

14. Spikelets usually dorsally flattened and falling entire, with 1 perfect terminal floret and usually 1 sterile lemma (which resembles a glume) or staminate floret below (1st glume sometimes minute) GROUP III

14. Spikelets usually flattened from the sides, the florets usually falling individually with the glumes persistent; spikelets with 1 or more perfect florets or the plants rarely dioecious; sterile or staminate florets, if any, above the perfect or with 2 below the perfect

15. Spikelets with 1 perfect terminal floret and 2 sterile lemmas or staminate florets below, the sterile often reduced to linear lemmas with long hairs

16. Lower florets staminate, well developed; spikelets brown and shiny; inflorescence an open panicle <u>Hierochloe</u>

16. Lower florets reduced to small scale-like or linear lemmas; spikelets green or yellow and dull; inflorescence a spike-like or contracted panicle <u>Phalaris</u>

15. Spikelets not as above, the sterile florets, if present, above the fertile, or rarely with 1 staminate floret below the 1 perfect floret

17. Spikelets mostly with 1 floret GROUP IV
17. Spikelets with 2 or more florets
  18. Glumes, or at least 1 glume, as long as or longer than the lowest floret, usually as long as the spikelet; lemmas awned from the back or from a bifid apex or awnless GROUP V
  18. Glumes mostly shorter than the lowest floret; lemmas awned from the tip or from a bifid apex or awnless GROUP VI

GROUP I

1. Spikelets with 1 or more modified florets above the perfect one, these sometimes merely awns *Bouteloua*
1. Spikelets without additional modified florets
  2. Plants with creeping rhizomes; glumes unequal in length; spikes erect *Spartina*
  2. Plants lacking rhizomes; glumes equal in length, or if not, the spikes spreading
    3. Glumes equal, broad and boat-shaped; leaf blades over 3mm wide; plants of wet places *Beckmannia*
    3. Glumes somewhat unequal, narrow, not boat-shaped; leaf blades mostly about 1mm wide; plants mostly of dry places *Schedonnardus*

GROUP II

1. Spikelets 1 flowered, the lemma much shorter than the glumes and awnless — <u>Phleum</u>

1. Spikelets 2 or more flowered, or if 1 flowered, the lemma conspicuously awned

 2. Spikelets mostly 1 per node

  3. Spikelets placed edgewise to rachis; 1st glume lacking except in terminal spikelet — <u>Lolium</u>

  3. Spikelets placed flatwise to rachis; glumes usually both present

   4. Plants annual; spikelets sunken into rachis — <u>Aegilops</u>

   4. Plants perennial (except <u>A</u>. <u>triticeum</u>); spikelets not sunken into rachis — <u>Agropyron</u>

 2. Spikelets mostly 2 or more per node, at least at middle of spike, the lateral ones sometimes reduced to awns

  5. Spikelets 3 per node, mostly 1 flowered, the lateral ones pediceled and usually reduced to awns — <u>Hordeum</u>

  5. Spikelets 2 or more per node, 2 or more flowered, the lateral ones like the central one and usually sessile

   6. Rachis usually continuous; glumes broad, or subulate and bristle-like; awns usually less than 3cm long — <u>Elymus</u>

   6. Rachis disarticulating when mature; glumes subulate, often bristle-like; awns usually over 3cm long — <u>Sitanion</u>

GROUP III

1. Inflorescence appearing like a simple spike; spikelets subtended by long bristles — Setaria
1. Inflorescence not appearing like a simple spike; spikelets not subtended by bristles
    2. Second glume awn-tipped; sterile lemma awned — Echinochloa
    2. Second glume and sterile lemma not awned
        3. Inflorescence of digitate racemes — Digitaria
        3. Inflorescence an open panicle — Panicum

GROUP IV

1. Disjointing below the glumes, the entire spikelet falling (most evident on mature plants but joints near tip of pedicels often apparent in younger plants)
    2. Glumes awned
        3. Awn of glumes 5mm or more long — Polypogon
        3. Awn of glumes 2mm or less long — Phleum
    2. Glumes not awned
        4. Panicle spike-like, cylindrical; keel of glumes long-ciliate — Alopecurus
        4. Panicle open; keel or midnerve of glumes glabrous or scabrous
            5. Plants annual; spikelets about as long as wide — Beckmannia
            5. Plants perennial; spikelets over twice as long as wide — Cinna

1. Disjointing above the glumes, the glumes not falling with the florets

6. Awn 3 parted — Aristida

6. Awn simple or lacking

7. Lemma hardened, much more so than glumes at maturity, closely enveloping grain, often without evident nerves, terminally awned or nearly so, the awn sometimes deciduous and sometimes over 2cm long

8. Awn persistent, strongly twisted and bent, often over 2cm long; callus often sharp pointed, usually acuminate; glumes 6-50mm long — Stipa

8. Awn often deciduous, not strongly twisted, sometimes bent, mostly 2cm or less long; callus usually obtuse; glumes 2.5-8mm long — Oryzopsis

7. Lemma not hardened, loose around the grain, usually with 1 or more evident nerves, awned or not, the awns, when present, usually less than 2cm long

9. Glumes (excluding awns) longer than the lemmas

10. Glumes strongly flattened and keeled, stiff ciliate on the keel, short awned; panicle spike-like; lemma awnless — Phleum

10. Glumes not as above; panicle open to spike-like; lemmas awned or not

11. Floret with a tuft of hairs at the base from the callus, the hairs usually 1/4 as long to as long as lemma; palea well developed

12. Ligules membranous
13. Lemma awned from back — Calamagrostis
13. Lemma awned from tip — Muhlenbergia
12. Ligules of hairs — Calamovilfa
11. Floret lacking hairs at base or with very short hairs; palea often small or lacking — Agrostis
9. Glumes (excluding awns) mostly shorter than the lemmas
14. Lemmas awned from tip (tip rarely bifid) or mucronate (at least some with the midnerve slightly prolonged beyond body); glumes often awned — Muhlenbergia
14. Lemmas awnless or awned from the back; glumes not awned
15. Floret with a tuft of hairs at the base from the callus — Calamovilfa
15. Floret lacking a tuft of hairs at the base
16. Lemma 1 nerved; ligules of hairs (at least the upper half) — Sporobolus
16. Lemma 3 or 5 nerved; ligules membranous
17. Plants alpine, 10cm or less high — Phippsia
17. Plants usually not alpine, usually over 10cm high
18. Plants annual — Muhlenbergia
18. Plants perennial — Catabrosa

GROUP V

1. Florets 2, the lower staminate, the upper perfect; lemmas awned — *Arrhenatherum*
1. Florets 2 or more, alike except sometimes the reduced upper ones; lemmas awned or awnless
  2. Spikelets 8mm long or less; rachilla often prolonged beyond terminal floret
    3. Glumes dissimilar, the 2nd much wider than the 1st; spikelets 2-4mm long — *Sphenopholis*
    3. Glumes usually relatively similar; spikelets 4-8mm long
      4. Lemmas awned from near or below middle — *Deschampsia*
      4. Lemmas awnless or awned from above the middle
        5. Lemmas with an exserted, usually geniculate awn — *Trisetum*
        5. Lemmas awnless or with a very short, straight awn
          6. Ligules mostly 2.5-4mm long; rachilla strongly bearded — *Trisetum*
          6. Ligules mostly 0.5-2.5mm long; rachilla glabrous to very short hairy
            7. Pedicels mostly less than 3mm long; upper floret rarely exceeding longest glume by more than 1mm — *Koeleria*
            7. Pedicels, or some of them, usually over 3mm long; upper floret usually exceeding longest glume by 1.5mm or more — *Poa*

2. Spikelets mostly 9mm or more long; rachilla usually not prolonged beyond terminal floret

8. Lemmas awnless or minutely awn-tipped — Scolochloa

8. Lemmas with awns 4mm or more long

9. Lemmas bifid at tip, awned from between the lobes — Danthonia

9. Lemmas mostly toothed, awned from the back

10. Plants annual; leaf blades 3-15mm wide — Avena

10. Plants perennial; leaf blades 1-3mm wide — Helictotrichon

GROUP VI

1. Plants dioecious, sometimes with rudiments of the opposite sex in a normally developed flower

2. Plants with long creeping rhizomes; sheaths usually long-hairy near throat; ligules usually with a fringe of hairs at tip; spikelets mostly 7-15 flowered — Distichlis

2. Plants with rhizomes or not; sheaths not long-hairy; ligules mostly membranous; spikelets mostly 3-7 flowered

3. Panicle narrow and congested; some leaf blades often over 3.5mm wide, often glaucous; plants short-rhizomatous — Leucopoa

3. Panicle often open; leaf blades rarely over 3.5mm wide, not glaucous; plants rhizomatous or not — Poa

1. Plants with perfect flowers

4. Plants annuals or tufted perennials; 1st glume much narrower than 2nd; spikelets mostly 2 flowered, 2-4mm long, awnless, falling entire (the glumes not persistent) — Sphenopholis

4. Plants without the above combination of characters

5. Lemmas with 3 prominent nerves
   6. Callus densely hairy — Redfieldia
   6. Callus glabrous
      7. Spikelets mostly 2 flowered; lemmas truncate — Catabrosa
      7. Spikelets 3 to many flowered; lemmas acute or obtuse — Eragrostis
5. Lemmas with 5 or more nerves or appearing nerveless
   8. Spikelets crowded in 1 sided clusters at the ends of stiff, naked panicle branches; glumes usually hispid-ciliate on the keel and sometimes on the margins and nerves, otherwise mostly glabrous — Dactylis
   8. Spikelets and glumes not as above
      9. Callus of florets bearded with straight hairs, the lemmas otherwise glabrous or scabrous
         10. Lemmas erose at apex, awnless — Scolochloa
         10. Lemmas bifid at apex, awned — Schizachne
      9. Callus of florets not bearded (lemmas sometimes cobwebby at base in Poa), the lemmas sometimes pubescent
         11. Stems usually bulbous at base in the soil; spikelets often tawny or purplish tinged; upper floret often sterile and lacking a palea; sheaths usually closed most of their length — Melica
         11. Stems usually not bulbous at base; spikelets often completely green; upper florets perfect, or if not, with paleas, or the floret reduced to a rudiment; sheaths often split most of their length

12. Lemmas mostly obtuse and scarious at apex, not awned, 5-9 nerved; glumes mostly 3mm or less long; leaves not boat-shaped at tip or slightly so in drying

13. Second glume 1 nerved; styles well developed; nerves of lemmas prominent; plants of fresh water shores or moist woods <u>Glyceria</u>

13. Second glume usually 3 nerved; styles lacking or nearly so; nerves of lemmas prominent or obscure; plants of alkaline soil or fresh water areas <u>Puccinellia</u>

12. Lemmas sometimes acute at the apex, scarious or not, often awned, if not awned, rarely over 5 nerved; glumes often over 3mm long; leaves sometimes boat-shaped at tip

14. Lemmas awned or awn-tipped from a minutely or strongly bifid apex (except sometimes <u>B</u>. <u>brizaeformis</u> which has inflated spikelets); spikelets usually 15mm or more long <u>Bromus</u>

14. Lemmas often entire, pointed or obtuse, awnless or awned usually from the tip; spikelets mostly less than 15mm long

15. Lemmas awned, or if awnless, the plants not rhizomatous nor with boat-shaped leaf tips, the lemmas mostly with slender pointed tips and mostly rounded on back <u>Festuca</u>

15. Lemmas awnless (midnerve rarely slightly extended), often keeled and blunt and scarious at tip; plants rhizomatous or not; leaves often with boat-shaped tips *Poa*

Aegilops L. Goatgrass

Aegilops cylindrica Host, Icon. Gram. Austr. 2:6, pl. 7. 1802. Annual to 7dm high; leaf blades 1.5-5mm wide; spikelets sessile, sunken into rachis, 2-5 flowered, 7-14mm long excluding awns; glumes with awns 5-35mm long; lemmas awnless or the upper ones awned like the glumes. Disturbed areas. NE.

Agropyron Gaertn. Wheatgrass

Perennials or rarely annual; spikelets several flowered, solitary or rarely in pairs, sessile; glumes mostly subequal, usually shorter than 1st lemma; lemmas 5-7 nerved, mostly acute or awned. This treatment largely follows that of C. L. Hitchcock in Vasc. Pl. Pac. N. W.

1. Plants with creeping rhizomes
  2. Glumes and lemmas blunt at tip — *A. intermedium*
  2. Glumes and lemmas sharp-pointed at tip
    3. Leaf blades mostly flat and some 5-10mm wide; awn, if present, straight — *A. repens*
    3. Leaf blades either involute or much less than 5mm wide; awn, if present, often divergent
      4. Glumes rigid, usually widest near base, often as long as 1st lemma, mostly 3-5 nerved, short-awned — *A. smithii*
      4. Glumes not rigid, widest at or above middle, shorter than 1st lemma, mostly 5-7 nerved, acute to awn-tipped — *A. dasystachyum*
1. Plants without creeping rhizomes
  5. Plants annual; spike less than 2cm long — *A. triticeum*
  5. Plants perennial; spike usually over 2cm long
    6. Spikelets strongly divergent, much compressed and crowded, some at least 4 times as long as internodes of rachis — *A. cristatum*
    6. Spikelets usually erect or ascending, not much compressed or crowded, mostly 3 times as long as internodes or less
      7. Lower internodes of rachis 14-30mm long, much longer than upper ones — *A. elongatum*
      7. Lower internodes of rachis shorter, about equal to the upper ones
        8. Glumes and lemmas blunt at tip — *A. intermedium*
        8. Glumes and lemmas sharp-pointed at tip

9. Anthers 4-6mm long; spikelets shorter to slightly longer than internodes of rachis; glumes acute or awn-tipped; lemmas often with a divergent awn
A. spicatum

9. Anthers 1-3mm long; spikelets mostly 2-3 times as long as internodes of rachis; glumes and lemmas various

10. Awn of lemma divergent; glumes narrowly lanceolate, long-awned; rachis tending to disarticulate when mature A. scribneri

10. Awn of lemma either lacking or mostly straight; glumes oblong-elliptic, awned or not; rachis not disarticulating A. caninum

Agropyron caninum (L.) Beauv., Ess. Agrost. 102. 1812. Plants caespitose; culms to 15dm high; leaf blades 1-8mm wide; spikelets 1-2cm long, 3-7 flowered; glumes 6-10mm long, often awned; lemmas 8-12mm long, awnless or awned, the awn to 3cm long. Woods and open areas. Statewide. A. bakeri E. Nels., A. gmelinii (Griseb.) Scribn. & Sm., A. latiglume (Scribn. & Sm.) Rydb., A. pseudorepens Scribn. & Sm., A. subsecundum (Link) Hitchc., A. tenerum Vasey, A. trachycaulum (Link) Malte ex Lewis, A. violaceum (Hornem.) Lange.

Agropyron cristatum (L.) Gaertn., Nov. Comm. Petrop. 14(1):540. 1770. Crested Wheatgrass. Plants tufted; culms to 1m high; leaf blades 2-10mm wide; spikelets 5-17mm long, mostly 5-8 flowered; glumes 4-8mm long, awned; lemmas 4-8mm long, the awns 2-4mm long. Disturbed areas. Statewide. *A. desertorum* (Fisch. ex Link) Schult., *A. sibiricum* (Willd.) Beauv.

Agropyron dasystachyum (Hook.) Scribn., Bull. Torrey Club 10:78. 1883. Plants rhizomatous; culms to 1m high; leaf blades 1-5mm wide; spikelets 9-17mm long, 4-10 flowered; glumes 6-12mm long, acute to awn-tipped; lemmas 6-12mm long, acute or awned, the awn to 15mm long. Open areas. Statewide. *A. albicans* Scribn. & Sm., *A. griffithsii* Scribn. & Sm., *A. pseudorepens* Scribn. & Sm., *A. riparium* Scribn. & Sm., *A. subvillosum* (Hook.) E. Nels.

Agropyron elongatum (Host) Beauv., Ess. Agrost. 102. 1812. Plants tufted; culms to 15dm high; leaf blades 3-7mm wide; spikelets 5-25mm long, 5-11 flowered; glumes 8-13mm long; lemmas 9-15mm long, mostly awnless. Disturbed areas. NW, SE.

Agropyron intermedium (Host) Beauv., Ess. Agrost. 102. 1812. Plants rhizomatous or not; culms to 1m high; leaf blades 2-8mm wide; spikelets 8-17mm long, 5-9 flowered; glumes 7-9mm long, blunt; lemmas 5-10mm long, blunt. Disturbed areas. NW, SW, SE.

<u>Agropyron</u> <u>repens</u> (L.) Beauv., Ess. Agrost. 102. 1812. Quackgrass. Plants rhizomatous; culms to 1m high; leaf blades 3-10mm wide; spikelets 8-15mm long, 4-8 flowered; glumes 6-9mm long, usually awn-tipped; lemmas 5-9mm long, awnless or awned, the awn to 1cm long. Disturbed areas. Statewide. <u>A</u>. <u>pseudorepens</u> Scribn. & Sm.

<u>Agropyron</u> <u>scribneri</u> Vasey, Bull. Torrey Club 10:128. 1883. Plants caespitose; culms to 6dm high; leaf blades 1-4mm wide; spikelets 10-16mm long, 4-6 flowered; glumes 6-9mm long, long-awned; lemmas 7-8mm long, the awns mostly 1-4cm long. Mostly alpine and subalpine. Statewide. <u>A</u>. <u>bakeri</u> E. Nels., <u>Sitanion</u> <u>marginatum</u> Scribn. & Merr. Plants of intergeneric hybrid origin which have been called <u>A</u>. <u>saxicola</u> (Scribn. & Sm.) Piper and <u>A</u>. <u>saundersii</u> (Vasey) Hitchc. will also key here.

<u>Agropyron</u> <u>smithii</u> Rydb., Mem. N. Y. Bot. Gard. 1:64. 1900. Western Wheatgrass. Plants rhizomatous; culms to 6dm high; leaf blades 1-4mm wide; spikelets 1-2.5cm long, 6-10 flowered; glumes 7-14mm long, short-awned; lemmas 7-11mm long, acuminate to short-awned. Plains and foothills. Statewide. <u>A</u>. <u>occidentale</u> Scribn.

<u>Agropyron</u> <u>spicatum</u> (Pursh) Scribn. & Smith, USDA Agrost. Bull. 4:33. 1897. Bluebunch Wheatgrass. Plants tufted; culms to 1m high; leaf blades 1-4mm wide; spikelets 1-2cm long, mostly 6-8 flowered; glumes 5-12mm long, acute to awn-tipped; lemmas about 1cm long, the awn to 2cm long. Hills and plains. Statewide. <u>A</u>. <u>inerme</u> (Scribn. & Sm.) Rydb.

Agropyron triticeum Gaertn., Nov. Comm. Petrop. 14(1):540. 1770. Annual; culms to 3dm high; leaf blades 2-3mm wide; spike only 1-1.5cm long; spikelets about 7mm long; glumes and lemmas acuminate. Disturbed areas. NW, SE. *Eremopyrum triticeum* (Gaertn.) Nevski.

Agrostis L. Bentgrass; Redtop

Perennials or rarely annual; spikelets 1 flowered; glumes usually equal or nearly so; lemma usually shorter than glumes, mostly 3 nerved, awned or not.

1. Palea evident, 2 nerved, over half as long as lemma
  2. Rachilla prolonged behind palea as a minute stub or bristle *A. thurberiana*
  2. Rachilla not prolonged
    3. Plants tufted, alpine, less than 18cm high *A. humilis*
    3. Plants with rhizomes, rarely alpine, usually over 25cm high *A. alba*
1. Palea lacking or a minute nerveless scale
  4. Plants annual, around hot springs and geysers *A. rossiae*
  4. Plants perennial, in various habitats
    5. Panicle contracted, at least some lower branches spikelet-bearing near base

6. Culms not over 20cm high, in dense tufts with many basal leaves; leaf blades less than 2mm wide *A. variabilis*

6. Culms over 20cm high, not in tufts with many basal leaves; leaf blades mostly over 2mm wide *A. exarata*

5. Panicle open at maturity, the lower branches spikelet-bearing near base

7. Lemmas awned; plants alpine *A. borealis*

7. Lemmas usually not awned; plants alpine or not

8. Panicle diffuse, the capillary branches branching toward the end *A. scabra*

8. Panicle not diffuse, the branches branching at or below the middle

9. Spikelets mostly 1.5-2mm long; plants less than 30cm high *A. idahoensis*

9. Spikelets 2-3mm long; plants usually over 30cm high *A. oregonensis*

Agrostis alba L., Sp. Pl. 63. 1753. Redtop. Culms to 1m high, with creeping rhizomes; leaf blades 3-10mm wide; panicle to 25cm long; spikelets 2-2.5mm long; glumes acute; lemma awnless or rarely awned; palea mostly 1/2-2/3 as long as lemma. Moist areas. Statewide. *A. palustris* Huds., *A. stolonifera* L.

Agrostis borealis Hartm., Handb. Skand. Fl. 3:17. 1838. Culms to 4dm high, tufted; leaf blades 0.5-3mm wide; panicle 5-15cm long; glumes 2-3mm long, acute; lemma slightly shorter than glumes, awned; palea obsolete or nearly so. Alpine. Albany Co.

Agrostis exarata Trin., Gram. Unifl. 207. 1824. Culms to 1m high, mostly tufted; leaf blades mostly 2-10mm wide; panicle narrow, 5-30cm long; glumes 2.5-4mm long, acuminate or awn-tipped; lemma 1.7-2mm long, awned or not; palea minute. Moist, open areas. Statewide. A. asperifolia Trin.

Agrostis humilis Vasey, Bull. Torrey Club 10:21. 1883. Culms mostly less than 18cm high, tufted; leaf blades 1mm or less wide; panicle 1-5cm long; spikelets about 2mm long; lemma nearly as long as glumes, awnless; palea about 2/3 as long as lemma. Alpine. NW, SE.

Agrostis idahoensis Nash, Bull. Torrey Club 24:42. 1897. Culms to 3dm high, tufted; leaf blades to 1.5mm wide; panicle 5-15cm long; spikelets 1.5-2.5mm long; lemma about 1.3mm long, awnless; palea minute. Mountain meadows. Yellowstone Park and Carbon Co.

Agrostis oregonensis Vasey, Bull. Torrey Club 13:55. 1886. Culms to 9dm high; leaf blades 1-4mm wide; panicle 1-3dm long; glumes 2.5-3mm long; lemma 1.5-2mm long, awnless; palea 0.5mm or less long. Wet areas. Teton Co.

Agrostis rossiae Vasey, Contr. U. S. Nat. Herb. 3:76. 1892. Annual to 2dm high; leaf blades 0.5-2mm wide; panicle 2-6cm long; spikelets 2-2.5mm long; lemma about 1.5mm long, awnless; palea minute. Around hot springs and geysers. Yellowstone Park.

Agrostis scabra Willd., Sp. Pl. 1:370. 1797. Culms to 8dm high, tufted; leaf blades 1-3mm wide; panicle 15-25cm long; spikelets 2-3mm long; glumes unequal; lemma about 1.5mm long; palea minute or obsolete. Moist areas. Statewide. *A. hiemalis* (Walt.) B. S. P.

Agrostis thurberiana Hitchc., USDA Pl. Indus. Bull. 68:23. 1905. Culms to 4dm high, tufted, sometimes short-rhizomatous; leaf blades 0.5-2mm wide; panicle 2-7cm long; spikelets 1.5-2mm long; lemma nearly as long as glumes; palea 2/3 or more as long as lemma. Moist areas in the mountains. Statewide.

Agrostis variabilis Rydb., Mem. N. Y. Bot. Gard. 1:32. 1900. Culms to 20cm high, tufted; leaf blades less than 2mm wide; panicle 2-6cm long; spikelets about 2.5mm long; lemma about 1.5mm long, awnless; palea minute. Moist areas in the high mountains. SW, SE.

Alopecurus L. Foxtail

Panicle spike-like; spikelets 1 flowered; glumes equal, usually united at base, ciliate on keel; lemma about half as long to as long as glumes, 5 nerved, awned from below middle; palea lacking.

1. Spikelets at middle of panicle 5-6mm long; leaf blades often over 5mm wide; introduced *A. pratensis*
1. Spikelets at middle of panicle 2-4mm long; leaf blades mostly less than 5mm wide; native
    2. Spikelets densely woolly; panicle about 1cm wide *A. alpinus*
    2. Spikelets often hairy but not woolly; panicle less than 7mm wide
        3. Plants annual *A. carolinianus*
        3. Plants perennial
            4. Awn scarcely exceeding the glumes *A. aequalis*
            4. Awn exserted 2mm or more beyond glumes *A. geniculatus*

Alopecurus aequalis Sobol., Fl. Petrop. 16. 1799. Culms to 6dm high, usually not rooting at nodes; leaf blades 1-4mm wide; panicle 2-7cm long, about 4mm thick; spikelets 2mm long; awn of lemma scarcely exserted. In water and wet areas. Statewide. *A. fulvus* Smith.

Alopecurus alpinus J. E. Smith in Sowerby, Engl. Bot. pl. 1126. 1803. Culms to 8dm high, with slender rhizomes; leaf blades 3-5mm wide; panicle 1-4cm long, about 1cm wide, woolly; glumes 3-4mm long; awn of lemma slightly exserted or as much as 5mm. Meadows and streambanks in the mountains. NW, SE. *A. occidentalis* Scribn. & Tweedy.

Alopecurus _carolinianus_ Walt., Fl. Carol. 74. 1788. Annual to 5dm high; leaf blades 1-4mm wide; panicle 3-6cm long, 3-6mm thick; spikelets 2-2.5mm long; awn long-exserted. Open areas. Campbell Co.

Alopecurus _geniculatus_ L., Sp. Pl. 60. 1753. Culms to 6dm high, rooting at nodes; leaf blades 1-4mm wide; panicle 2-7cm long, about 4mm thick; spikelets 2-2.5mm long; awn exserted 2-3mm. In water and wet areas. NE, SE.

Alopecurus _pratensis_ L., Sp. Pl. 60. 1753. Culms to 8dm high; leaf blades 2-8mm wide; panicle 3-7cm long, 7-10mm thick; glumes mostly 5-6mm long; awn exserted 2-5mm. Fields. SW, NW.

Andropogon L. Bluestem

Perennials; spikelets in pairs at each node, 1 sessile and perfect, the other pediceled and either staminate, neuter, or reduced to the pedicel; fertile lemma usually awned; palea small or lacking.

Racemes solitary on each peduncle; sheaths strongly keeled
_A_. _scoparius_

Racemes 2 or more on each peduncle; sheaths not strongly keeled

Rhizomes short or lacking; awns of sessile spikelets mostly 1-2cm long *A. gerardii*

Rhizomes well developed; awns of sessile spikelets less than 7mm long *A. hallii*

Andropogon gerardii Vitman, Summa Pl. 6:16. 1792. Big Bluestem. Culms to 2m high, sometimes with short rhizomes; leaf blades 3-10mm wide; racemes mostly 3-6 on terminal peduncle, fewer on branches, 5-10cm long; sessile spikelet 7-10mm long, the awn 1-2cm long; pediceled spikelet staminate, awnless. Prairies and hills. NE, SE. *A. furcatus* Muhl.

Andropogon hallii Hack., Sitzungsb. Akad. Naturw. 89(1):127. 1884. Sand Bluestem. Similar to *A. gerardii* but with creeping rhizomes and awns of sessile spikelets rarely over 7mm long. Prairies and hills especially where sandy. NE, SE.

Andropogon scoparius Michx., Fl. Bor. Am. 1:57. 1803. Little Bluestem. Culms to 1.5m high, tufted; leaf blades 2-6mm wide; racemes solitary on each peduncle, 3-6cm long; sessile spikelets mostly 6-8mm long, the awn 8-15mm long; pediceled spikelet usually reduced, short-awned. Prairies and hills. NW, NE, SE.

Aristida L. Three-awn

Spikelets 1 flowered, in a panicle; glumes mostly unequal, acute to awn-tipped; lemma terete, with a trifid awn.

Plants annual; lateral branches of awns 4mm or less long *A. curtissii*

Plants perennial; lateral branches of awns well over 1cm long

Leaves mostly in a short cluster at base of plant; awns 1.5-5cm long *A. fendleriana*

Leaves not conspicuously basal; awns mostly 5-8cm long *A. longiseta*

Aristida curtissii (Gray) Nash in Britt., Man. 94. 1901. Plants annual, to 4dm high; leaf blades about 1mm wide or less; 1st glume 5-8mm long; 2nd glume 8-11mm long; lemma about 9mm long, the trifid awn about 1cm long. Open, dry areas. Reported from Wyoming.

Aristida fendleriana Steud., Syn. Pl. Glum. 1:420. 1854. Culms tufted, to 3dm high; leaf blades involute; 1st glume about 7mm long, the 2nd nearly twice as long; lemma 10-15mm long, the trifid awn 1.5-5cm long. Plains and hills. Statewide.

Aristida longiseta Steud., Syn. Pl. Glum. 1:420. 1854. Culms tufted, to 5dm high; leaf blades involute; 1st glume 8-12mm long, the 2nd about twice as long or more; lemma 12-15mm long, the trifid awn 5-8cm long. Plains and hills. NW, NE, SE.

Arrhenatherum Beauv. Tall Oatgrass

Arrhenatherum elatius (L.) J. & K. Presl, Fl. Cech. 17. 1819. Perennial to 15dm high; leaf blades 4-10mm wide; spikelets in a narrow panicle, 2 flowered, the lower staminate, the other perfect; glumes unequal, 4-10mm long; lemmas 7-9mm long, the 1st with an awn 1-2cm long, the 2nd with one 1-9(15)mm long. Open places. Statewide.

Avena L. Oats

Avena fatua L., Sp. Pl. 80. 1753. Wild Oat. Annual to 8dm high; leaf blades 3-15mm wide; spikelets in an open panicle, usually 3 flowered; glumes 15-27mm long; lemmas mostly 7-18mm long, with geniculate and twisted awns 1-4cm long. Disturbed areas. NE, SE.

Beckmannia Host Sloughgrass

Beckmannia syzigachne (Steud.) Fern., Rhodora 30:27. 1928. Plants annual; culms to 1m high; leaf blades 2-9mm wide; panicle 10-25cm long, each spike 1-2cm long; spikelets 1 flowered, about 3mm long, sessile or nearly so, in 2 rows along 1 side of rachis; lemma about as long as glumes, 5 nerved, acuminate. Ditches and wet areas. Statewide. *B. erucaeformis* (L.) Host of authors.

Bouteloua Lag. Grama

Perennials or rarely annual; inflorescence a raceme of spikes or a single spike; spikelets 1 flowered, with rudiments of a floret(s) above, sessile, in 2 rows along 1 side of rachis; glumes unequal; lemma about as long as 2nd glume, usually 1-3 awned; palea sometimes 2 awned; rudiment usually 3 awned.

Reference: Gould, F. W. & Z. J. Kapadia. 1964. Brittonia 16:182-207.

1. Spikes of raceme many, to 20mm long, pendulous; rhizomes present ... *B. curtipendula*
1. Spikes 1-4, often over 20mm long, spreading or ascending; rhizomes lacking
   2. Rachis of spikes prolonged beyond the spikelets as a naked point mostly 4-8mm long; largest glume with long, tubercle-based hairs ... *B. hirsuta*

2. Rachis of spikes not prolonged beyond the spikelets, or if so, not naked but bearing rudimentary spikelets; largest glume with or without tubercle-based hairs
   3. Plants perennial — *B. gracilis*
   3. Plants annual — *B. simplex*

Bouteloua curtipendula (Michx.) Torr., Bot. Marcy Exp. 300. 1853. Side-oats Grama. Culms to 8dm high, tufted, with scaly rhizomes; leaf blades 2-4mm wide; spikes mostly 35-50, 1-2cm long; spikelets 5-8, 5-10mm long; fertile lemma mucronate; rudiment 3 awned or inconspicuous. Plains and hills. NE, SE. *Atheropogon curtipendula* (Michx.) Fourn. The above combination by Torrey was not valid in Emory, Notes Mil. Recon., 1848.

Bouteloua gracilis (H. B. K.) Lag. ex Griffiths, Contr. U. S. Nat. Herb. 14:375. 1912. Blue Grama. Culms to 5dm high, tufted; leaf blades 1-2mm wide; spikes mostly 1-3, 2.5-5cm long; spikelets many, about 5mm long at middle of spike; fertile lemma awned; rudiment bearded at summit of rachilla, awned. Plains and hills. NW, NE, SE. This name was not validated by Steudel, Nom. Bot. 2(1):219. 1840, since it was merely listed as a synonym. *B. oligostachya* (Nutt.) Torr.

Bouteloua hirsuta Lag., Var. Cienc. 2(4):141. 1805. Culms to 6dm high, tufted; leaf blades 1-2mm wide; spikes 1-4, 1-3.5cm long, the rachis extending beyond spikelets 4-8mm; spikelets 10-45, about 5mm long including awn; fertile lemma awn-tipped; rudiment bearded and awned. Plains and hills. Laramie Co.

Bouteloua simplex Lag., Var. Cienc. 2(4):141. 1805. Tufted annual to 2dm high; leaf blades 0.3-1.5mm wide; spike solitary on each culm, 6-25mm long; spikelets mostly 5-30, about 5mm long; fertile lemma awned; rudiment bearded and awned. Plains and hills. Laramie Co. B. prostrata Lag.

Bromus L. Bromegrass; Chess

Spikelets several to many flowered, in a panicle; glumes unequal to subequal; lemmas 5-9 nerved, 2 toothed, awned from between the teeth or awnless.

Reference: Wagnon, H. K. 1952. Brittonia 7:415-480.

1. Spikelets strongly flattened, the lemmas compressed-keeled; plants tufted perennials B. carinatus
1. Spikelets terete or somewhat flattened but the lemmas not compressed-keeled; plants annual or perennial
  2. Plants perennial
    3. Creeping rhizomes present B. inermis
    3. Creeping rhizomes lacking
      4. Lemmas pubescent along the margin and on lower part of back, the upper part glabrous or nearly so
        5. Ligule 3-5mm long; awn usually over 5mm long B. vulgaris

5. Ligule about 1mm long; awn 3-5mm long *B. ciliatus*

4. Lemmas pubescent evenly over back, usually more densely so along the lower part of margin

6. Leaf blades 5-15mm wide; 1st glume 1 nerved, 2nd glume 3 nerved *B. pubescens*

6. Leaf blades less than 5mm wide; 1st glume 3 nerved, 2nd glume 5 nerved or nearly so *B. anomalus*

2. Plants annual

7. Teeth of lemmas 2-3mm long; awns mostly 10-15mm long; 1st glume 1 nerved, 2nd glume usually 3 nerved *B. tectorum*

7. Teeth of lemmas mostly less than 1mm long; awns often less than 10mm long; 1st glume 3-5 nerved, 2nd glume 5-9 nerved

8. Panicle contracted, dense, the branches erect or ascending

9. Sheaths usually densely pilose; lemma not curling around edges of mature fruit *B. mollis*

9. Sheaths glabrous to sparsely pubescent; lemma tending to curl around edges of mature fruit *B. secalinus*

8. Panicle open, the branches mostly spreading

10. Lemmas awnless or with awns mostly less than 1mm long *B. brizaeformis*

10. Lemmas with awns mostly over 2mm long

11. Foliage glabrous to sparsely pubescent; lemma tending to curl around edges of mature fruit *B. secalinus*

11. Foliage mostly densely pilose; lemma not curling around edges of mature fruit
    12. Awn straight; panicle branches stiffly spreading or ascending, usually not flexuous — *B. commutatus*
    12. Awn flexuous, usually divergent when dry; panicle branches lax or flexuous — *B. japonicus*

Bromus anomalus Rupr. ex Fourn., Mex. Pl. 2:126. 1886. Culms to 6dm high, tufted; leaf blades mostly 2-4mm wide; panicle drooping; spikelets 15-35mm long; lemmas pubescent on back; awn 2-4mm long. Woods and slopes. Statewide. *B. porteri* (Coult.) Nash, *B. scabratus* Scribn.

Bromus brizaeformis Fisch. & Mey., Ind. Sem. Hort. Petrop. 3:30. 1837. Annual to 6dm high; leaf blades 1-5mm wide; panicle somewhat drooping; spikelets 15-25mm long; lemmas glabrous, awnless or the awn to 1.5mm long. Sandy and disturbed areas. Sheridan and Albany Cos.

Bromus carinatus Hook. & Arn., Bot. Beechey Voy. 403. 1840. Culms to 9dm high, tufted; leaf blades 1-12mm wide; panicle erect; spikelets 2-4cm long; lemmas puberulent or pubescent; awn 3-10mm long. Woods, meadows, and disturbed areas. Statewide. *B. breviaristatus* Buckl., *B. marginatus* Nees.

Bromus _ciliatus_ L., Sp. Pl. 76. 1753. Culms to 12dm high, tufted; leaf blades to 1cm wide; panicle drooping; spikelets 15-25mm long; lemmas pubescent near margin, much less so on back; awn 3-5mm long. Moist areas. Statewide. _B_. _richardsonii_ Link.

Bromus _commutatus_ Schrad., Fl. Germ. 353. 1806. Annual to 6dm high; leaf blades 1-5mm wide; panicle mostly erect; spikelets 1-2.5cm long; lemmas glabrous or scaberulous; awns 4-10mm long. Disturbed areas. NW, NE, SE.

Bromus _inermis_ Leyss., Fl. Hal. 16. 1761. Culms to 12dm high, from creeping rhizomes; leaf blades 3-15mm wide; panicle erect; spikelets 1.5-3cm long; lemmas glabrous to villous or scaberulous; awn lacking or to 6mm long. Disturbed areas, woods, and meadows. Statewide. _B_. _polyanthus_ Scribn., _B_. _pumpellianus_ Scribn., _B_. _multiflorus_ Scribn.

Bromus _japonicus_ Thunb., Fl. Japon. 52. 1784. Annual to 7dm high; leaf blades 1-7mm wide; panicle somewhat drooping; spikelets 15-25mm long; lemmas glabrous; awn 8-15mm long. Disturbed areas. NW, NE, SE.

Bromus mollis L., Sp. Pl. 2:112. 1762. Annual to 8dm high; leaf blades 1-5mm wide; panicle erect; spikelets 8-18mm long; lemmas glabrous or scabrous; awn 4-9mm long. Disturbed areas. Hot Springs Co. *B. racemosus* L.

Bromus pubescens Muhl. ex Willd., Enum. Pl. 120. 1809. Culms to 12dm high, tufted; leaf blades 5-15mm wide; panicle drooping; spikelets 15-25mm long; lemmas pubescent; awn 2-5mm long. Woods and slopes. Converse Co. *B. purgans* L. of authors.

Bromus secalinus L., Sp. Pl. 76. 1753. Annual to 6dm high; leaf blades 1-4mm wide; panicle nodding to erect; spikelets 1-3cm long; lemmas glabrous or scaberulous; awn 1-5mm long. Disturbed areas. NE, SE.

Bromus tectorum L., Sp. Pl. 77. 1753. Cheatgrass. Annual to 6dm high; leaf blades 1-5mm wide; panicle drooping; spikelets 12-20mm long; lemmas villous or pilose; awn 10-15mm long. Disturbed areas, especially rangeland. Statewide.

Bromus vulgaris (Hook.) Shear, USDA Agrost. Bull. 23:43. 1900. Culms to 12dm high, tufted; leaf blades to 12mm wide; panicle drooping; spikelets about 2.5cm long; lemmas pubescent to nearly glabrous; awn 4-8mm long. Woods and thickets. Teton Co.

Buchloe Engelm. Buffalo Grass

Buchloe dactyloides (Nutt.) Engelm., Trans. Acad. Sci. St. Louis 1:432. 1859. Dioecious or monoecious, stoloniferous perennial forming a dense sod, to 2dm high; leaf blades 0.5-2mm wide; staminate spikelets 2 flowered, sessile, in 2 rows on 1 side of rachis, lemmas longer than the unequal glumes; pistillate spikelets mostly 2-5 in a spike or head, the thickened rachis and 2nd glume forming a rigid white or yellowish globular structure crowned by the green-toothed summits of the glumes. Plains and hills. NE, SE.

Calamagrostis Adans. Reedgrass

Perennials; spikelets 1 flowered, in open to spike-like panicles; glumes about equal; lemma shorter than glumes, awned from the back, the callus bearing a tuft of hairs mostly 1/4 as long to as long as lemma.

1. Awn exserted 1-4mm beyond glumes, geniculate; glumes mostly 6-8mm long _C. purpurascens_
1. Awn either included or scarcely longer than the glumes, straight or geniculate; glumes mostly shorter
  2. Culms less than 40cm high, scabrous just below panicle; leaf blades 1-3mm wide, involute; awn geniculate; panicle mostly spike-like _C. montanensis_

2. Culms mostly over 40cm high, scabrous or not below panicle; leaf blades often over 3mm wide, often flat; awn geniculate or not; panicle open to spike-like

3. Callus hairs rarely over half as long as lemma (hairs of rachilla sometimes longer); awn geniculate or straight

4. Sheaths, or some of them, pubescent on the collar, the hairs much longer than foliage hairs if present — *C. rubescens*

4. Sheaths glabrous on the collar or the hairs the same as the foliage hairs

5. Awn attached toward base of lemma — *C. koelerioides*

5. Awn attached at or above middle of lemma — *C. scopulorum*

3. Callus hairs mostly 2/3 as long to as long as lemma; awn straight

6. Panicle loose and usually open, mostly over 2cm wide; awn delicate; leaf blades often over 4mm wide, usually flat — *C. canadensis*

6. Panicle contracted or spike-like, rarely over 3cm wide; awn somewhat stout; leaf blades 1-4mm wide, sometimes involute

7. Ligules of upper leaves 4-8mm long — *C. inexpansa*

7. Ligules of upper leaves 1-3.5mm long — *C. stricta*

<u>Calamagrostis</u> <u>canadensis</u> (Michx.) Beauv., Ess. Agrost. 15, 152. 1812. Bluejoint. Culms to 15dm high, with creeping rhizomes; leaf blades 2-8mm wide; panicle mostly loose and open; glumes mostly 3-5mm long; awn extending to or slightly beyond lemma tip; callus hairs about half as long to as long as lemma. Moist areas. Statewide. <u>C</u>. <u>scribneri</u> Beal.

<u>Calamagrostis</u> <u>inexpansa</u> Gray, N. A. Gram. Cyp. 1:No. 20. 1834. Culms to 12dm high, with slender rhizomes; leaf blades mostly 2-4mm wide; panicle dense; glumes 3-4mm long; awn extending to or slightly beyond lemma tip; callus hairs about 3/4 as long as lemma. Moist areas. Statewide. <u>C</u>. <u>hyperborea</u> Lange, <u>C</u>. <u>wyomingensis</u> Gandog.

<u>Calamagrostis</u> <u>koelerioides</u> Vasey, Bot. Gaz. 16:147. 1891. Culms to 14dm high, with stout rhizomes; leaf blades 2-8mm wide; panicle spike-like; glumes 4.5-6.5mm long; awn about as long as lemma; callus hairs about 1-2mm long. Slopes and meadows. Teton Co. and Yellowstone Park.

<u>Calamagrostis</u> <u>montanensis</u> Scribn. ex Vasey, Contr. U. S. Nat. Herb. 3:82. 1892. Culms mostly to 4dm high, with creeping rhizomes; leaf blades mostly less than 2mm wide; panicle mostly spike-like; glumes 4-5mm long; awn about equaling lemma; callus hairs about half as long as lemma. Plains and hills. NE, SE, SW.

Calamagrostis purpurascens R. Br. in Richards. in Frankl., Narr. 1st Journ. 731. 1823. Culms to 6 or sometimes 10dm high, tufted, sometimes with short rhizomes; leaf blades 2-5mm wide; panicle spike-like; glumes 6-8mm long; awn 2-4mm longer than lemma; callus hairs about 1/4 as long as lemma. Rocky areas in the mountains. Statewide.

Calamagrostis rubescens Buckl., Proc. Acad. Phila. 1862:92. 1862. Pinegrass. Culms to 1m high, with creeping rhizomes; leaf blades 1.5-5mm wide; panicle spike-like or loose; glumes 4-5mm long; awn slightly longer than lemma, exserted from glumes; callus hairs about 1/3 as long as lemma. Woods or occasionally in open areas. NW, SW, SE. C. suksdorfii Vasey.

Calamagrostis scopulorum Jones, Proc. Calif. Acad. Sci. II, 5:722. 1895. Culms to 8dm high, with short rhizomes; leaf blades 3-7mm wide; panicle spike-like or somewhat loose; glumes 4-6mm long; awn about as long as lemma or shorter; callus hairs about 1/2 as long as lemma. Moist areas. NW, NE.

Calamagrostis stricta (Timm) Koeler, Descr. Gram. 105. 1802. Culms to 1m high, with rhizomes; leaf blades 1-4mm wide; panicle dense; glumes 2-4mm long; awn about as long as lemma; callus hairs about 3/4 as long as lemma. Wet areas. NE, SE, SW. C. neglecta (Ehrh.) Gaertn. et al. (see Taxon 19:299, 1970, concerning nomenclature).

Calamovilfa Hack.

Calamovilfa longifolia (Hook.) Scribn. in Hack., True Grasses 113. 1890. Culms to 18dm high, with creeping rhizomes; leaf blades 4-8mm wide near base, narrower above; panicle usually narrow; spikelets 1 flowered, 6-7mm long; glumes unequal; lemma slightly shorter than 2nd glume, awnless; callus hairs over half as long as lemma. Plains, hills, and open woods. NE, SE, SW.

Catabrosa Beauv. Brookgrass

Catabrosa aquatica (L.) Beauv., Ess. Agrost. 97. 1812. Culms to 4dm high, creeping at base; leaf blades 2-8mm wide; panicle mostly 1-2dm long; spikelets mostly 2 flowered, about 3mm long; glumes unequal, shorter than lower floret, truncate or erose at tip; lemmas 2-3mm long. Mostly in shallow water. Statewide.

Cenchrus L. Sandbur

Cenchrus longispinus (Hack.) Fern., Rhodora 45:388. 1943. Culms spreading, 2-9dm long; leaf blades 2-7mm wide; spikelets solitary or few in a raceme, enclosed by a spiny bur, the bur mostly 4-6mm wide excluding spines. Sandy areas. NW, SE. *C. tribuloides* L. of authors, *C. pauciflorus* Benth.

<u>Cinna</u> L. Woodreed

<u>Cinna</u> <u>latifolia</u> (Trevir. ex Goepp.) Griseb. in Ledeb., Fl. Ross. 4:435. 1853. Culms to 15dm high, solitary or tufted; leaf blades to 15mm wide; panicle loose; spikelets 1 flowered, about 4mm long; glumes subequal; lemma about as long as glumes, awned or not. Moist woods and streambanks. NW, NE, SE.

<u>Dactylis</u> L. Orchard Grass

<u>Dactylis</u> <u>glomerata</u> L., Sp. Pl. 71. 1753. Culms to 12dm high, tufted; leaf blades 2-8mm wide; panicle mostly 5-20cm long; spikelets several flowered, nearly sessile in dense 1 sided fascicles; glumes unequal; lemmas 5-8mm long, mucronate or short-awned, ciliate on keel. Meadows and disturbed areas. Statewide.

<u>Danthonia</u> DC. Oatgrass

Plants tufted; spikelets several flowered, in open or spike-like panicles; glumes subequal, mostly exceeding uppermost floret; lemmas with a flat, twisted, geniculate awn arising from between 2 terminal teeth.

1. Lemmas pilose on back, sometimes sparsely so
   2. Glumes mostly 19mm or more long — D. parryi
   2. Glumes mostly less than 15mm long — D. spicata
1. Lemmas glabrous on back, pilose on margin only
   3. Panicle narrow, the pedicels mostly appressed to rachis; spikelets mostly 4-10 per panicle — D. intermedia
   3. Panicle open, the pedicels mostly spreading or reflexed; spikelets mostly 1-5 per panicle
      4. Panicle usually with a single spikelet, rarely 2 or 3; plants less than 3dm high — D. unispicata
      4. Panicle with few to several spikelets; plants usually over 3dm high — D. californica

Danthonia californica Bolander, Proc. Calif. Acad. Sci. 2:182. 1863. Culms to 1m high; leaf blades 0.5-5mm wide; panicle with mostly 2-5 spikelets; glumes mostly 15-20mm long; lemmas 8-12mm long; awn 5-10mm long. Meadows and open woods. Statewide.

Danthonia intermedia Vasey, Bull. Torrey Club 10:52. 1883. Culms to 5dm high; leaf blades 0.5-3mm wide; panicle with mostly 4-10 spikelets; glumes 12-18mm long; lemmas 7-8mm long; awn 6-11mm long. Woods and meadows in the mountains. Statewide.

Danthonia parryi Scribn., Bot. Gaz. 21:133. 1896. Culms to 8dm high; leaf blades 0.5-3mm wide; panicle with mostly 3-9 spikelets; glumes 19-24mm long; lemmas 9-13mm long; awns mostly 10-18mm long. Open woods and slopes. Albany Co.

Danthonia spicata (L.) Beauv. ex Roem. & Schult., Syst. Veg. 2:690. 1817. Culms to 7dm high; leaf blades 0.5-3mm wide; panicle with mostly 4-10 spikelets; glumes 9-12mm long; lemmas 3.5-5mm long; awn 5-9mm long. Woods and open areas. NW, NE.

Danthonia unispicata (Thurb.) Munro ex Macoun, Can. Pl. Cat. 2(4):215. 1888. Culms to 3dm high; leaf blades 0.5-3mm wide; panicle with a single spikelet or rarely with 2 or 3; glumes 1-2cm long; lemmas 7-12mm long; awn 4-8mm long. Wooded and open areas. Statewide.

Deschampsia Beauv. Hairgrass

Annuals or tufted perennials; spikelets 2 flowered, in narrow or open panicles; glumes subequal, exceeding the florets or nearly so; lemmas awned from near or below the middle; rachilla prolonged beyond upper floret.

Reference: Kawano, S. 1963. Canad. Journ. Bot. 41:719-742.

Several alpine and subalpine specimens with contracted panicles match specimens which have been called *D. brevifolia* R. Br. and *D. cespitosa* ssp. *orientalis* Hult. There is little agreement on the limits of these two taxa, so until the types are studied, our plants are best considered as conspecific with *D. cespitosa*.

1. Plants annual; leaves few — *D. danthonioides*

1. Plants perennial; leaves usually numerous

   2. Glumes 3.5mm or less long; panicle usually contracted; plants alpine or subalpine — *D. cespitosa*

   2. Glumes mostly over 3.5mm long; panicle open or contracted; plants alpine or not

      3. Panicle narrow, the branches mostly appressed to rachis or nearly so; leaf blades 0.5-1.5mm wide — *D. elongata*

      3. Panicle open, the branches spreading or ascending; leaf blades often over 1.5mm wide

         4. Awns mostly 2.5-3mm long, from near middle of lemma; anthers 0.8-1.2mm long — *D. atropurpurea*

         4. Awns mostly 3-4mm long, from near base of lemma; anthers 1.2-2.2mm long — *D. cespitosa*

Deschampsia atropurpurea (Wahl.) Scheele, Flora 27:56. 1844. Culms loosely tufted, to 8dm high; leaf blades flat, 2-6mm wide; panicle somewhat open; glumes mostly 4-6mm long, exceeding florets; lemmas about 2.5mm long; awn about 2.5-3mm long. Woods and moist meadows in the mountains. NW, SE.

Deschampsia cespitosa (L.) Beauv., Ess. Agrost. 91, 149. 1812. Culms densely tufted, to 12dm high; leaf blades flat or folded, 1-4mm wide; panicle open or rarely contracted; glumes 2-5mm long, about as long as florets; lemmas 2.5-3.5mm long; awns mostly 3-4mm long. Moist areas. Statewide.

Deschampsia danthonioides (Trin.) Munro ex Benth., Pl. Hartw. 342. 1857. Plants annual; culms to 6dm high; leaf blades 0.5-1.5mm wide; panicle open; glumes 5-8mm long, exceeding florets; lemmas 2-3mm long; awns 4-6mm long. Open areas. Park Co.

Deschampsia elongata (Hook.) Munro ex Benth., Pl. Hartw. 342. 1857. Culms densely tufted, to 12dm high; leaf blades flat or folded, 1-1.5mm wide; panicle narrow; glumes 4-6mm long, equaling or slightly exceeding florets; lemmas 1.5-2.5mm long; awns 3-5mm long. Open areas in the mountains. Yellowstone Park.

Digitaria Haller Crabgrass

Annuals; spikelets flattened from the back, in digitate racemes, 2 flowered, the lower one reduced to a sterile lemma which appears like a 2nd glume; 1st glume minute, 2nd glume similar to sterile lemma.

Sheaths glabrous; sterile lemma with minutely glandular hairs
D. ischaemum

Sheaths pilose; sterile lemma lacking glandular hairs
D. sanguinalis

Digitaria ischaemum (Schreb.) Schreb. ex Muhl., Descr. Gram. 131. 1817. Erect or decumbent annual; leaf blades 1-4mm wide; racemes 3-10cm long; 1st glume hyaline, obscure; 2nd glume and sterile lemma as long as fertile lemma, about 2mm long. Weed in disturbed areas. Natrona and Albany Cos.

Digitaria sanguinalis (L.) Scop., Fl. Carn. 2(1):52. 1771. Spreading annual, rooting at decumbent base; leaf blades 3-10mm wide; racemes 3-15cm long; 1st glume minute; 2nd glume about 1.5mm long, half as long as spikelet. Weed in disturbed areas. SW, SE, NE. *Syntherisma sanguinalis* (L.) Dulac.

Distichlis Raf. Saltgrass

Distichlis stricta (Torr.) Rydb., Bull. Torrey Club 32:602. 1905. Plants dioecious, with creeping rhizomes, to 35cm high; leaf blades 0.5-2.5mm wide; panicles with 7-15 flowered spikelets; glumes 4-6mm long; lemmas 3-7mm long. Mostly alkaline areas. Statewide.

Echinochloa Beauv. Barnyard Grass

Reference: Gould, F. W., et al. 1972. Am. Midl. Nat. 87:36-59.

Echinochloa muricata (Beauv.) Fern., Rhodora 17:106. 1915. Erect to decumbent annual to 1.5m high; leaf blades 3-15mm wide; panicle of several spike-like branches; 1st glume about half as long as spikelet or less; 2nd glume about equal to sterile lemma, mostly 3-4mm long; sterile lemma awned and tuberculate-hispid; fertile lemma acuminate and smooth. Disturbed areas. SW, SE, NE. E. crusgalli (L.) Beauv. of authors.

Elymus L. Wild Rye

Spikelets mostly 2-6 flowered, usually in pairs, mostly sessile; glumes subequal, sometimes bristle-like; lemmas obscurely 5 nerved, acute or awned. Two intergeneric hybrids commonly key to Elymus. They are tufted and have the rachis tardily disjointing. They have been called E. macounii Vasey and E. aristatus Merr.

1. Plants with slender creeping rhizomes
   2. Ligules mostly 2mm or more long; leaf blades mostly flat, some often over 6mm wide E. cinereus
   2. Ligules mostly less than 2mm long; leaf blades sometimes involute, less than 6mm wide

3. Spikelets glabrous to scabrous *E. triticoides*

3. Spikelets densely villous to coarsely, sometimes sparsely, pubescent *E. innovatus*

1. Plants lacking rhizomes

4. Glumes subulate to subsetaceous, not broadened above the base, the nerves obscure except in *E. villosus*

5. Lemmas awnless or awn-tipped, the awn shorter than the lemma body

6. Lemmas hairy; spikelets mostly about 7-8mm long; leaf sheaths fibrillose at maturity *E. junceus*

6. Lemmas glabrous or hairy; spikelets mostly 10mm or more long; leaf sheaths not fibrillose at maturity

7. Ligules mostly 2mm or more long; leaf blades flat, often over 6mm wide *E. cinereus*

7. Ligules mostly 0.5-1.5mm long; leaf blades involute, 1-5mm wide *E. ambiguus*

5. Lemmas awned, the awn as long as the lemma body or longer

8. Awns straight; lemmas about 1.5mm wide across the back *E. villosus*

8. Awns mostly flexuous-divergent; lemmas about 2mm wide across the back *E. interruptus*

4. Glumes lanceolate or narrower, broadened above the base, strongly nerved

9. Glumes relatively thin, not indurate at base *E. glaucus*

9. Glumes firm, usually strongly indurate at base

10. Awns divergently curved when dry, mostly 2-4cm long; base of glumes not terete *E. canadensis*

10. Awns straight, mostly less than 1.5cm long; base of glumes somewhat terete ... *E. virginicus*

Elymus ambiguus Vasey & Scribn. in Vasey, Contr. U. S. Nat. Herb. 1:280. 1893. Culms loosely tufted, to 8dm high; leaf blades 1-5mm wide; glumes 3-15mm long, somewhat bristle-like; lemmas 7-12mm long, pubescent or glabrous, awnless or with an awn to 5mm long. Hills and plains. Sweetwater and Albany Cos. *E. salinus* Jones, *E. strigosus* Rydb.

Elymus canadensis L., Sp. Pl. 83. 1753. Culms tufted, to 15dm high; leaf blades mostly 5-15mm wide; glumes mostly 10-15mm long excluding awn; lemmas about 1cm long, mostly pubescent, the awns 2-4cm long. Open areas especially where moist. Statewide.

Elymus cinereus Scribn. & Merr., Bull. Torrey Club 29:467. 1902. Culms mostly tufted, sometimes short-rhizomatous, to 2m high; leaf blades 4-15mm wide; glumes 1-2cm long, somewhat bristle-like; lemmas 7-12mm long, pubescent, mostly awn-tipped. Plains and foothills. Statewide. *E. condensatus* Presl of authors.

Elymus glaucus Buckl., Proc. Acad. Phila. 1862:99. 1862. Culms tufted, to 12dm high; leaf blades mostly 5-15mm wide; glumes 8-14mm long; lemmas 8-14mm long, the awns mostly 1-3cm long. Open woods and hills. Statewide.

Elymus innovatus Beal, Grasses N. Am. 2:650. 1896. Culms tufted, with slender rhizomes, to 1m high; leaf blades mostly 1-5mm wide; glumes 4-10mm long, villous, somewhat bristle-like; lemmas mostly 7-9mm long, mostly villous, the awns 1-10mm long. Open woods and stream banks. Sweetwater Co. *E. hirtiflorus* Hitchc.

Elymus interruptus Buckl., Proc. Acad. Phila. 1862:99. 1862. Culms tufted, to 13dm high; leaf blades 3-12mm wide; glumes 1-3cm long, somewhat bristle-like; lemmas about 1cm long, glabrous or not, the awns 1-3cm long. Open areas. Albany Co. *E. occidentalis* Scribn., *E. diversiglumis* Scribn. & Ball.

Elymus junceus Fisch., Mem. Soc. Nat. Mosc. 1:25. 1811. Culms tufted, to 15dm high; leaf blades 0.5-4mm wide; glumes 3-6mm long, bristle-like; lemmas about 6mm long, awn-tipped. Disturbed areas. Albany Co.

Elymus triticoides Buckl., Proc. Acad. Phila. 1862:99. 1862. Plants rhizomatous, to 12dm high; leaf blades mostly 2-6mm wide, usually involute; glumes mostly 5-20mm long, awn-tipped or nearly completely bristle-like; lemmas 6-10mm long, glabrous, the awn minute or to 14mm long. Moist, sandy, or alkaline areas. NW, SW, SE. *E. simplex* Scribn. & Williams.

Elymus villosus Muhl. ex Willd., Enum. Pl. 1:131. 1809. Culms tufted, to 1m high; leaf blades 2-12mm wide; glumes 12-20mm long, bristle-like, hirsute; lemmas 6-9mm long, mostly hirsute, the awns mostly 1-3cm long. Woods and canyons. Crook Co.

Elymus virginicus L., Sp. Pl. 84. 1753. Culms tufted, to 12dm high; leaf blades mostly 3-15mm wide; glumes 6-12mm long, scabrous, awn-tipped; lemmas 6-9mm long, scabrous or hirsute, the awns 4-15mm long. Moist areas. NE, SE.

Eragrostis Beauv. Lovegrass

Plants annual; spikelets few to many flowered, in a panicle; glumes somewhat unequal, shorter than 1st lemma; lemmas obtuse to acuminate, 3 nerved; palea half as long to about as long as lemma.

1. Plants creeping, rooting at nodes, forming mats _E. hypnoides_
1. Plants often decumbent at base but not creeping and forming mats
    2. Plants not glandular _E. pectinacea_
    2. Plants with glandular depressions on panicle branches, or on keel of the lemmas, or on margins of leaf blades or keel of sheaths

3. Spikelets mostly 2.5mm or more wide; glands prominent on keel of most lemmas *E. cilianensis*

3. Spikelets 2mm wide or less; glandular depressions mostly on panicle branches and leaves *E. poaeoides*

Eragrostis cilianensis (All.) Lutati, Malpighia 18:386. 1904. Culms ascending or spreading, to 5dm high, with a ring of glands below the nodes; leaf blades 1-7mm wide; spikelets 3-15mm long, 3-40 flowered; glumes 1.5-2.5mm long; lemmas about 2.5mm long. Disturbed areas. Statewide. *E. major* Host.

Eragrostis hypnoides (Lam.) B. S. P., Prel. Cat. N. Y. Pl. 69. 1888. Plants creeping and matlike, to 1dm high; leaf blades 0.5-1.5mm wide; spikelets mostly 3-10mm long, several to many flowered; glumes 0.5-1.5mm long; lemmas 1.5-2mm long. Moist, usually sandy areas. Sheridan Co.

Eragrostis pectinacea (Michx.) Nees, Fl. Afr. Austr. 406. 1841. Culms erect or ascending from a decumbent base, to 5dm high; leaf blades 1-6mm wide; spikelets 3-8mm long, 3-15 flowered; glumes 1-2mm long; lemmas 1.5-2mm long. Disturbed areas. SE. *E. diffusa* Buckl.

Eragrostis poaeoides Beauv. ex Roem. & Schult., Syst. Veg. 2:574. 1817. Culms ascending or spreading, to 4dm high; leaf blades 2-7mm wide; spikelets 3-10mm long, 3-20 flowered; glumes 1-2mm long; lemmas about 2mm long. Disturbed areas. Albany Co.

<u>Festuca</u> L. Fescue

Perennials or rarely annuals; spikelets few to several flowered, or rarely 1 flowered in some spikelets, in a panicle; glumes unequal, mostly shorter than or subequal to 1st lemma; lemmas awned from tip or awnless, 5 nerved, the nerves often obscure.

References: Terrell, E. E. 1968. Rhodora 70:564-568.
Lonard, R. I. & F. W. Gould. 1974. Madroño 22:217-230.

1. Plants annual — *F. octoflora*
1. Plants perennial
  2. Leaf blades flat, mostly over 3mm wide
    3. Lemmas with awns over 4mm long — *F. subulata*
    3. Lemmas awnless or with awns less than 2mm long — *F. pratensis*
  2. Leaf blades involute, or if flat, less than 3mm wide
    4. Ligule over 2mm long; lemmas awnless or awn-tipped — *F. thurberi*
    4. Ligule less than 2mm long; lemmas awnless or awned
      5. Lower lemmas 7mm or more long, acute or awn-tipped — *F. scabrella*
      5. Lower lemmas mostly less than 7mm long, awned
        6. Awns, or some of them, as long as or longer than body of lemma; NW Wyo. — *F. occidentalis*

6. Awns usually shorter than body of lemma; widespread
  7. Culms either decumbent at the usually red or purple, fibrillose base or from rhizomes *F. rubra*
  7. Culms erect, without rhizomes, often not red, purple, or fibrillose at base
    8. Culms mostly over 30cm high; panicles 10-20cm long, mostly open; anthers 2-4mm long *F. idahoensis*
    8. Culms mostly less than 30cm high; panicles mostly less than 10cm long, mostly narrow; anthers 0.3-2mm long
      9. Culms glabrous or scabrous below the panicle *F. ovina*
      9. Culms hairy below the panicle *F. baffinensis*

Festuca baffinensis Polunin, Bull. Nat. Mus. Can. 92:91. 1940. Plants caespitose; culms to 2dm high, puberulent or tomentulose below the panicle; leaf blades about 1mm wide or less; spikelets 5-8mm long, mostly 3-4 flowered; glumes 2-5mm long; lemmas 3-5.5mm long, the awn 1-3mm long. Alpine. Park Co.

Festuca idahoensis Elmer, Bot. Gaz. 36:53. 1903. Culms tufted, to 1m high; leaf blades 0.2-0.5mm wide; spikelets 6-12mm long, mostly 4-7 flowered; glumes 3-5mm long; lemmas 4-7mm long, the awn usually 2-5mm long. Open woods and hills. Statewide.

Festuca occidentalis Hook., Fl. Bor. Am. 2:249. 1840. Culms tufted, to 1m high; leaf blades 0.2-1mm wide; spikelets 6-10mm long, 3-6 flowered; glumes 3-5mm long; lemmas 5-7mm long, the awn often as long or longer. Dry slopes. Yellowstone Park.

Festuca octoflora Walt., Fl. Carol. 81. 1788. Annual to 3dm or rarely 6dm high; leaf blades 0.2-0.7mm wide; spikelets 6-10mm long, 5-15 flowered; glumes 3-4.5mm long; lemmas 3-5mm long, the awn 1-7mm long. Plains and disturbed areas. NW, NE, SE. Vulpia octoflora (Walt.) Rydb.

Festuca ovina L., Sp. Pl. 73. 1753. Culms tufted, to 4dm high; leaf blades 0.2-1mm wide; spikelets 5-10mm long, mostly 4 or 5 flowered; glumes 2-4mm long; lemmas 3-5mm long, the awns 1-4mm long. Open woods and slopes from the lowlands to alpine. Statewide. F. brachyphylla Schult.

Festuca pratensis Huds., Fl. Angl. 37. 1762. Culms tufted, to 12dm high; leaf blades 2-8mm wide; spikelets 8-15mm long, 6-10 flowered; glumes 2-4mm long; lemmas 5-7mm long, acute or rarely short-awned. Meadows and disturbed areas. Statewide. F. elatior L. of authors.

Festuca rubra L., Sp. Pl. 74. 1753. Culms mostly loosely tufted, from rhizomes, or bent or decumbent at base, to 1m high; leaf blades 0.5-2mm wide; spikelets mostly 7-10mm long, 4-6 flowered; glumes 2-5mm long; lemmas 3-7mm long, the awns mostly 1-3mm long. Meadows, hills, and swamps. NW, SW, NE.

<u>Festuca</u> <u>scabrella</u> Torr. ex Hook., Fl. Bor. Am. 2:252. 1840. Culms mostly tufted, to 9dm high; leaf blades 0.5-2mm wide; spikelets 8-12mm long, 4-6 flowered; glumes 4-9mm long; lemmas 5-10mm long, acute to awn-tipped. Plains, hills, and open woods. Johnson Co.

<u>Festuca</u> <u>subulata</u> Trin. in Bong., Mem. Acad. St. Petersb. VI, 2:173. 1832. Culms tufted, to 1m high; leaf blades 3-10mm wide; spikelets mostly 7-10mm long, 3-5 flowered; glumes 2.5-5mm long; lemmas 3-6mm long, the awn 4-20mm long. Moist thickets and shaded banks. Teton Co. <u>F</u>. <u>jonesii</u> Vasey.

<u>Festuca</u> <u>thurberi</u> Vasey in Wats., Cat. Pl. Surv. 100th Merid. 56. 1874. Culms tufted, to 9dm high; leaf blades 0.5-2mm wide; spikelets 8-12mm long, 3-6 flowered; glumes 4-6mm long; lemmas 5-8mm long, acute to awn-tipped. Dry slopes. Albany Co.

<u>Glyceria</u> R. Br. Mannagrass

Plants usually rhizomatous; spikelets few to many flowered, in panicles; glumes unequal, 1 nerved; lemmas scarious at apex, usually obtuse, 5-9 nerved.

1. Spikelets linear, nearly terete, usually 1cm long or more; panicle narrow and erect ..... <u>G</u>. <u>borealis</u>
1. Spikelets ovate or oblong, somewhat compressed, usually 6mm or less long; panicle usually nodding

2. Leaf blades mostly 2-5mm wide; 1st glume 0.5-1mm long; upper ligules closed in front (sometimes opened by drying) *G. striata*

2. Leaf blades mostly 6mm or more wide; 1st glume about 1mm or more long; upper ligules usually open in front

   3. Ligules pubescent-scabridulous; 1st glumes averaging about 1mm long *G. elata*

   3. Ligules glabrous; 1st glumes averaging about 1.5mm long *G. grandis*

Glyceria borealis (Nash) Batchelder, Proc. Manchester Inst. 1:74, 106. 1900. Culms to 1m high; leaf blades 2-5mm wide; spikelets 1-1.5cm long, mostly 6-12 flowered; glumes about 1.5-3mm long; lemmas 3-4mm long. Moist areas. NW, SW, SE.

Glyceria elata (Nash ex Rydb.) Jones, Bull. U. Mont. Biol. 15:17. 1910. Culms to 2m high; leaf blades mostly 6-12mm wide; spikelets 3-6mm long, 3-8 flowered; glumes 0.8-1.5mm long; lemmas 1.5-2.5mm long. Moist areas. NW, SE.

Glyceria grandis Wats. in Gray, Man. 6:667. 1890. Culms to 15dm high; leaf blades 6-12mm wide; spikelets 4-6mm long, 4-7 flowered; glumes about 1.5-2mm long; lemmas 2-2.5mm long. Wet areas. Statewide.

Glyceria striata (Lam.) Hitchc., Proc. Biol. Soc. Wash. 41:157. 1928. Culms to 1m high; leaf blades usually 2-6mm wide; spikelets 3-4mm long, 3-7 flowered; glumes about 0.5-1.2mm long; lemmas 1.5-2mm long. Moist areas. Statewide. *G. nervata* Trin.

Helictotrichon Besser Spike Oat

Helictotrichon hookeri (Scribn.) Henr., Blumea 3:429. 1940. Culms tufted, to 4dm high; leaf blades 1-3mm wide; panicle narrow; spikelets about 1.5cm long, 3-6 flowered; glumes very thin, slightly shorter than spikelet; lemmas mostly 1-1.2cm long, the awn 1-2cm long. Woods and slopes. Park Co.

Hierochloe R. Br. Sweetgrass

Reference: Weimarck, G. 1971. Bot. Notiser 124:129-175.

Hierochloe odorata (L.) Beauv., Ess. Agrost. 62, 164. 1812. Plants rhizomatous; culms to 6dm high; leaf blades 1-5mm wide; panicle pyramidal; spikelets 4-5mm long, with 1 terminal perfect floret and 2 staminate florets below; glumes 4-5mm long; lemmas about as long as glumes, awnless or nearly so. Moist meadows in the mountains. Statewide.

<u>Hilaria</u> H. B. K.

<u>Hilaria jamesii</u> (Torr.) Benth., Journ. Linn. Bot. Soc. 19:62. 1881. Galleta. Plants rhizomatous; culms to 4dm high; leaf blades 1-4mm wide; spike terminal; spikelets mostly sessile, in groups of 3, 6-8mm long, villous at base, the central one fertile and 1 flowered or occasionally 2 flowered, the lateral ones staminate and 2 flowered or occasionally 3 flowered; glumes awned or awn-tipped. Dry areas. SW, SE.

<u>Hordeum</u> L. Barley

Spikelets usually 1 flowered, mostly 3 at each node, the central one sessile, the lateral ones usually pedicelled; lateral spikelets usually imperfect, sometimes reduced to bristles; spikes dense and bristly; glumes awn-like or awned; lemmas awned or awn-tipped. Perennial plants with the lateral spikelets sessile are intergeneric hybrids and have been called <u>H. montanense</u> Scribn.

1. Plants perennial
   2. Awns 1.8-8cm long .... <u>H. jubatum</u>
   2. Awns 1.5cm long or less .... <u>H. brachyantherum</u>
1. Plants annual

3. Leaf blades with prominent auricles at base; awn of central lemma 3-4cm long ..... H. leporinum

3. Leaf blades lacking auricles at base; awn of central lemma 1.5cm or less long ..... H. pusillum

Hordeum brachyantherum Nevski, Acta Inst. Bot. Acad. Sci. URSS I, 2:61. 1936. Tufted perennial; culms to 1m high; leaf blades 2-8mm wide; floret of central spikelet usually 5-10mm long, the awn 5-10mm long, the glumes awn-like, 5-10mm long; florets of lateral spikelets well-developed and staminate, or much reduced and empty, the awns mostly 1-5mm long, the glumes awn-like, 5-15mm long. Open, mostly moist areas. Statewide. H. nodosum L. of authors.

Hordeum jubatum L., Sp. Pl. 85. 1753. Tufted perennial; culms to 6dm high; leaf blades 1-5mm wide; lateral spikelets mostly reduced to 1-3 awns; glumes of perfect spikelet awn-like, 2-7cm long; lemma 6-8mm long, the awns mostly 1.8-8cm long. Plains and disturbed areas. Statewide. H. caespitosum Scribn.

Hordeum leporinum Link, Linnaea 9:133. 1834. Annual; culms to 5dm high; leaf blades 2-5mm wide; glumes of central spikelet long-ciliate on both margins, the awns 1.5-2.5cm long; lemma 9-12mm long, the awn 3-4cm long; lateral spikelets usually staminate, the glumes much shorter than central ones and dissimilar, the lemma 1-2cm long, the awn 2-4cm long. Disturbed areas. Hot Springs Co.

Hordeum pusillum Nutt., Gen. Pl. 1:87. 1818. Annual; culms to 35cm high; leaf blades 1-3mm wide; 1st glume of lateral spikelets and both glumes of fertile spikelet dilated above base, the awn 5-15mm long; lemma of central spikelet awned, of lateral spikelets awn-tipped. Plains and meadows, especially where alkaline. NW, SE.

Koeleria Pers. Junegrass

Koeleria macrantha (Ledeb.) Schultes, Mant. 2:345. 1824. Tufted perennial to 6dm high; leaf blades 1-3mm wide; panicle spike-like or somewhat loose at anthesis; spikelets mostly 4-6mm long, 2-4 flowered; glumes 3-5mm long, the 2nd wider than the 1st; lemmas 3-6mm long, the lowest usually slightly longer than glumes, acute or short-awned. Open woods, plains, and hills. Statewide. K. cristata (L.) Pers. of authors. See Greuter, 1968, Candollea 23:81-108, concerning nomenclature.

Leersia Sw.

Plants rhizomatous; spikelets 1 flowered, compressed laterally, in a panicle; glumes lacking; lemma boat-shaped, usually 5 nerved.

Reference: Pyrah, G. L. 1969. Iowa St. Journ. Sci. 44:215-270.

Spikelets about 3mm long and 1mm wide — *L. virginica*

Spikelets about 5mm long and 1.5-2mm wide — *L. oryzoides*

Leersia oryzoides (L.) Sw., Nov. Gen. Ind. Occ. 21. 1788. Rice Cutgrass. Culms to 1.5m high; leaf blades mostly 4-10mm wide; spikelets about 5mm long; glumes lacking; lemma hispidulous. Wet areas. Goshen Co. *Homalocenchrus oryzoides* (L.) Poll.

Leersia virginica Willd., Sp. Pl. 1:325. 1797. Culms to 12dm high; leaf blades mostly 5-12mm wide; spikelets about 3mm long; glumes lacking; lemma sparsely hispidulous. Woods and moist areas. Uinta Co.

Leucopoa Griseb. Western Grass

Leucopoa kingii (Wats.) Weber, Univ. Colo. Stud. (Biol.) No. 23:2. 1966. Plants dioecious and short-rhizomatous; culms to 6dm high; leaf blades 2-6mm wide; panicle narrow; spikelets 7-12mm long, 3-5 flowered; 1st glume 3-4mm long; 2nd glume 4-6mm long; lemmas 5-8mm long, acute or acuminate. Open woods and hills. Statewide. *Festuca confinis* Vasey, *Hesperochloa kingii* (Wats.) Rydb.

<u>Lolium</u> L. Ryegrass

Annuals to perennials; spikelets several to many flowered, sessile and turned edgewise to rachis; glume next to rachis lacking except on terminal spikelet; lemmas awned or not.

Glume as long as spikelet or nearly so ... <u>L</u>. <u>persicum</u>

Glume about half as long as spikelet or less ... <u>L</u>. <u>perenne</u>

<u>Lolium</u> <u>perenne</u> L., Sp. Pl. 83. 1753. Short-lived perennial to 8dm high, usually with many basal innovations; leaf blades 1-5mm wide; spikelets 8-25mm long, 3-15 flowered; glumes 6-11mm long; lemmas 4-7mm long, awnless. Disturbed areas. Statewide.

<u>Lolium</u> <u>persicum</u> Boiss. & Hohen ex Boiss., Diagn. Pl. Orient. Nov. I, 2(13):66. 1854. Annual to 6dm high; leaf blades 2-6mm wide; spikelets 10-25mm long, 5-8 flowered; glumes 7-20mm long; lemmas 5-10mm long, with awns 4-14mm long. Disturbed areas. NE.

<u>Melica</u> L. Melicgrass; Oniongrass

Perennials with the base of culm often swollen into a corm; spikelets 2 to several flowered, in a panicle; glumes somewhat unequal, scarious-margined, shorter than lowest floret, 3-7 nerved; lemmas scarious-margined, awned or not.

1. Lemmas awned or long-tapering to a pointed tip
   2. Lemmas awned from a bifid tip ... *M. smithii*
   2. Lemmas acute or acuminate ... *M. subulata*
1. Lemmas mostly obtuse, awnless
   3. Pedicels mostly flexuous or recurved; 1st glume 3.5-5.5mm long ... *M. spectabilis*
   3. Pedicels mostly straight; 1st glume 6-9mm long ... *M. bulbosa*

Melica bulbosa Geyer ex Porter & Coult., Syn. Fl. Colo. 149. 1874. Culms to 6dm high, bulbous at base, usually densely tufted; leaf blades 1-5mm wide; spikelets mostly 7-15mm long; glumes 6-12mm long; lemmas 6-12mm long, somewhat obtuse. Meadows and hills. Statewide.

Melica smithii (Porter ex Gray) Vasey, Bull. Torrey Club 15:294. 1888. Culms to 12dm high; leaf blades 6-12mm wide; spikelets 12-20mm long including awns; glumes mostly 4-6mm long; lemmas 4-10mm long, the awn 2-5mm long. Moist woods. Teton Co.

Melica spectabilis Scribn., Proc. Acad. Phila. 1885:45. 1885. Culms to 1m high, bulbous at base, often short-rhizomatous; leaf blades 1-4mm wide; spikelets 10-15mm long; glumes 3.5-7mm long; lemmas 3-8mm long, obtuse. Meadows, hills, and woods. Statewide.

Melica subulata (Griseb.) Scribn., Proc. Acad. Phila. 1885:47. 1885. Culms to 13dm high, mostly bulbous at base; leaf blades usually 2-5mm wide; spikelets 1-2cm long; glumes 4-8mm long; lemmas 7-12mm long, acute to acuminate. Meadows and shaded areas. Teton Co. *M. geyeri* Munro of authors.

<u>Muhlenbergia</u> Schreb. Muhly

Spikelets 1 flowered or occasionally 2 flowered, in a panicle; glumes usually shorter than lemma but sometimes longer; lemmas usually awned from tip, the awn sometimes minute or rarely lacking.

Reference: Pohl, R. W. 1969. Am. Midl. Nat. 82:512-542.

1. Plants annual, the culms rarely decumbent and rooting at the nodes and appearing perennial; glumes 1mm long or less
  2. Panicle open; pedicels mostly over 3 times as long as spikelets — <u>M</u>. <u>minutissima</u>
  2. Panicle narrow; pedicels mostly less than 3 times as long as spikelets — <u>M</u>. <u>filiformis</u>
1. Plants perennial; glumes often over 1mm long
  3. Plants with creeping rhizomes
    4. Panicle diffuse, the spikelets very remote on long pedicels or panicle branches
      5. Spikelets awned, 2.5-5mm long; leaf blades involute
        6. Lemmas 4-5mm long, the awns about 1mm long — <u>M</u>. <u>pungens</u>
        6. Lemmas about 3mm long, the awns mostly 2-3mm long — <u>M</u>. <u>torreyi</u>
      5. Spikelets awnless, 1-2mm long; leaf blades flat — <u>M</u>. <u>asperifolia</u>
    4. Panicle narrow and condensed, the spikelets crowded on short pedicels

7. Hairs at base of floret about as long as body of lemma *M. andina*

7. Hairs at base of floret not more than half as long as lemma

8. Leaf blades mostly involute or sometimes flat, 2mm wide or less

9. Ligules 0.5-1mm long; culms not nodulose-roughened below the nodes *M. cuspidata*

9. Ligules usually 1-3mm long; culms minutely nodulose-roughened below nodes *M. richardsonis*

8. Leaf blades flat, over 2mm wide

10. Glumes including awns over 4mm long, much exceeding the lemma

11. Culms mostly simple or branching at base; internodes minutely puberulent; sheaths not or scarcely keeled *M. glomerata*

11. Culms mostly branching from the middle nodes; internodes smooth and glossy except at summit; sheaths keeled *M. racemosa*

10. Glumes including awns usually less than 4mm long, not exceeding the lemma or barely so *M. mexicana*

3. Plants densely tufted, without rhizomes

12. Panicle diffuse, the spikelets remote on long panicle branches *M. torreyi*

12. Panicle narrow, the spikelets crowded

13. Second glume 1 nerved, acute or short-awned
M. cuspidata

13. Second glume 3 nerved, usually 3 toothed or 3 awned

14. Awns mostly over 9mm long; culms often over 30cm high *M. montana*

14. Awns mostly 0.5-4mm long; culms 5-30cm high
M. filiculmis

Muhlenbergia andina (Nutt.) Hitchc., USDA Bull. 772:145. 1920. Plants rhizomatous, to 1m high; leaf blades 1-6mm wide; panicle usually spike-like; glumes 3-4mm long; lemma 2.5-3mm long, the awn 4-9mm long, hairs at base about as long as lemma. Moist areas. NW, SW, SE. *M. comata* Thurb. ex Benth.

Muhlenbergia asperifolia (Mey.) Parodi, Univ. Nac. Buenos Aires Rev. Agron. 6:117. 1928. Plants rhizomatous, to 5dm high; leaf blades 1-2mm wide; panicle diffuse; glumes 0.8-2mm long; lemma 1.5-2mm long, minutely mucronate. Moist, especially alkaline areas. Statewide. *Sporobolus asperifolius* (Mey.) Thurb.

Muhlenbergia cuspidata (Torr. ex Hook.) Rydb., Bull. Torrey Club 32:599. 1905. Culms to 5dm high, tufted with bulb-like, scaly bases; leaf blades 1-2mm wide; panicle somewhat spike-like; glumes 1.5-2.5mm long; lemma about 3mm long, acuminate-cuspidate. Plains and foothills. NE, SE. *Sporobolus brevifolius* (Nutt.) Scribn.

Muhlenbergia _filiculmis_ Vasey, Contr. U. S. Nat. Herb. 1:267. 1893. Culms tufted, to 3dm high; leaves in a basal cluster, the blades 1mm or less wide; panicle slender; glumes 1.5-3mm long; lemmas 2.5-4mm long, the awn 0.5-4mm long. Sandy or rocky soil. NE, SE.

Muhlenbergia _filiformis_ (Thurb. ex Wats.) Rydb., Bull. Torrey Club 32:600. 1905. Plants annual or rarely appearing perennial, tufted, to 25cm high; leaf blades less than 1.5mm wide; panicle narrow; glumes less than 1mm long; lemma 1.5-2.5mm long, mucronate. Open woods and meadows. Statewide. _Sporobolus simplex_ Scribn.

Muhlenbergia _glomerata_ (Willd.) Trin., Gram. Unifl. 191. 1824. Plants rhizomatous, to 9dm high; leaf blades mostly 2-5mm wide; panicle narrow; glumes 2-2.5mm long, awned; lemma 2-3mm long, awn-tipped. Moist areas. Goshen and Sublette Cos.

Muhlenbergia _mexicana_ (L.) Trin., Gram. Unifl. 189. 1824. Plants rhizomatous, to 1m high; leaf blades mostly 2-4mm wide; panicle narrow; glumes 2-3.5mm long; lemma 2-3mm long, awn-tipped. Moist areas. NE, SE.

Muhlenbergia _minutissima_ (Steud.) Swallen, Contr. U. S. Nat. Herb. 29:207. 1947. Plants annual, to 35cm high; leaf blades about 1mm wide or less; panicle open; glumes 0.6-1mm long; lemma 1-1.5mm long, acute to obtuse. Moist areas. SE, SW.

Muhlenbergia montana (Nutt.) Hitchc., USDA Bull. 772:145, 147. 1920. Culms tufted, to 6dm high; leaf blades 1-2mm wide; panicle narrow; 1st glume acute, 1.5-3mm long; 2nd glume longer and broader, 3 toothed or 3 awned; lemma 3-4mm long, the awns mostly 1-1.5cm long. Dry hills. Carbon and Laramie Cos. *M. gracilis* Trin.

Muhlenbergia pungens Thurb. in Gray, Proc. Acad. Phila. 1863:78. 1863. Plants rhizomatous, to 4dm high; leaf blades 0.3-1mm wide, sharp-pointed; panicle open; glumes 1.5-2.5mm long; lemma 4-5mm long, the awn about 1mm long. Plains and foothills. Albany and Goshen Cos.

Muhlenbergia racemosa (Michx.) B. S. P., Prel. Cat. N. Y. Pl. 67. 1888. Plants rhizomatous, to 1m high; leaf blades mostly 2-7mm wide; panicle narrow; glumes mostly 4-5mm long including awns; lemma 2.5-3.5mm long, acuminate to short-awned. Moist or dry areas. NW, NE, SE.

Muhlenbergia richardsonis (Trin.) Rydb., Bull. Torrey Club 32:600. 1905. Plants rhizomatous, to 6dm high; leaf blades 0.4-1.2mm wide; panicle narrow; glumes 1-1.5mm long; lemma 2-3mm long, mucronate. Moist or dry areas. Statewide. *M. squarrosa* (Trin.) Rydb.

Muhlenbergia torreyi (Kunth) Hitchc. ex Bush, Am. Midl. Nat. 6:84. 1919. Plants tufted and/or rhizomatous, sometimes forming "fairy rings," to 3dm high; leaves in a basal cluster, the blades 0.2-0.8mm wide, sharp-pointed; panicle open; glumes 1.5-2mm long; lemma about 3mm long, the awns mostly 2-3mm long. Plains and hills. Reported from SE Wyo. *M. gracillima* Torr.

Munroa Torr. False Buffalo Grass

Munroa squarrosa (Nutt.) Torr., Rep. Bot. Pac. R. R. Exp. 4(4):158. 1857. Plants annual but mat forming by branching and spreading, to 1dm high; leaf blades 1-3mm wide, stiff, fascicled; spikelets in fascicles of 2 or 3, 2-5 flowered, the fascicle about 7mm long and enclosed in leaf sheaths; glumes of lower spikelets equal, those of upper unequal or the 1st glume sometimes obsolete; lemmas with a tuft of hair on each margin at about the middle, awn-tipped. Plains, canyons, and hills. NW, NE, SE.

Oryzopsis Michx. Ricegrass

Plants perennial; spikelets 1 flowered, in a panicle; glumes subequal; lemma about as long as glumes or shorter, usually awned, the awn deciduous.

1. Lemmas glabrous or rarely puberulent; leaf blades somewhat involute, less than 2mm wide; panicle branches spreading or reflexed at maturity; awn 5-10mm long *O. micrantha*
1. Lemmas pubescent; leaf blades, panicle branches, and awns various
  2. Pubescence on lemma long and silky
    3. Panicle diffuse, the pedicels widely divergent from the panicle branches; awns mostly 3-8mm long *O. hymenoides*
    3. Panicle mostly contracted, the pedicels parallel to panicle branches or nearly so; awns mostly 6-12mm long *O. contracta*
  2. Pubescence on lemma short and appressed
    4. Spikelets, excluding awn, 6-9mm long; leaf blades flat *O. asperifolia*
    4. Spikelets, excluding awn, 5.5mm long or less; leaf blades involute or subinvolute
      5. Awns mostly 1-2mm long, straight *O. pungens*
      5. Awns about 5mm long or more, bent
        6. Awns about 5mm long, strongly once geniculate *O. exigua*
        6. Awns 7-20mm long, weakly twice geniculate *O. canadensis*

Oryzopsis asperifolia Michx., Fl. Bor. Am. 1:51. 1803. Culms tufted, the fertile spreading or prostrate, to 7dm long; leaf blades 2-8mm wide; glumes 6-8mm long; lemma 6-8mm long, sparsely pubescent, the awn 5-10mm long. Woods and slopes. NW, NE, SE.

Oryzopsis canadensis (Poir.) Torr., Fl. N. Y. 2:433. 1843. Culms tufted, to 7dm high; leaf blades 0.3-0.8mm wide; glumes 3-5mm long; lemma about 3mm long, appressed-pilose, the awn 7-20mm long. Woods and thickets. Reported from Wyo. *O. juncea* (Michx.) B. S. P.

Oryzopsis contracta (Johnson) Shechter in Shechter & Johnson, Brittonia 18:342. 1966. Culms tufted, to 7dm high; leaf blades 0.5-3mm wide; glumes 5-7mm long; lemma about 3mm long, long-pilose, the awn 6-12mm long. Dry plains and hills. SE, SW. An apparent intergeneric hybrid which has been called *O. bloomeri* (Boland.) Ricker will key here. It has glumes 7-10mm long and lemmas about 5mm long.

Oryzopsis exigua Thurb. in Torr., Bot. Wilkes Exp. 17(2):481. 1874. Culms tufted, to 3dm high; leaf blades 0.5-1.5mm wide; glumes 4-5mm long; lemma 4-5.5mm long, appressed-pilose, the awn about 5mm long, bent. Dry woods and slopes. NW, SW, SE.

Oryzopsis hymenoides (Roem. & Schult.) Ricker ex Piper, Contr. U. S. Nat. Herb. 11:109. 1906. Indian Ricegrass. Culms tufted, to 6dm high; leaf blades 0.5-2mm wide; glumes 5-7mm long; lemma about 3mm long, densely long-pilose, the awn 3-8mm long. Deserts and plains. Statewide. *Eriocoma cuspidata* Nutt., *E. caduca* (Scribn.) Rydb.

Oryzopsis micrantha (Trin. & Rupr.) Thurb. in Gray, Proc. Acad. Phila. 1863:78. 1863. Culms tufted, to 7dm high; leaf blades 0.5-2mm wide; glumes 2.5-3.5mm long; lemma 2-2.5mm long, glabrous or rarely puberulent, the awn 5-10mm long. Dry woods and rocky slopes. SE, NE.

Oryzopsis pungens (Torr. ex Spreng.) Hitchc., Contr. U. S. Nat. Herb. 12:151. 1908. Culms tufted, to 5dm high; leaf blades less than 2mm wide; glumes 3-4mm long; lemma about 3mm long, densely pubescent, the awn usually 1-2mm long, straight. Sandy or rocky areas. Crook Co.

Panicum L.

Spikelets compressed dorsiventrally, mostly in panicles; glumes usually very unequal, the 1st often minute; sterile lemma simulating a 3rd glume, sometimes with a staminate flower in its axil; fertile lemma nerveless or the nerves obscure, the margins inrolled over an enclosed palea.

Reference: Spellenberg, R. 1975. Madroño 23:134-153.

1. Plants annual

2. Spikelets 3.5-5mm long; 1st glume 2-3.5mm long P. miliaceum

2. Spikelets 2.5-3.5mm long; 1st glume 1-1.5mm long P. capillare

1. Plants perennial

3. Rhizomes present; spikelets 3.5mm or more long P. virgatum

3. Rhizomes lacking; spikelets mostly less than 3.5mm long

4. Spikelets 3-3.5mm long — P. oligosanthes

4. Spikelets 1-2mm long — P. lanuginosum

Panicum capillare L., Sp. Pl. 58. 1753. Plants annual, to 8dm high; leaf blades mostly 5-15mm wide; 1st glume 1-1.5mm long; 2nd glume 2.5-3.5mm long; fertile lemma 1.5-2.5mm long. Disturbed areas. NW, NE, SE.

Panicum lanuginosum Ell., Sk. Bot. S. C. & Ga. 1:123. 1816. Plants perennial, tufted, to 6dm high; leaf blades 3-12mm wide; 1st glume 0.2-0.5mm long; 2nd glume about 1.5-2mm long, pubescent; fertile lemma about 1.5-2mm long. Open places, often near hot springs and geysers. Yellowstone Park and Teton Co. P. huachucae Ashe, P. thermale Boland., P. ferventicola Schmoll, P. occidentale Scribn., P. pacificum Hitchc. & Chase, Dichanthelium lanuginosum (Ell.) Gould.

Panicum miliaceum L., Sp. Pl. 58. 1753. Annual to 1m high; leaf blades 7-17mm wide; 1st glume 2-3.5mm long; 2nd glume 3.5-5mm long; fertile lemma 3-4mm long. Disturbed areas. Albany Co.

Panicum oligosanthes Schult., Mant. 2:256. 1824. Plants perennial, tufted, to 5dm high; leaf blades mostly 6-12mm wide; spikelets sparsely pubescent to glabrous; 1st glume 1-1.5mm long; 2nd glume about 3-3.3mm long, the fertile lemma about as long. Dry areas. NE, SE. P. scribnerianum Nash, Dichanthelium oligosanthes (Schult.) Gould.

Panicum virgatum L., Sp. Pl. 59. 1753. Plants rhizomatous, to 2m high; leaf blades mostly 3-15mm wide; 1st glume 2-4mm long; 2nd glume 3.5-5mm long; fertile lemma 3-4.5mm long. Plains and bottomlands. SE.

Phalaris L. Canary Grass

Spikelets laterally compressed, with 1 terminal perfect floret and 2 much smaller, scale-like or linear, sterile lemmas below, in a narrow or spike-like panicle; glumes equal, boat-shaped; fertile lemma shorter than glumes.

Plants annual; panicle 2-4cm long *P. canariensis*
Plants perennial; panicle 6cm or more long *P. arundinacea*

Phalaris arundinacea L., Sp. Pl. 55. 1753. Plants rhizomatous, to 15dm high; leaf blades 5-20mm wide; glumes about 5mm long; fertile lemma about 4mm long; sterile lemmas mostly linear, 1-2mm long, long-hairy. Wet areas. Statewide.

Phalaris canariensis L., Sp. Pl. 54. 1753. Plants annual, to 6dm high; leaf blades 2-10mm wide; glumes 7-9mm long; fertile lemma 5-6mm long; sterile lemmas usually about half as long as fertile. Moist disturbed areas. NW, NE, SE.

<u>Phippsia</u> (Trin.) R. Br. Icegrass

<u>Phippsia</u> <u>algida</u> (Phipps) R. Br., Chlor. Melv. cclxxxv. 1823. Culms densely tufted, 2-10cm high; leaf blades 1-2.5mm wide; spikelets 1 flowered, in a panicle; glumes unequal, minute or the 1st sometimes lacking; lemma about 1.5mm long. Alpine. Park Co.

<u>Phleum</u> L. Timothy

Spikelets 1 flowered, strongly laterally compressed, in a spike-like panicle; glumes equal, awned, ciliate on keel; lemma shorter than glumes, awnless.

Panicle long-cylindric, usually over 5 times as long as wide; culms usually bulbous at base ..... <u>P</u>. <u>pratense</u>

Panicle ovoid or oblong, usually not over 4 times as long as wide; culms not bulbous at base ..... <u>P</u>. <u>alpinum</u>

<u>Phleum</u> <u>alpinum</u> L., Sp. Pl. 59. 1753. Culms tufted, sometimes with short rhizomes, to 5dm high; leaf blades 2-7mm wide; panicle ovoid or oblong; glumes about 5mm long including awns, the awns about 2mm long; lemma about 2mm long. Mountain meadows and slopes. Statewide.

<u>Phleum</u> <u>pratense</u> L., Sp. Pl. 59. 1753. Culms tufted, usually bulbous at base, to 1m high; leaf blades mostly 2-8mm wide; panicle long-cylindric; glumes about 2.5-3mm long, the awns about 1mm long; lemma about 2mm long. Disturbed areas and meadows. Statewide.

<u>Phragmites</u> Trin. Common Reed

<u>Phragmites</u> <u>australis</u> (Cav.) Trin. ex Steud., Nom. Bot. II, 2:324. 1841. Plants rhizomatous, to 4m high; leaf blades mostly 1-4cm wide; panicle plume-like at maturity; spikelets 12-15mm long, several flowered; florets exceeded by hairs of rachilla; 1st glume 4-7mm long; 2nd glume 6-10mm long; 1st lemmas mostly 9-14mm long, the others shorter, long-acuminate or awned. Wet areas, often in water. Statewide. <u>P</u>. <u>communis</u> Trin. (See Clayton, Taxon 17:168, 1968, concerning nomenclature).

<u>Poa</u> L. Bluegrass

Spikelets 2 to several flowered, the uppermost floret usually reduced or rudimentary, in a panicle; glumes somewhat unequal, the 1st usually 1 nerved, the 2nd usually 3 nerved; lemmas awnless, mostly 5 nerved. The key presented below will not work for all specimens, as is the case with most keys for this very difficult genus. The user is urged to also try other keys covering Wyoming.

1. Plants annual but often densely clustered, lacking remains of old culms, mostly 25cm or less high *P. annua*

1. Plants perennial, usually with remains of old culms, often over 25cm high

  2. Creeping rhizomes present, the culms often densely tufted also GROUP I

  2. Creeping rhizomes lacking (culms sometimes decumbent and rooting)

    3. Florets usually converted into bulblets with a dark purple base; culms bulbous at base *P. bulbosa*

    3. Florets normal; culms not bulbous at base

      4. Spikelets little compressed, the lemmas convex on back, the keels lacking or mostly obscure GROUP II

      4. Spikelets distinctly compressed, the glumes and lemmas usually strongly keeled

        5. Lemmas with tangled cobwebby hairs at base (sometimes scant or obscure in *P. interior*) GROUP III

        5. Lemmas not cobwebby at base GROUP IV

GROUP I

1. Culms strongly flattened, 2 edged *P. compressa*

1. Culms terete or slightly flattened, not 2 edged

  2. Plants often dioecious, mostly pistillate; lower sheaths minutely retrorsely pubescent and usually purplish *P. nervosa*

  2. Plants usually perfect flowered; lower sheaths not retrorsely pubescent, usually green

3. Lemmas with tangled cobwebby hairs at base
    4. Lemmas glabrous or scabrous on internerves, 3-4mm long; mostly middle and lower elevations *P. pratensis*
    4. Lemmas pubescent on internerves at least on lower half, 4-5mm long; mostly alpine and subalpine, rarely lower *P. grayana*
3. Lemmas lacking cobwebby hairs at base
    5. Plants alpine or subalpine *P. grayana*
    5. Plants below subalpine
        6. Panicle contracted, the branches ascending or appressed
            7. First glume 2.5-3.5mm long, 1 nerved; anthers mostly about 1.5mm long *P. arida*
            7. First glume 3.5-5mm long, 3-5 nerved; anthers mostly 1.8-2.3mm long *P. glaucifolia*
        6. Panicle open, the branches spreading or reflexed
            8. Lemmas puberulent to glabrous; anthers 2.5-3mm long; panicle branches mostly nodding *P. curta*
            8. Lemmas villous at least on lower nerves; anthers 1.4-2.5mm long; panicle branches spreading to ascending
                9. Plants usually glaucous; lemmas 3-4mm long *P. glaucifolia*
                9. Plants not glaucous; lemmas 4-5mm long *P. grayana*

GROUP II

1. Lemmas crisp-puberulent on the back especially on nerves (sometimes obscure or only at very base)
  2. Panicle open, the lower branches naked at base, mostly spreading ... *P. gracillima*
  2. Panicle contracted, the branches appressed or somewhat spreading at anthesis, the lower branches often with spikelets near base
    3. Culms slender, averaging under 30cm high, with numerous short innovations at base; basal leaf blades mostly less than 1.5mm wide and 5cm long; spikelets often purplish tinged
      4. Plants usually of deserts or dry foothills, flowering in spring, often strongly purplish tinged all over ... *P. sandbergii*
      4. Plants montane to alpine, often in open rocky areas, flowering mostly in summer, rarely purplish tinged all over ... *P. incurva*
    3. Culms stout, averaging over 40cm high; innovations usually not numerous; basal leaf blades mostly 1-3mm wide and over 5cm long; spikelets sometimes purple banded but usually not purplish tinged
      5. Ligules of upper leaves mostly 3-7mm long ... *P. scabrella*
      5. Ligules of upper leaves 1-3mm long ... *P. juncifolia*
1. Lemmas glabrous or minutely scabrous, not crisp-puberulent

6. Sheaths scaberulous; ligules on upper leaves about 4mm long, decurrent P. nevadensis

6. Sheaths glabrous; ligules shorter, usually not decurrent P. juncifolia

GROUP III

1. Lemmas glabrous or only the keel pubescent (some of the cobwebby hairs may arise from base of marginal nerves); spikelets mostly 3mm or less long P. trivialis

1. Lemmas pubescent on keel and marginal nerves; spikelets often over 3mm long

2. Lower panicle branches distinctly reflexed at maturity, mostly capillary with the spikelets borne near the ends; anthers less than 1mm long P. reflexa

2. Lower panicle branches not reflexed, capillary or not, sometimes with spikelets borne near the base; anther length variable

3. Lower panicle branches in pairs, elongate, capillary, with a few spikelets near the ends; anthers less than 1mm long

4. Sheaths retrorsely scabrous to glabrous; spikelets usually green; plants not alpine P. tracyi

4. Sheaths glabrous or scabridulous; spikelets often purplish; plants often alpine or subalpine P. leptocoma

3. Lower panicle branches often more than 2, or if not, not capillary and elongate; anthers often over 1mm long

5. Culms 2-5dm high, densely tufted; ligule 0.5-1.5mm long; panicle 5-15cm long

6. Second glume usually 2-3mm long; mostly below subalpine — *P. interior*

6. Second glume usually 3-4.5mm long; mostly alpine and subalpine — *P. pattersonii*

5. Culms 3-12dm high, loosely tufted; ligule 1.5-5mm long; panicle 12-30cm long

7. Ligules 1.5-3mm long; lemmas 3-5mm long — *P. tracyi*

7. Ligules 3-5mm long; lemmas 1.5-3mm long — *P. palustris*

GROUP IV

1. Lemmas conspicuously pubescent on keel or marginal nerves or both, sometimes also on the internerves but these hairs shorter

2. Leaf blades folded or involute, firm, rather stiff; usually lower than subalpine — *P. fendleriana*

2. Leaf blades flat, or if involute, rather lax or soft; usually alpine or subalpine

3. Panicle usually about as broad as long, pyramidal; leaf blades mostly over 1.5mm wide — *P. alpina*

3. Panicle longer than broad, oblong; leaf blades mostly 1.5mm or less wide

4. Second glume 3-4.5mm long, extending to about tip of 1st lemma — *P. pattersonii*

4. Second glume 2.5-3.5mm long, exceeded by 1st lemma — *P. rupicola*

1. Lemmas glabrous or sometimes uniformly puberulent
   5. Lemmas 2-3mm long; plants alpine, mostly less than 1dm high *P. lettermanii*
   5. Lemmas 4-6mm long; plants alpine or not, often over 1dm high *P. cusickii*

Poa alpina L., Sp. Pl. 67. 1753. Culms tufted, to 3dm high; leaf blades mostly 1.5-4mm wide; glumes 2.5-4mm long; lemmas 3-4mm long, villous on keel and nerves, pubescent on internerves below. Mountain meadows or talus. Statewide.

Poa annua L., Sp. Pl. 68. 1753. Plants annual, mostly to 25cm high, forming mats; leaf blades mostly 1-3mm wide; glumes 1.5-3mm long; lemmas 2-4mm long, somewhat pubescent on lower half of nerves. Disturbed areas, open woods, and meadows. NW, SW, SE.

Poa arida Vasey, Contr. U. S. Nat. Herb. 1:270. 1893. Plants rhizomatous, to 6dm high; leaf blades usually 0.5-3mm wide; glumes 2.5-4mm long; lemmas 3-4mm long, pubescent. Plains and meadows. Statewide. *P. sheldonii* Vasey.

Poa bulbosa L., Sp. Pl. 70. 1753. Culms tufted, to 6dm high; leaf blades 0.5-3mm wide; florets converted into bulblets with a purple base about 2-3mm long, the bracts extending into slender green tips 4-15mm long; bulblets usually subtended by several empty lemmas. Disturbed areas and meadows. NW, NE, SE.

Poa compressa L., Sp. Pl. 69. 1753. Plants rhizomatous, to 7dm high; culms strongly flattened; leaf blades 1-4mm wide; glumes 2-3mm long; lemmas 2-3mm long, not webbed at base or barely so, keel and marginal nerves usually slightly pubescent toward base. Disturbed areas and meadows. NE, SE.

Poa curta Rydb., Bull. Torrey Club 36:534. 1909. Plants rhizomatous, to 8dm high; leaf blades 2-6mm wide; glumes 2.5-4.5mm long; lemmas 4-5.5mm long, without a web, usually slightly scaberulous, sometimes puberulent at base. Moist, shaded areas. Teton Co.

Poa cusickii Vasey, Contr. U. S. Nat. Herb. 1:271. 1893. Culms tufted, to 6dm high; leaf blades 0.5-3mm wide; glumes 3-5mm long; lemmas 4-6mm long, glabrous or puberulent. Plains, meadows, and slopes. Statewide. *P. subaristata* Scribn., *P. epilis* Scribn., *P. subpurpurea* Rydb.

Poa fendleriana (Steud.) Vasey, USDA Div. Bot. Bull. 13(2):No. 74. 1893. Plants incompletely dioecious; culms tufted, to 5dm high; leaf blades 0.5-3mm wide; glumes 3-5mm long; lemmas 4-5mm long, villous on lower keel and marginal nerves, not webbed. Hills and meadows. Statewide. *P. longiligula* Scribn. & Wms., *P. longipedunculata* Scribn.

Poa glaucifolia Scribn. & Williams ex Williams, USDA Div. Agrost. Cir. 10:6. 1899. Plants rhizomatous, usually glaucous, to 1m high; leaf blades 1-3mm wide; glumes 3-5mm long; lemmas 3-4mm long, usually villous on lower half of keel and marginal nerves, sometimes also on internerves. Moist areas. Albany and Washakie Cos.

Poa gracillima Vasey, Contr. U. S. Nat. Herb. 1:272. 1893. Culms loosely tufted, to 6dm high, usually decumbent at base; leaf blades to 1.5mm wide; glumes 3-5mm long; lemmas 3-5mm long, minutely scabrous, pubescent near base especially on nerves. Rocky areas. Reported from NW Wyo.

Poa grayana Vasey, Contr. U. S. Nat. Herb. 1:272. 1893. Plants rhizomatous, to 6dm high; leaf blades mostly 1-3mm wide; glumes 3-5mm long; lemmas 4-5mm long, villous on keel and marginal nerves, pubescent on lower internerves, often webbed at base. Mountain meadows mostly above timberline. Statewide. P. longipila Nash, P. arctica R. Br. of authors, P. laxa Haenke.

Poa incurva Scribn. & Williams ex Scribn., USDA Div. Agrost. Cir. 9:6. 1899. Culms tufted, to 6dm high; leaf blades 0.5-2mm wide; glumes 2-4mm long; lemmas 3-4mm long, minutely scabrous, pubescent near base especially on nerves. Alpine slopes to open woods. NW, SW.

Poa interior Rydb., Bull. Torrey Club 32:604. 1905. Culms tufted, to 5dm high; leaf blades 0.5-2mm wide; glumes 2-3mm long; lemmas 2-3.5mm long, villous on lower keel and marginal nerves, usually webbed at base. Woods, slopes, and meadows. Statewide.

Poa juncifolia Scribn., USDA Agrost. Bull. 11:52. 1898. Culms tufted, to 12dm high; leaf blades 1-3mm wide; glumes 3-5mm long; lemmas 3-6mm long, usually glabrous or nearly so, keel lacking at least below. Plains, woods, and meadows. Statewide. P. ampla Merr., P. confusa Rydb.

Poa leptocoma Trin., Mem. Acad. St. Petersb. VI, 1:374. 1830. Culms solitary or few to many in a tuft, to 5dm high; leaf blades mostly 1-4mm wide; glumes 1.5-4mm long; lemmas 1.5-4mm long, usually pubescent on keel and marginal nerves, webbed at base. Bogs and thickets. NW, SE. P. paucispicula Scribn. & Merr.

Poa lettermanii Vasey, Contr. U. S. Nat. Herb. 1:273. 1893. Culms tufted, mostly less than 1dm high; leaf blades mostly 1.5mm or less wide; glumes 2-3mm long; lemmas 2-3mm long, glabrous. Alpine. Park Co.

Poa nervosa (Hook.) Vasey, USDA Div. Bot. Bull. 13(2):No. 81. 1893. Plants rhizomatous, to 6dm high; leaf blades mostly 1-4mm wide; glumes 3-4mm long; lemmas 3-5mm long, glabrous or pubescent on lower part of nerves. Open woods and slopes. Statewide. P. wheeleri Vasey.

Poa _nevadensis_ Vasey ex Scribn., Bull. Torrey Club 10:66. 1883. Culms tufted, to 1m high; leaf blades 1-4mm wide; glumes 3-5.5mm long; lemmas 3.5-5mm long, glabrous or scabrous, keel obscure. Moist areas. NW, SW, SE.

Poa _palustris_ L., Syst. Nat. 10th ed. 874. 1759. Culms loosely tufted, to 15dm high; leaf blades 1-4mm wide; glumes 2-3mm long; lemmas 1.5-3mm long, villous on keel and marginal nerves, webbed at base. Moist areas. Statewide.

Poa _pattersonii_ Vasey, Contr. U. S. Nat. Herb. 1:275. 1893. Culms loosely tufted, to 3dm high; leaf blades about 1mm wide; glumes 2.5-4.5mm long; lemmas 3-4.5mm long, strongly pubescent on keel and marginal nerves, less so on internerves, sparsely webbed at base or not webbed. Alpine or subalpine. Statewide.

Poa _pratensis_ L., Sp. Pl. 67. 1753. Kentucky Bluegrass. Plants rhizomatous, to 1m high; leaf blades mostly 1-4mm wide; glumes 2-4mm long; lemmas 3-4mm long, pubescent on lower part of keel and marginal nerves, webbed at base. Disturbed and other open areas. Statewide.

Poa _reflexa_ Vasey & Scribn. ex Vasey, Contr. U. S. Nat. Herb. 1:276. 1893. Culms solitary or in small tufts, to 5dm high; leaf blades 1-4mm wide; glumes 2-3.5mm long; lemmas 2-3.5mm long, villous on keel and marginal nerves, webbed at base. Moist areas in the mountains. NW, SW, SE.

Poa rupicola Nash ex Rydb., Mem. N. Y. Bot. Gard. 1:49. 1900. Culms tufted, to 2dm high; leaf blades 0.5-1.5mm wide; glumes 2-3.5mm long; lemmas 2-3.5mm long, villous below on keel and marginal nerves and sometimes on internerves, not webbed. Rocky slopes in the high mountains. NW, SE.

Poa sandbergii Vasey, Contr. U. S. Nat. Herb. 1:276. 1893. Culms tufted, to 6dm high; leaf blades 0.5-1.5mm wide; glumes 2-4mm long; lemmas 3-4mm long, pubescent near base, not webbed. Plains and hills. Statewide. *P. buckleyana* Nash, *P. secunda* Presl.

Poa scabrella (Thurb.) Benth. ex Vasey, Grasses U. S. 42. 1883. Culms tufted, to 12dm high; leaf blades 1-3mm wide; glumes 3-5mm long; lemmas 4-5mm long, finely pubescent on back, not webbed. Plains, hills, and mountain slopes. Statewide. *P. canbyi* (Scribn.) Piper, *P. laevigata* Scribn., *P. lucida* Vasey.

Poa tracyi Vasey, Bull. Torrey Club 15:49. 1888. Culms loosely tufted, to 15dm high; leaf blades mostly 2-6mm wide; glumes 2.5-3.5mm long; lemmas 3-4.5mm long, villous on keel and marginal nerves, webbed at base. Moist areas. NW, NE, SE. *P. occidentalis* Vasey, *P. platyphylla* Nash & Rydb.

Poa trivialis L., Sp. Pl. 67. 1753. Culms tufted, often decumbent, to 1m high; leaf blades 1-4mm wide; glumes 1.5-3mm long; lemmas 2-3mm long, glabrous except slightly pubescent on keel, webbed at base. Moist areas. Sublette Co.

Polypogon Desf.

Polypogon monspeliensis (L.) Desf., Fl. Atlant. 1:67. 1798. Rabbitfoot Grass. Plants annual, mostly to 5dm high; leaf blades mostly 2-6mm wide; panicle spike-like; spikelets 1 flowered; glumes equal, about 2mm long, the awns 5-9mm long; lemma about 1mm long, the awn slightly longer than lemma, deciduous. Disturbed areas. Statewide.

Puccinellia Parl. Alkali Grass; False Manna

Spikelets several flowered, in a panicle; glumes unequal, shorter than 1st lemma, 1 or 3 nerved; lemmas 5-7 nerved, usually scarious at tip.

1. Nerves of lemmas prominent, usually raised from surface; ligules 3mm or more long; plants mostly of fresh water areas
   2. Leaf blades 1-3mm wide; culms weak and decumbent P. fernaldii
   2. Leaf blades, or some of them, over 4mm wide; culms stout and erect to ascending P. pauciflora
1. Nerves of lemmas mostly obscure; ligules rarely over 3mm long; plants mostly of alkaline areas

3. Lower panicle branches usually reflexed at maturity; lemmas mostly 2mm or less long; anthers 0.5-0.8mm long ... P. distans

3. Lower panicle branches erect to spreading at maturity; lemmas mostly 2mm or more long; anthers 0.7-1.8mm long

4. Second glumes mostly 2-2.8mm long; anthers mostly 1.5mm or more long ... P. cusickii

4. Second glumes mostly 1.5-2.2mm long; anthers mostly 1mm long or less ... P. nuttalliana

Puccinellia cusickii Weatherby, Rhodora 18:182. 1916. Culms tufted, to 8dm high; leaf blades mostly 1-2mm wide; 2nd glume 2-2.8mm long; lemmas 2.5-3mm long; anthers mostly 1.5-1.8mm long. Alkaline areas. Reported from SE Wyo.

Puccinellia distans (L.) Parl., Fl. Ital. 1:367. 1850. Culms tufted, to 4dm high; leaf blades mostly 1-3mm wide; 2nd glume 1.5-2mm long; lemmas 1.5-2mm long; anthers 0.5-0.8mm long. Moist or alkaline areas. NW, SW, SE.

Puccinellia fernaldii (Hitchc.) Voss, Rhodora 68:445. 1966. Culms decumbent and rooting at base, to 4dm high; leaf blades 1-3mm wide; 2nd glume 1.5-2.5mm long; lemmas 2-3mm long; anthers 0.2-0.5mm long. Moist areas or in water. Teton Park. Torreyochloa fernaldii (Hitchc.) Church, Glyceria fernaldii (Hitchc.) St. John.

<u>Puccinellia</u> <u>nuttalliana</u> (Schult.) Hitchc. in Jeps., Fl. Calif. 3:162. 1912. Culms tufted, mostly to 6dm high; leaf blades 1-4mm wide; glumes 1.5-2.5mm long; lemmas 2-3mm long; anthers about 0.7mm long. Moist, usually alkaline areas. Statewide. <u>P</u>. <u>lucida</u> Fern. & Weatherby for Wyo. material only, <u>P</u>. <u>airoides</u> (Nutt.) Wats. & Coult.

<u>Puccinellia</u> <u>pauciflora</u> (Presl) Munz, Aliso 4:87. 1958. Culms erect to ascending, to 12dm high; leaf blades mostly 3-15mm wide; 2nd glume about 1.5mm long; lemmas 1.5-2.5mm long; anthers 0.5-0.7mm long. Moist areas or in water. NW, NE, SE. <u>Torreyochloa</u> <u>pauciflora</u> (Presl) Church, <u>Glyceria</u> <u>pauciflora</u> Presl.

<u>Redfieldia</u> Vasey Blowout Grass

<u>Redfieldia</u> <u>flexuosa</u> (Thurb.) Vasey, Bull. Torrey Club 14:133. 1887. Plants rhizomatous, to 1m high; leaf blades 0.5-3mm wide; spikelets mostly 2-4 flowered, 5-7mm long, in an open panicle; glumes slightly unequal, 2-3.5mm long; lemmas 4-5mm long, acute to mucronate. Sandy areas. Carbon and Natrona Cos.

Schedonnardus Steud. Tumblegrass

Schedonnardus paniculatus (Nutt.) Trel. in Branner & Cov., Rep. Geol. Surv. Ark. 1888(4):236. 1891. Culms tufted, to 4dm high; leaf blades about 1mm wide; spikelets 1 flowered, sessile and somewhat distant in 2 rows on 1 side of a 3 angled rachis, appressed to rachis; spikes divergent along a common axis; glumes somewhat unequal, 2-4mm long; lemma 3-4mm long, acuminate. Dry, mostly open areas. NW, NE, SE.

Schizachne Hack. False Melic

Schizachne purpurascens (Torr.) Swallen, Journ. Wash. Acad. Sci. 18:204. 1928. Culms loosely tufted, to 1m high; leaf blades 1-5mm wide; spikelets several flowered, in an open panicle; glumes unequal, 5-9mm long; lemmas about 1cm long, the awn as long or longer. Open woods. NW, NE, SE. Avena striata Michx.

Scolochloa Link Prickle Grass

Scolochloa festucacea (Willd.) Link, Hort. Berol. Desc. 1:137. 1827. Plants rhizomatous, to 1.5m high; leaf blades mostly 4-10mm wide; spikelets 3-5 flowered, about 9mm long, in a panicle; glumes slightly unequal, 7-10mm long; lemmas 4-8mm long, erose at tip. Shallow water and moist areas. Yellowstone Park.

__Setaria__ Beauv. Bristlegrass

Plants annual; spikelets dorsally flattened, subtended by 1 to several bristles, awnless, in a spike-like panicle; 1st glume usually about half the spikelet length or less; 2nd glume longer than 1st; sterile lemma present below the fertile; fertile lemma usually transversely rugose.

Bristles subtending spikelets retrorsely scabrous __S. verticillata__
Bristles subtending spikelets antrorsely scabrous
  Bristles below each spikelet 5 or more __S. glauca__
  Bristles below each spikelet 1-3, or 4-6 by abortion of spikelet __S. viridis__

__Setaria glauca__ (L.) Beauv., Ess. Agrost. 51, 178. 1812. Culms to 1m high; leaf blades 3-10mm wide; spikelets about 3mm long; bristles 5-20 below each spikelet. Disturbed areas. Big Horn Co. __S. lutescens__ (Weigel) Hubb. of authors.

__Setaria verticillata__ (L.) Beauv., Ess. Agrost. 51, 171, 178. 1812. Culms to 1m high; leaf blades 3-10mm wide; spikelets 2-2.5mm long; bristles 1(2) below each spikelet. Disturbed areas. Hot Springs Co.

<u>Setaria</u> <u>viridis</u> (L.) Beauv., Ess. Agrost. 51, 178. 1812. Culms mostly to 5dm high; leaf blades 2-12mm wide; spikelets 2-2.5mm long; bristles 1-3 below each spikelet. Disturbed areas. Statewide. <u>Chaetochloa</u> <u>viridis</u> (L.) Scribn.

<u>Sitanion</u> Raf. Squirreltail

<u>Sitanion</u> <u>hystrix</u> (Nutt.) Smith, USDA Agrost. Bull. 18:15. 1899. Culms tufted, to 5dm high; leaf blades 1-5mm wide; spikelets mostly 2 to several flowered, the uppermost floret usually reduced, usually 2 spikelets per node; spike bristly from 1 or more awns on each glume and lemma, the rachis disarticulating when mature; glumes 3-10cm long including awns; lemmas mostly 3-10cm long including awns. Plains, woods, slopes, and disturbed areas. Statewide. <u>S</u>. <u>longifolium</u> Smith, <u>S</u>. <u>brevifolium</u> Smith, <u>S</u>. <u>montanum</u> Smith. An intergeneric hybrid which has been called <u>S</u>. <u>hansenii</u> (Scribn.) Smith will key here. It differs from this species in that the glumes are 2-4 nerved and lanceolate as opposed to 1 or 2 nerved and bristle-like.

Sorghastrum Nash Indian Grass

Sorghastrum nutans (L.) Nash in Small, Fl. S. E. U. S. 66, 1326. 1903. Plants rhizomatous, to 2.5m high; leaf blades mostly 3-10mm wide; spikelets in pairs, 1 sessile and perfect, 6-8mm long excluding awn, the other reduced to a hairy pedicel, in a panicle; lemma with a twisted awn to 2cm long. Moist plains. Goshen Co.

Spartina Schreb. Cordgrass

Plants rhizomatous; spikelets 1 flowered, sessile and usually closely imbricate on 1 side of rachis; spikes racemose on main axis; glumes keeled, unequal, awned or not; lemma keeled, awnless.

Reference: Mobberley, D. G. 1956. Iowa St. Coll. Journ. Sci. 30:471-574.

Ligules about 1mm long; 2nd glume awnless, the midnerve sometimes slightly extended; culms mostly 5-10dm high *S. gracilis*
Ligules 1.5-3mm long; 2nd glume awned or awn-tipped; culms often over 1m high *S. pectinata*

Spartina gracilis Trin., Mem. Acad. St. Petersb. VI, 4(1):110. 1840. Culms to 1m high; leaf blades mostly less than 5mm wide; spikes mostly 4-8; spikelets 6-10mm long; glumes acute, 1st about half as long as 2nd; lemma nearly as long as 2nd glume. Alkaline areas. Statewide.

Spartina pectinata Link, Jahrb. Gewächsk. 1(3):92. 1820. Culms to 2m high; leaf blades to 15mm wide; spikes mostly 10-20; spikelets 10-16mm long including awns; 1st glume acuminate or short-awned, about as long as lemma; 2nd glume exceeding lemma, with an awn to 7mm long. Wet areas. NE, SE.

Sphenopholis Scribn. Wedgegrass

Tufted perennials or rarely annual; spikelets mostly 2 flowered, in a panicle; glumes unlike in shape, the 1st narrow and 1 nerved, the 2nd obovate and 3-5 nerved; lemmas awnless.

Panicle dense, usually spike-like; 2nd glume about 1½ times as long as wide or less S. obtusata

Panicle loose, not spike-like; 2nd glume almost 3 times as long as wide S. intermedia

Sphenopholis intermedia (Rydb.) Rydb., Bull. Torrey Club 36:533. 1909. Culms to 12dm high; leaf blades mostly 1.5-6mm wide; spikelets 3-4mm long; 2nd glume about 2.5mm long; lemmas mostly 2.5-3mm long. Moist areas. NW, NE, SE. Eatonia pennsylvanica (DC.) Gray.

Sphenopholis _obtusata_ (Michx.) Scribn., Rhodora 8:144. 1906. Culms to 1m high; leaf blades mostly 1-5mm wide; spikelets 2.5-3.5mm long; 2nd glume 2-2.5mm long; lemmas 2-3mm long. Moist areas. NW, SW, SE. _Eatonia_ _obtusata_ (Michx.) Gray.

Sporobolus R. Br. Dropseed

Spikelets 1 flowered, in a panicle; glumes usually unequal, 1 nerved; lemma 1 nerved, awnless.

1. Plants annual, the mature panicle often partly concealed by swollen sheaths _S_. _neglectus_
1. Plants perennial, the panicle not concealed by sheaths when mature
   2. Spikelets mostly over 3mm long _S_. _heterolepis_
   2. Spikelets 2-2.5mm long
      3. Margins of lower sheaths long-hairy or conspicuously ciliate, the collars usually long-hairy also; spikelets tending to be appressed to panicle branches, appearing crowded _S_. _cryptandrus_
      3. Margins of lower sheaths not long-hairy or conspicuously ciliate except sometimes at very summit, the collars glabrous or hairy only on the margins; spikelets tending to diverge from panicle branches, appearing scattered _S_. _airoides_

Sporobolus airoides (Torr.) Torr., Rep. Bot. Pac. R. R. Exp. 7:21. 1858. Culms tufted, to 1m high; leaf blades usually less than 4mm wide; spikelets 2-2.5mm long; 1st glume about half as long as spikelet; 2nd glume nearly as long as lemma. Moist, usually alkaline areas. Statewide.

Sporobolus cryptandrus (Torr.) Gray, Man. 1:576. 1848. Culms tufted, to 1m high; leaf blades 1-5mm wide; spikelets 2-2.5mm long; 1st glume 1/3-1/2 as long as spikelet; 2nd glume about as long as lemma. Dry, usually sandy areas. NW, SW, SE.

Sporobolus heterolepis (Gray) Gray, Man. 1:576. 1848. Culms tufted, to 7dm high; leaf blades 2mm or less wide; spikelets 3-5mm long; 1st glume 2-3mm long; 2nd glume 3-5mm long; lemma slightly shorter than 2nd glume. Plains and hills. Crook and Weston Cos.

Sporobolus neglectus Nash, Bull. Torrey Club 22:464. 1895. Plants annual, mostly to 4dm high; leaf blades 1-2mm wide; spikelets 2-3mm long; glumes subequal, 1.5-3mm long; lemma 2-3mm long. Disturbed areas. Park and Big Horn Cos.

Stipa L. Needlegrass

Tufted perennials; spikelets 1 flowered, in a panicle; glumes acute to aristate; lemma usually strongly convolute, hardened, awned, the awn twisted below.

1. Glumes 15mm or more long; lemmas 8-25mm long; awns mostly over 10cm long
   2. Terminal segment of awn plumose — _S. neomexicana_
   2. Terminal segment of awn not plumose
      3. Glumes 30-40mm long; lemmas 16-25mm long — _S. spartea_
      3. Glumes 15-25mm long; lemmas 8-12mm long — _S. comata_
1. Glumes 13mm or less long; lemmas 4-9mm long; awns mostly less than 6cm long
   4. Awns plumose below
      5. Ligules 2-6mm long — _S. thurberiana_
      5. Ligules mostly less than 1mm long — _S. occidentalis_
   4. Awns not plumose below
      6. Panicle open, the branches spreading or ascending and spikelet-bearing near tip — _S. richardsonii_
      6. Panicle narrow, the branches appressed and often spikelet-bearing near base
         7. Hairs on lemma usually 1.5-3mm long; awn arising from between 2 terminal teeth of lemma, these sometimes appressed to awn — _S. pinetorum_
         7. Hairs on lemma usually not as much as 1.5mm long or lemma not toothed

8. Sheaths, at least the lowermost, pubescent *S. williamsii*

8. Sheaths glabrous or villous only at the throat or on margins

9. Sheaths villous at the throat; callus broad and blunt; lower nodes of panicle villous

10. Plants mostly less than 1m high; panicle mostly 10-25cm long; glumes thin and papery *S. viridula*

10. Plants mostly over 1m high; panicle mostly 25-40cm long; glumes relatively firm *S. robusta*

9. Sheaths not villous at the throat or only slightly so; callus narrow and sharp-pointed; nodes of panicle glabrous or nearly so

11. Awns mostly over 2cm long; lemmas mostly 6-7mm long *S. columbiana*

11. Awns mostly less than 2cm long; lemmas mostly 4-5mm long *S. lettermanii*

Stipa columbiana Macoun, Cat. Can. Pl. 2(4):191. 1888. Culms to 1m high; leaf blades 0.5-3mm wide; glumes 8-10mm long; lemma 6-7mm long, the awn 2-4cm long. Plains, meadows, and woods. Statewide. *S. nelsonii* Scribn.

Stipa comata Trin. & Rupr., Mem. Acad. St. Petersb. VI, 5:75. 1842. Culms mostly to 6dm high; leaf blades 0.5-3mm wide; glumes 15-25mm long; lemma 8-12mm long, the awns mostly 7-15cm long. Plains and hills. Statewide.

Stipa lettermanii Vasey, Bull. Torrey Club 13:53. 1886. Culms to 6dm high; leaf blades 0.5-3mm wide; glumes 6-9mm long; lemma 4-5mm long, the awns mostly 1.5-2cm long. Open slopes and woods. NW, SW, SE.

Stipa neomexicana (Thurb. ex Vasey) Scribn., USDA Agrost. Bull. 17:132. 1899. Culms mostly to 8dm high; leaf blades 1mm or less wide; glumes 2.5-5cm long; lemma about 15mm long, the awn 12-18cm long, plumose throughout. Dry slopes and canyons. Platte Co.

Stipa occidentalis Thurb. ex Wats., Bot. King Exp. 380. 1871. Culms to 5dm high; leaf blades 0.5-2mm wide; glumes 9-13mm long; lemma 5-7mm long, the awn 2.5-4cm long and plumose below. Plains, hills, and open woods. NW, SE.

Stipa pinetorum Jones, Proc. Calif. Acad. Sci. II, 5:724. 1895. Culms to 5dm high; leaf blades less than 1mm wide; glumes 7-10mm long; lemma 4-5mm long, awn about 2cm long and nearly glabrous. Open woods and slopes. Sublette and Sweetwater Cos.

Stipa _richardsonii_ Link, Hort. Berol. Desc. 2:245. 1833. Culms to 1m high; leaf blades 0.5-2mm wide; glumes 8-9mm long; lemma 4-5mm long, the awn 1.5-3cm long. Woods and meadows. NW, NE, SE.

Stipa _robusta_ (Vasey) Scribn. in Rydb. & Shear, USDA Agrost. Bull. 5:23. 1897. Culms to 1.5m high; leaf blades to 8mm wide; glumes 9-12mm long; lemma 6-8mm long, the awn 2-4cm long. Plains, hills, and roadsides. Albany and Laramie Cos. _S_. _vaseyi_ Scribn.

Stipa _spartea_ Trin., Mem. Acad. St. Petersb. VI, 1:82. 1830. Culms about 1m high; leaf blades 2-5mm wide; glumes 3-4cm long; lemma 16-25mm long, the awn 12-20cm long. Plains and hills. Crook Co.

Stipa _thurberiana_ Piper in Scribn., USDA Div. Agrost. Cir. 27:10. 1900. Culms to 6dm high; leaf blades 0.5-2mm wide; glumes 10-13mm long; lemma 7-9mm long, the awn 3-5cm long and plumose below. Dry slopes. Reported from Wyoming.

Stipa _viridula_ Trin., Mem. Acad. St. Petersb. VI, 2:39. 1836. Culms to 1m high; leaf blades 1-5mm wide; glumes 6-10mm long; lemma 5-6mm long, the awn 2-3cm long. Plains and hills. Statewide.

<u>Stipa</u> <u>williamsii</u> Scribn., USDA Agrost. Bull. 11:45. 1898. Culms to 1m high; leaf blades 0.5-3mm wide; glumes 8-11mm long; lemma 5-7mm long, the awn 2-4cm long. Plains, hills, and woods. SW, NW, NE.

<u>Trisetum</u> Pers.

Tufted perennials; spikelets usually 2 or 3 flowered, sometimes 4 or 5 flowered, in an open or spike-like panicle; glumes somewhat unequal, acute, the 2nd usually longer than 1st floret; lemmas 2 toothed at tip, awned from the back or awnless.

Awns included in glumes or nearly so, 2mm long or less, straight, or awns lacking ... <u>T</u>. <u>wolfii</u>
Awns exserted from glumes, 3mm long or more, usually bent ... <u>T</u>. <u>spicatum</u>

<u>Trisetum</u> <u>spicatum</u> (L.) Richt., Pl. Eur. 1:59. 1890. Culms to 5dm high; leaf blades 1-5mm wide; panicle usually spike-like; spikelets 4-6mm long; lemmas 4-5mm long, the awn 3-6mm long, usually geniculate and exserted. Forests and alpine slopes. Statewide. <u>T</u>. <u>subspicatum</u> (L.) Beauv.

Trisetum wolfii Vasey, USDA Monthly Rep. Feb.-Mar. 156. 1874. Culms to 1m high; leaf blades 2-8mm wide; panicle dense but not spike-like; spikelets 5-7mm long; lemmas 3-6mm long, awnless or with straight awns to 2mm long. Moist meadows. NW, NE, SE. *Graphephorum wolfii* (Vasey) Vasey ex Coult.

GROSSULARIACEAE Currant Family

Shrubs with simple, alternate, usually 3-5 lobed and toothed leaves; flowers bisexual, regular, in racemes; sepals usually 5, united; petals usually 5, separate; stamens usually 5, alternate with the petals; ovary 1, at least partly inferior; hypanthium usually well developed; carpels usually 2; styles 1 or 2; locule 1; ovules several to many; placentation parietal; fruit berry-like.

Ribes L. Currant; Gooseberry

1. Plants with spines or prickles at least at the nodes, often also on the internodes
   2. Hypanthium shallowly cup-shaped or saucer-shaped; pedicels often jointed below ovary
      3. Leaf blades pubescent and somewhat glandular *R. montigenum*
      3. Leaf blades glabrous or sparsely pubescent, not glandular *R. lacustre*
   2. Hypanthium tubular-campanulate or cylindric; pedicels not jointed below ovary

4. Stamens about twice as long as petals or longer, conspicuously exserted — R. inerme

4. Stamens about equaling the petals, not conspicuously exserted — R. setosum

1. Plants lacking spines or prickles

5. Flowers bright yellow or the petals sometimes reddish, glabrous, not glandular — R. aureum

5. Flowers not bright yellow, often hairy or glandular

6. Hypanthium tubular-campanulate or narrowly cylindric, usually longer than wide

7. Leaf lobes sharply pointed; leaf blades with sessile yellow glands beneath — R. americanum

7. Leaf lobes rounded; leaf blades usually lacking sessile yellow glands beneath

8. Hypanthium usually 2 or more times as long as calyx lobes; calyx lobes 1.5-3mm long; petals 1-2mm long — R. cereum

8. Hypanthium less than twice as long as calyx lobes; calyx lobes 3-7mm long; petals 2-4mm long — R. viscosissimum

6. Hypanthium saucer-shaped or shallowly cup-shaped, mostly wider than long

9. Ovary with sessile yellow glands on outer surface; native — R. hudsonianum

9. Ovary without glands (pedicels sometimes glandular); introduced and escaped — R. rubrum

Ribes americanum Mill., Gard. Dict. 8th ed. Ribes No. 4. 1768. Unarmed shrub to about 1m high; leaf blades ovate to cordate, 1-9cm long and about as wide, mostly 3 lobed, with sessile glands at least beneath; calyx lobes 4-7mm long; hypanthium tubular-campanulate; petals whitish, 2.5-4mm long; berry glabrous. Moist areas. NE, SE.

Ribes aureum Pursh, Fl. Am. Sept. 164. 1814. Unarmed shrub to 3m high; leaf blades deltoid-ovate to ovate, 1-5cm long, slightly wider, with mostly 3 rounded lobes almost half the length, glandular or not; calyx lobes 3-7mm long; hypanthium narrowly cylindric; petals yellow to reddish, about 2mm long; berry glabrous. Thickets. Statewide. R. longiflorum Nutt. of authors.

Ribes cereum Dougl., Trans. Hort. Soc. Lond. 7:512. 1830. Unarmed shrub to 2m high; leaf blades reniform to cordate-orbicular, 0.5-3cm long and about as wide, glandular or not, shallowly 3 or 5 rounded-lobed; calyx lobes 1.5-3mm long; hypanthium narrowly cylindric; petals whitish, 1-2mm long; berry glabrous or glandular. Woods, thickets, and rocky areas. Statewide.

Ribes hudsonianum Richards. in Franklin, Narr. 1st Journ. 2:6. 1824. Unarmed shrub to 2m high, glandular; leaf blades mostly cordate, 1-11cm long and about as wide, with mostly 3 lobes less than half the length; calyx lobes 3-5mm long; hypanthium saucer- or cup-shaped; petals white, about 1.5mm long; berry glabrous or glandular. Moist areas. NW, SW. R. petiolare Dougl.

Ribes inerme Rydb., Mem. N. Y. Bot. Gard. 1:202. 1900. Sparingly armed shrub to 2m high; leaf blades ovate to cordate, 1-5cm long and about as wide, 3-5 lobed, not glandular; calyx lobes 3-5mm long; hypanthium tubular-campanulate; petals white or pink, 1-2mm long; berry glabrous. Streambanks, swamps, and thickets. Statewide. *R. saxosum* Hook. of authors.

Ribes lacustre (Pers.) Poir. in Lam., Encyc. Meth. Suppl. 2:856. 1812. Armed shrub to 2m high; leaf blades cordate, 1-7cm long and about as wide, mostly 5 lobed about half or more the length, not glandular; calyx lobes 2-3mm long; hypanthium saucer-shaped; petals pinkish, 1-2mm long; berry usually glandular. Woods, slopes, and streambanks. Statewide. *R. parvulum* (Gray) Rydb.

Ribes montigenum McClatchie, Erythea 5:38. 1897. Armed shrub to 1m high, glandular; leaf blades mostly cordate, 0.5-4cm long and about as wide, deeply 5 cleft and usually further lobed; calyx lobes 2.5-4mm long; hypanthium saucer-shaped; petals pink or purple, 1-2mm long; berry glandular or glabrous. Mostly alpine or subalpine, occasionally slightly lower. Statewide. *R. lentum* (Jones) Cov. & Rose.

Ribes rubrum L., Sp. Pl. 200. 1753. Unarmed shrub to 1.5m high; leaf blades cordate or deltoid-orbicular, 1-8cm long and about as wide, 3-5 lobed about half or less the length, usually not glandular; calyx lobes about 2mm long; hypanthium saucer-shaped; petals pinkish or cream, about 1mm or less long; berry usually glabrous. Disturbed areas. Teton Co. *R. sativum* (Reichb.) Syme.

Ribes setosum Lindl., Trans. Hort. Soc. Lond. 7:243. 1828. Armed shrub to 1m high; leaf blades cordate to orbicular, 1-5cm long and about as wide, 3-5 lobed about half the length or less, sometimes glandular beneath; calyx lobes 3-5mm long; hypanthium cylindric; petals white to pink, 2-3.5mm long; berry glabrous. Hills and streambanks. Statewide. *R. saximontanum* E. Nels.

Ribes viscosissimum Pursh, Fl. Am. Sept. 163. 1814. Unarmed shrub to 2m high; leaf blades mostly cordate, 1-8cm long and about as wide, with 3-5 rounded lobes less than half the length, glandular-pubescent; calyx lobes 3-7mm long; hypanthium tubular-campanulate; petals white, 2-4mm long; berry glandular. Moist areas. SW, NW, NE.